Agraria

Agrarian policies and rural poverty in Africa

Edited by Dharam Ghai
and Samir Radwan

Prepared with the financial support
of the Swedish Agency
for Research Co-operation
with Developing Countries

International Labour Office   Geneva

Copyright © International Labour Organisation 1983

Publications of the International Labour Office enjoy copyright under Protocol 2 of the Universal Copyright Convention. Nevertheless, short excerpts from them may be reproduced without authorisation, on condition that the source is indicated. For rights of reproduction or translation, application should be made to the Publications Branch (Rights and Permissions), International Labour Office, CH-1211 Geneva 22, Switzerland. The International Labour Office welcomes such applications.

ISBN 92-2-103100-4 (limp cover)
ISBN 92-2-103109-8 (hard cover)

First published 1983
Second impression 1985

The designations employed in ILO publications, which are in conformity with United Nations practice, and the presentation of material therein do not imply the expression of any opinion whatsoever on the part of the International Labour Office concerning the legal status of any country or territory or of its authorities, or concerning the delimitation of its frontiers.
The responsibility for opinions expressed in signed articles, studies and other contributions rests solely with their authors, and publication does not constitute an endorsement by the International Labour Office of the opinions expressed in them.

ILO publications can be obtained through major booksellers or ILO local offices in many countries, or direct from ILO Publications, International Labour Office, CH-1211 Geneva 22, Switzerland. A catalogue or list of new publications will be sent free of charge from the above address.

Printed in Switzerland

PREFACE

This volume presents some results of the continuing work on rural poverty, employment and agrarian systems being carried out by the International Labour Office within the framework of the World Employment Programme. In contrast with the situation in Asia and Latin America, relatively little comparative work has been done on the issues of rural poverty, income distribution and agrarian change in Africa. This is at least in part due to the paucity of statistical and other data on this subject. One of the objectives of the present volume is to help to fill this gap in the knowledge and understanding of the process and results of rural development in African countries since independence. The country studies included here attempt to provide a quantitative and qualitative profile of rural poverty and income distribution and to evaluate the impact of state policies on them, especially with respect to land tenure, public expenditure and resource extraction from the rural areas.

Given the large number of countries in sub-Saharan Africa, it was difficult to select studies which would be completely representative of their diversity. Nevertheless, in choosing the case studies, an attempt was made to cover a wide range of production systems, agrarian relations, rates and patterns of growth and strategies of development. The results of these studies should therefore be of interest and relevance to most African countries. The contributors to this volume are acutely aware of the constraints imposed by data limitations on analysis of the pattern of, and changes in, rural poverty and income distribution. While the availability of additional data would no doubt lead to revision of some of the results presented here, evidence from a number of sources confirms the validity of the main conclusions of the study.

In preparing this volume, the editors have accumulated a debt of gratitude to many institutions and individuals. We should like to extend our thanks to the Swedish Agency for Research Co-operation with Developing Countries (SAREC), whose grant financed much of the external collaboration and travel costs. Our thanks are due also to the contributors to this volume for commenting on various draft chapters. Iqbal Ahmed, Ajit Ghose and Rick Szal deserve our special thanks for their generous assistance in editorial work. We are grateful to Enyinna Chuta, Jacques Gaude, Wouter van Ginneken, Rolf van der Hoeven, Willem Keddeman, Frank Lisk, Jean Majeres, Thandica Mkandawire, Peter Richards and Lawrence Smith for their comments on earlier drafts of this manuscript. We acknowledge with gratitude the help on documentation received from Evelyn Schaad at all stages of this work. Finally, we owe a special debt of gratitude to Sandra Johnson, Frances Kaufmann and Lesley Brooks for typing successive drafts of the manuscript against exacting deadlines.

Dharam Ghai
Samir Radwan

CONTENTS

Preface	v
1. Agrarian change, differentiation and rural poverty in Africa: a general survey *(Dharam Ghai and Samir Radwan)*	1
Growth: rates, patterns and sources	3
Income distribution	9
Poverty	12
Mechanisms generating differentiation and perpetuating poverty	15
Processes behind differentiation	16
Persistence of rural poverty	21
Conclusion	27
2. Social justice and development policy in Kenya's rural economy *(William J. House and Tony Killick)*	31
An introduction to the rural economy	31
Nature of the rural economy	31
Agriculture in the national economy	33
Extent, nature and causes of poverty in rural Kenya	34
A national survey	34
Inequalities within the rural economy	35
Poverty in rural areas	39
Deficiencies in other basic needs	42
Impact of government policies	44
Evolution of rural development policy	44
Distributive effects of land policies	46
Government economic services to agriculture	53
Questions of rural-urban balance	58
Planning rural development	62
Summary of main findings	64
Extent and nature of the problems	64
Impact of government policies	64

3. **Growth and inequality: rural development in Malawi, 1964-78** *(Dharam Ghai and Samir Radwan)* 71
   Growth and structural changes, 1964-78 71
      Background 71
      Growth, exports and employment 73
      Rural economy 78
   Growth, distribution and poverty 82
      Some features of income distribution in 1968 83
      Agricultural estate workers 84
      Smallholders 86
      Some concluding remarks on growth and distribution 90
   State policies 93
   Conclusion 96

4. **Export-led rural development: the Ivory Coast** *(Eddy Lee)* 99
   Sources and pattern of growth 100
   Rural income and its distribution 108
   Impact of government policies 120
   Conclusion 124

5. **Rural poverty in Botswana: dimensions, causes and constraints** *(Christopher Colclough and Peter Fallon)* 129
   Economic trends since independence 129
   Rural income distribution: statistical analysis 133
   Over-all distribution and measures of inequality 133
   Sources of income and characteristics associated with poverty .. 138
   Rural income distribution: dynamic analysis 143
      Cattle 144
      Crops 146
      Employment 146
      Transfers 147
      The role of government 149
   Conclusion 151

6. **Equity and growth: an unresolved conflict in Zambian rural development policy** *(Charles Elliott)* 155
   Equity and growth 160
   Rural poverty 164
   Poverty and policy 167
   Limits of policy 179
   Conclusion 186

7. **Oil and inequality in rural Nigeria** *(Paul Collier)* 191
   National growth and rural stagnation 192
   Rural inequality 193
   Oil and rural inequality: mechanisms of transmission 206
   Conclusion 214

8. **Stagnation and inequality in Ghana** *(Assefa Bequele)* 219
   Characteristics of the growth process in Ghana 220
   Poverty and inequality 225
   Towards a perspective on Ghana's economic problems 230
   Conclusion 245

9. **Growth and distribution: the case of Mozambique** *(R. K. Srivastava and I. Livingstone)* 249
   The colonial economy 249
      The setting 249
      The rural sector 253
      Employment and basic needs 257
      Conclusion 262
   Changes since independence 263
      The economic situation 263
      Socio-economic objectives and priorities 267
      Rural institutions 269
   Growth and distribution 274
   Conclusion 277

10. **Nomads and farmers: incomes and poverty in rural Somalia** *(Vali Jamal)* 281
    Structure of the economy 282
       Livestock 284
       Agriculture 288
       Contribution of nomads and farmers 291
    Incomes and poverty 293
    Trends in inequality 297
    Government policy 299
       Livestock 299
       Agriculture 300
    Conclusion 302
    Appendix A. Poverty line for the nomads 303
    Appendix B. Previous studies of income distribution 307

AGRARIAN CHANGE, DIFFERENTIATION AND RURAL POVERTY IN AFRICA: A GENERAL SURVEY

Dharam Ghai and Samir Radwan

The purpose of this volume is to study the conditions of agrarian change in Africa. The emphasis is placed on the impact of economic growth on income distribution and poverty in rural areas. Within the limits of the data available, the studies contained in this volume have attempted to provide evidence on the nature and extent of rural differentiation and poverty and to evaluate the impact on them of certain processes and policies.[1] The present chapter draws on these studies and other sources to present a general survey of growth, inequalities and poverty in rural areas, as well as of the underlying processes and mechanisms. In particular, it addresses itself to the question whether the economic growth experienced in the countries covered by this volume has led to a widespread improvement in the living standards of the great mass of the rural population or whether the benefits of growth have accrued only to a small minority, with the rest of the population continuing to live in poverty or experiencing further impoverishment. While, as indicated below, the inadequate existing data cannot be used to test this hypothesis rigorously, some tentative conclusions are suggested by the available evidence in a number of countries. These conclusions are reinforced by an analysis of the mechanisms and policies which affect rural differentiation and poverty.

Before coming to these issues, it is necessary to comment briefly on the selection of case studies and the availability of data. Sub-Saharan Africa is a vast region comprising 45 independent countries with enormous differences in ecology and economic structure, as well as in the evolution of their respective agrarian systems, inherited colonial traditions, and development strategies and policies pursued since independence. It would be quite impossible to capture this diversity in a limited number of case studies. Nevertheless, the countries included in this volume—which comprise 131 million people (mid-1976) or 42 per cent of the total population in independent sub-Sahara—illustrate a sufficiently broad range of experience in terms of structure and organisation of agricultural production, evolution of relations of production, patterns of growth and models of development, for their conclusions to be of interest and relevance to most African countries. Some of these features may be highlighted here.

In terms of agrarian structures, the case studies represent a wide spectrum with a dominance of small- and medium-scale peasant agriculture in Ghana, the Ivory Coast and Nigeria, and a significant role being played by plantations and large farms in Kenya, Malawi, Mozambique and to a lesser extent in Zambia. Historically, the former group of countries depended largely on African peasant cultivation for the generation of export earnings, while the latter were characterised by alienation of land for the White settlers. While in most countries private or traditional communal ownership of land forms the basis of the agrarian system, Mozambique represents a case of the transformation of agriculture into co-operative and collective forms of production. The countries studied also represent a wide spectrum in terms of transition from subsistence peasant agriculture to highly modernised, commercialised production. While crop cultivation remains the main rural activity in most countries, Botswana and Somalia illustrate the operation of livestock-dominated economies.

In terms of relations of production, substantial wage employment is a feature of the rural economies of Ghana, the Ivory Coast, Kenya and Malawi; it remains relatively unimportant, however, in Botswana, Nigeria, Somalia and Zambia. Landlessness is emerging as a serious problem in Kenya and Malawi. Outmigration on a large scale has figured significantly in the history of Botswana, Malawi and Mozambique, and more recently in that of Somalia. On the other hand, Ghana and the Ivory Coast have received immigrants from their neighbouring countries.

The countries studied also show considerable diversity in growth experience and development strategies. Botswana, the Ivory Coast, Kenya, Malawi and Nigeria have experienced high growth rates, while Ghana and Somalia have suffered stagnation or decline in per head incomes. The driving force behind growth in the Ivory Coast, Malawi and, to a lesser extent, Kenya was the rapid expansion of agricultural exports, while the growth in Botswana and Nigeria was fuelled by the emergence of mineral exports. In terms of export dependence, Mozambique and Somalia represent one end of the spectrum and Botswana and the Ivory Coast the other, with Ghana and Zambia showing a marked decline in export-orientation over the period.

In summary, the countries studied represent a wide range of production systems and development experiences.

The studies included in this volume make extensive use of the existing data on growth, inequalities and rural poverty. A word of caution is therefore in order regarding the range and quality of the data available for an analysis of this kind. While there is a great deal of country variation in this respect, hardly any African countries offer sufficient data to permit a rigorous, quantitative analysis of the dimensions and trends in rural poverty and inequality.

The studies presented here draw upon the available evidence—quantitative and qualitative, macro and micro—to build up a general picture, if not of trends, then at least of the broad magnitude and characteristics of rural poverty and inequality. It has not been possible in all cases to check the reliability and consistency of the data used in analysis; the conclusions presented here must thus be interpreted in the light of such data limitations.

This chapter begins with an examination of evidence on growth, equity and poverty. This is followed by an analysis of the main mechanisms and forces contributing to differentiation and perpetuation or intensification of rural poverty. The concluding section presents some general conclusions suggested by the studies.

GROWTH: RATES, PATTERNS AND SOURCES

Since the rate and pattern of growth are primary determinants of changes in poverty and income distribution, it is useful to begin with a summary of the growth experience in the nine countries. Table 1 presents data on the level of GNP, rates of growth of GNP and agricultural and food production, covering for the most part the period 1960 to 1978.[2] Taking the period as a whole, the countries may be divided into three categories of fast growth (annual growth of income per head in excess of 2 per cent), moderate growth (increase per head between 0 and 2 per cent) and stagnation or decline (zero growth or decline in income per head). According to this criterion, Botswana, the Ivory Coast, Kenya, Malawi and Nigeria fall into the first category; Mozambique and Zambia into the second, and Ghana and Somalia into the third.

Turning now from over-all growth to agricultural growth, we find as expected that countries with low or negative growth of GDP such as Ghana, Mozambique and Somalia also experienced poor agricultural performance, while high over-all growth rates in the Ivory Coast, Kenya and Malawi were accompanied by good agricultural performance. There are, however, some exceptions to this correlation between over-all growth and agricultural growth. Nigeria experienced a strong over-all growth despite an absolute decline in agricultural production in both the 1960s and the 1970s. In Zambia, on the other hand, despite reasonable agricultural performance, the over-all growth was low.

Data on food production are not available over the period between 1960 and 1978 but the indices of food production per head between 1969 and 1971 and 1976 and 1978 draw attention to the food crisis which is confronting an increasing number of African countries. Only two countries—the Ivory Coast and Zambia—managed to secure increases in food production per head, and two others—Botswana and Malawi—were able to prevent a decline. Ghana and Mozambique represent the other extreme: they experienced declines of nearly 20 per cent, compared with those in Kenya and Nigeria of about 10 per cent.

Agrarian policies and rural poverty in Africa

Table 1. Growth in GDP and agriculture

Country	GDP growth		GNP per head (in $), 1978	GNP growth per head, 1960-78	Agriculture		Food production per head (1969-71 = 100), 1976-78
	1960-70	1970-78			1960-70	1970-78	
Somalia	1.0	3.1	130	−0.5	−1.5	2.7	87
Mozambique	4.6	−3.2	140	0.4	2.1	−1.8	81
Malawi	4.9	6.5	180	2.9	.	4.2	99
Kenya	6.0	6.7	330	2.2	.	5.5	91
Ghana	2.1	0.4	390	−0.5	.	−1.2	79
Zambia	5.0	2.3	480	1.2	.	3.1	109
Nigeria	3.1	6.2	560	3.6	−0.4	−1.5	89
Botswana	15[1]	.	620	.	.	.	100
Ivory Coast	8.0	6.8	840	2.5	4.2	3.9	104

. = not available.
[1] 1966-76; data from the country chapter.
Source. World Bank: *World Development Report, 1980.*

The pattern and sources of growth experienced by these countries provide some insights on changes in income distribution. Among the fast-growing economies, a distinction may be made between the Ivory Coast, Malawi and to a lesser extent Kenya, on the one hand, and Botswana and Nigeria on the other. The driving force behind the first group of economies was the rapid expansion of agricultural exports, while in the latter case it was the mining sector. In the Ivory Coast exports increased in volume by slightly less than 9 per cent per annum over the period 1960-78 (table 2). Rapid expansion of coffee, cocoa, palm oil, timber and pineapple cultivation fuelled the growth process. The export orientation of the economy, already very high in 1960, increased to some extent over the next 18 years. Such growth was achieved largely by bringing additional land under cash crop cultivation. This in turn required a growing labour force supplied by migration from the northern region and from the neighbouring countries, principally the Niger and the Upper Volta. The combination of cheap, plentiful labour and untapped reserves of fertile soil in the southern forest zone enabled an increasing number of medium- and large-scale farmers to increase their incomes through the rapid expansion of coffee and cocoa cultivation.

Malawi achieved a high growth rate through essentially the same pattern. In value terms, exports increased by more than six and a half times between 1964-65 and 1977-78. The export orientation of the economy was intensified, exports as a proportion of GDP rising from 15 per cent in 1964-65 to 20 per cent in 1977-78. The main crops were tea and tobacco. However, in contrast with the Ivory Coast, where the bulk of the additional production came from

Table 2. Exports, terms of trade and gross domestic investment

Country	Exports of goods and services as % of GDP		Annual growth rate in quantity of exports		Terms of trade (1970 = 100)		Annual growth rate in gross domestic investment	
	1960	1978	1960-70	1970-78	1960	1978	1960-70	1970-78
Somalia	11	12	2.3	7.8	107	72	4.3	8.5
Mozambique	14	13	6.0	−15.9	103	96	8.3	−9.6
Malawi	21	21	11.6	2.4	116	112	15.4	1.1
Kenya	31	27	7.2	0.8	112	104	7.0	2.3
Ghana	28	10	0.1	−0.1	92	80	−3.2	−8.3
Zambia	56	32	2.2	−4.7	50	56	10.6	−2.9
Nigeria	15	31	6.1	0.5	97	290	7.4	23.3
Botswana	.	.	8.9	29.5	.	.	.	.
Ivory Coast	37	38	8.8	8.5	89	94	12.7	14.0

. = not available.
Sources. World Bank: *World Development Report, 1980*. For Botswana, idem: *Accelerated development in sub-Saharan Africa: An agenda for action* (Washington DC, 1981), Statistical annex.

farmers with 5- to 20-hectare holdings, in Malawi it was the estates which spearheaded the export drive. Between 1964-65 and 1977-78, the share of estates in total agriculture rose from 7 to 13.2 per cent, and in monetary agriculture from just over 25 to 37 per cent. However, as in the Ivory Coast, the expansion of exports both from estates and from "progressive" farmers was made possible by the availability of cheap and plentiful labour. Whereas in the Ivory Coast the labour was drawn largely from the neighbouring countries, in Malawi it was the peasant households which were the source of supply.

In Kenya, although the rapid expansion of agricultural exports (especially in the 1960s) was a significant contributory factor to high over-all growth rates, the export orientation of the economy—particularly when one takes into account merchandise exports only—declined to some extent over the period 1960-78. The export of primary products did not play the same predominant role in generating growth as in the case of the Ivory Coast and Malawi. There were also significant differences in the pattern of agricultural growth: whereas between 1960 and 1964, smallholders provided less than 38 per cent of marketed agricultural output, between 1975 and 1978 their share had risen to over 53 per cent. This is in sharp contrast with the pattern in the Ivory Coast and Malawi and is explained in large part by the land registration, consolidation and settlement schemes for smallholders initiated in the years preceding and following independence. The bulk of the increased production came from smallholders in high-potential areas. Thus, as a summary oversimplification, it may be stated that, whereas in Malawi it was the expansion of estate production for exports and, in the Ivory Coast, increased production by medium- and large-scale planters which were the critical factors in the rapid growth of these economies, in Kenya the dynamic element was provided by smallholder production.

Botswana and Nigeria illustrate the dramatic impact on the growth and structure of the economy made by the discovery and large-scale exploitation of minerals. In Nigeria, between 1960 and 1977, the share of mining in GDP rose from 5.2 to over 37 per cent, and that of agriculture declined from 54 to 22 per cent. While in the early 1960s agriculture provided between 75 and 80 per cent of total exports, by the late 1970s, with the exception of cocoa, agricultural exports had virtually disappeared, their share declining to a mere 3-4 per cent. An equally dramatic transformation took place in Botswana, where the share of mining in GDP rose from less than 1 per cent in 1966 to 15 per cent in 1975-77, while that of agriculture declined from 40 to 25 per cent.

However, the two experiences also display a number of differences, notably concerning the performance of the agricultural sector. In Nigeria, as noted earlier, agricultural output over the period underwent a decline, which intensified significantly in the 1970s. In Botswana, on the other hand, the agricultural sector achieved a good performance: livestock holdings, which generate nearly three-quarters of the agricultural output, rose from between 1 and 1.5 million to 3.5 million heads between 1966 and 1979. A further major difference is that whereas in Botswana the major impulse to growth came from increases in the volume of production of diamonds, copper and nickel, in Nigeria—where the index of mining production rose from 40 in 1973 to 123 in 1978—the increase in petrol prices was a dominant factor, at least in the 1970s, resulting in a large improvement in the terms of trade.

The Zambian experience offers yet another illustration of the impact of the mining sector on the over-all economy. With 90 per cent of exports and two-thirds of government revenue in good years, copper has been the dominant influence on the growth of the economy. However, unlike the situation in Botswana and Nigeria, mining was already an established sector in the early 1960s and did not experience dramatic increases in production. On the contrary, with rising costs and fluctuating and highly adverse prices for copper in the 1970s, there was a relative shrinkage of the sector, with its share in GDP declining from nearly 40 to less than 20 per cent between 1965-67 and 1974-76.

Ghana and Somalia are characterised by declining output per head over the period 1960-78. Apart from the fact that they both suffered from political instability and armed conflict—a characteristic also shared by Mozambique and Nigeria, and which obviously must have strongly inhibited growth—these economies have little in common which can explain such a poor economic performance. The outcome in Somalia is easier to understand. The Somali economy typifies other countries in the Saharan belt which are handicapped in their development drive by the poverty of their natural resources. Much of Somalia is desert and semi-desert land with low rainfall and poor soils. Thus, in many parts of the country, agriculture is only possible through costly and complex irrigation schemes. Unlike some other countries in the desert zone, Somalia has not yet discovered and exploited any mineral resources which could provide the basis for rapid growth. Livestock is the key sector in the economy and its sustained expansion is dependent upon the creation of ecologically sound grazing facilities. Although the country has benefited from rapidly rising prices for its livestock exports, this has not prevented a fairly substantial deterioration in its over-all terms of trade since the mid-1960s. Thus, given the difficult nature of basic natural resources and taking into account the military conflicts with neighbouring countries, it is not surprising that domestic production failed to keep pace with population growth.

The Ghanaian performance is nearly unique and presents many puzzling aspects. By the mid-1950s Ghana had very substantially overtaken practically all sub-Saharan African countries in terms of income per head, human resource development and potential for growth. After achieving substantial growth in the 1950s, the country entered a phase in the early 1960s of steadily declining output per head so that in 1978 incomes per head were little different from those in 1950. In the 1970s there was a dramatic decline in food and, notably, cocoa production—a most surprising performance in a country which had had Ghana's output per head in the mid-1950s. The existing data are not adequate to provide a satisfactory interpretation of this phenomenon. What appears certain is that it was the stagnation and subsequent decline in agricultural production which brought the economy to a halt. Stagnation in agriculture must have resulted from a massive deterioration of the terms of trade faced by cocoa growers, through excessive taxation and other deductions from the proceeds of cocoa production, the huge overvaluation of the currency and sharp domestic inflation provoked in the late 1960s by skyrocketing food prices.

The situation in Mozambique is different from others in that the country became independent only in 1974 after a prolonged struggle. Economic growth accelerated sharply between 1965 and 1970 after a period of low growth between 1960 and 1965. The basis of this growth was a rise in primary exports stimulated in turn by the existence of cheap labour, together with a major expansion of the country's transit role and establishment of new processing and manufacturing enterprises. The sharp decline in GDP, agricultural exports, food production and employment since 1973 was the result of the structural upheaval associated with independence. Some of the relevant factors were the damage caused during the final phase of the liberation struggle, the mass exodus of most of the skilled people, the closure of the Rhodesian border, the sharp decline in recruitment of workers by South Africa and the coincidence of a series of natural disasters. At the same time the independent Government carried out a far-reaching reorganisation of industry, agriculture and the services sector. Under the circumstances it would have been a miracle if aggregate production and incomes had not declined.

At this stage it may be useful to comment briefly on the trends in the external barter terms of trade experienced by these countries. Movements in the terms of trade provide an indication of how developments in the world economy have affected a given country, as well as of the extent to which changes in domestic output are reflected in changes in real incomes for such country. A study of the changes in the terms of trade over the period 1960-78 indicates unfavourable movement for agricultural exporters in most years, particularly in the 1960s.[3] Of the countries studied here, Kenya, Malawi and Mozambique have experienced relatively more favourable trends. Ghana has suffered more from adverse trends in terms of trade than other agricultural exporters. The terms of trade for Nigeria declined until 1968 but have been improving subsequently, especially in 1974. Zambia, on the other hand, benefited from improving terms of trade in the 1960s but suffered major declines in the 1970s, which led to sharp falls in real GDP per head between 1972 and 1978. Likewise, Somalia experienced favourable trends until 1964, but suffered from almost continuously declining terms of trade in the subsequent period. However, in view of the relatively much smaller importance of the trade sector in the Somali economy, the adverse impact on real incomes has been much lower than in Zambia.

The diverse patterns of development noted above have wide-ranging implications for rural poverty and differentiation; these are discussed in the subsequent sections. Here it may be useful to consider a few broad implications of the main patterns of development experienced by these countries. The first point which must be made concerns the role played by different sectors in spearheading the growth process. Although in all countries the non-agricultural sectors expanded more rapidly than the farming sector, in none of the countries did the manufacturing sector play a key role in income generation and labour absorption. Among the leading growth-generating sectors, it is illuminating to distinguish between crops, livestock and mining. Generally speaking, the direct benefits of a given growth in incomes are likely to be most widespread in crops and least felt in mining. The two main reasons for this are to be found in the degree of concentration of asset ownership and of the labour intensity of production in the three sectors. Even where the mining sectors are in the hands of the State, payments to foreigners for management, technology, marketing and compensation may absorb a substantial part of the revenues generated, and the benefits to the rural population are ultimately dependent on the allocation of these revenues. Furthermore, through the impact on the exchange rate of the foreign exchange earned by exports, the mining sector may exercise a powerful negative effect on the level and growth of incomes in the agricultural sector.

According to this reasoning, it may therefore be expected that the effects of a given increase in domestic incomes in countries such as Botswana (copper, diamonds, livestock), Nigeria (petroleum), Somalia (livestock) and Zambia (copper) would have much less impact on rural incomes and poverty than in countries such as the Ivory Coast, Kenya and Malawi, where increases in crop production were the main source of increase in incomes. Within the latter group of countries, the diffusion of benefits from growth would further depend on the pattern of organisation of production. Thus it may be recalled that the major benificiaries of agricultural growth in the Ivory Coast were identified as medium- to large-scale planters, in Malawi the plantation owners and a relatively small group of progressive farmers, and in Kenya the smallholders and some large-scale farmers located in areas benefiting from land settlement and consolidation schemes. However, the increase in wage employment generated on these farming units enabled a much larger number of rural households to enhance their family incomes, albeit by modest amounts.

A notable feature of the pattern of agricultural development in many countries has been a shift of production in favour of cash crops destined primarily for exports. A variety of government policies have reinforced this trend, which has resulted in growing food shortages and increasing food imports in recent years in a number of African countries. Given the direct link between food and poverty, the food production and price trends have accentuated poverty since the rural poor often tend to be net food purchasers.

INCOME DISTRIBUTION

Any discussion on income distribution in Africa must start with a recognition of the constraints imposed by the paucity and deficiency of data. There are few countries in sub-Saharan Africa which have reliable national household income or expenditure surveys at a given period, let alone comparable time series data. Nevertheless, in many countries there exist partial surveys and indirect empirical evidence on income distribution. We have relied upon the available data and evidence to build up a picture of a general kind on the pattern and trends of rural income distribution. While the availability of additional and more reliable data would no doubt lead to modification of the results presented here, there seems to be adequate evidence from a number of different sources to support a judgement that the main conclusions broadly reflect the underlying reality.

Many observers of the African scene have been impressed by the relatively high inequality of over-all income distribution in most African countries. One of the contributing factors is the sharp inequality between the rural and urban areas. The latter contain the predominant share of modern sector activities, in which the salaries and incomes of skilled employees, both in the public and in the private sector, as well as of professionals and business and entrepreneurial groups, have historically been linked to the levels prevailing in the erstwhile colonial Powers in Europe. The existence of significant expatriate communities in countries such as Bostwana, the Ivory Coast, Kenya, Malawi and Zambia and before independence in Mozambique, has reinforced this tendency. In the rural sector, on the other hand, the great majority of peasant households have incomes and living standards close to that represented by subsistence production. A recent World Bank study has pointed out that "non-agricultural incomes are 4-9 times agricultural incomes. All other countries in the world have ratios which are lower, 2.0 and 2.5".[4] This characteristic feature of most African economies partly explains the high degree of income inequality for the country as a whole.[5]

While such sharp disparities in over-all income distribution are not too unexpected, it is often believed that incomes and consumption in the rural areas are relatively evenly distributed. This belief is founded on the assumption of land abundance, the role of the customary land tenure system in preventing landlessness and the widespread prevalence of subsistence production based on family labour. One of the conclusions of this study is that this picture is no longer valid in the African countryside. The available evidence points both to the existence of very significant inequalities in income distribution and to a sharpening of these inequalities over the past two decades. In Botswana in the mid-1970s the bottom 40 per cent of the rural households received less than 12 per cent of the total rural income, while the top 10 per cent received nearly 40 per cent. The Gini coefficient of income distribution was estimated at 0.49. The basis for such disparities rests in the highly unequal distribution of livestock ownership. In Kenya, also in the mid-1970s, the Gini coefficient of rural income distribution was estimated at 0.50.[6] In Ghana over-all figures for the rural sector are not available, but among the cocoa growers (who in 1970 formed about 22 per cent of the rural labour force) the share in total income of the bottom 20 per cent was 5 per cent, compared with 50 per cent for the highest 20 per cent. In Nigeria, likewise, no estimates of over-all rural income distribution are available. However, on the basis of a large number of village studies carried out over different time periods, the income differential between the top 10 and the bottom 40 per cent of cocoa growers ranged from 7.1 to 20.1; for food producers, the differentials were of the order of 4.1. In Zambia the income differentials between the 850,000 poor and subsistence farmers and the 20,000 prosperous farmers may have been of the order of 1.10 or 15 in the early 1970s.

Some indirect evidence lends support to the existence of substantial inequalities in rural income distribution. Table 3 provides estimates of Gini coefficients for land distribution in various countries. The land distribution data are taken either from agricultural censuses or from samples for various categories of land. The data are subject to the usual qualifications: *(a)* apart from errors in data collection and aggregation, lands of widely different quality are pooled together; *(b)* no attempt been made to allow for different household sizes; *(c)* the landless are excluded from the data; *(d)* in view of the differences in definition and coverage in different countries, it is not possible to make inter-country comparisons; and *(e)* as is well known, income differentials tend to be less than land differentials. Even making allowance for all these factors, it would appear that there are very considerable inequalities in

Table 3. Land distribution

Country	Year	Coverage	Gini coefficient
Botswana	1968-69	Traditional holdings	0.50
Ivory Coast	1973-75	Traditional sector	0.42
Kenya	1969	Registered smallholdings	0.55
	1960	African holdings (sample survey)	0.50
Malawi	1968-69	Smallholdings	0.41
Mozambique	1970	Traditional sector	0.42
	1970	Modern sector	0.81
	1970	Total	0.71
Nigeria	1963-64	Sample survey (North farm crops only)	0.43
	1963-64	Sample survey (Eastern farm/tree crops)	0.56
		Sample survey (Western including mid-West: farm and tree crops)	0.40
Somalia	1968	Sample survey (5 districts)	0.55
Zambia	1970-71	Commercial sector	0.76
Ghana	1970	All holdings	0.64

Note. The term "smallholders" is used in many African countries to refer to small- and medium-scale peasant farmers to differentiate them from plantations and large private farms.
Source. Gini coefficients for distribution of landholdings based on data from the 1960 and 1970 census of agriculture in the respective countries as reported in FAO: *Report on the 1970 World Census of Agriculture*, Country bulletins (Rome, 1975).

land distribution. Even within the traditional or smallholder sector, the value of the Gini coefficient varies between 0.40 and 0.55.

The distribution of livestock often tends to be even more skewed. In Botswana, for example, where cattle constitute the backbone of the rural economy, it was estimated that in 1974 about 5 per cent of rural households owned nearly 50 per cent of the national herd, while 90 per cent of households owned a mere 20 per cent. Similar nation-wide data are not available for Somalia, but pilot surveys in two districts in 1973-74 indicate an extremely high degree of concentration of cattle and camel ownership.

Data are even less satisfactory for analysing changes in income distribution. However, the limited evidence available points to increasing inequality in incomes and assets, as represented for example by the shares of the bottom 20 to 40 and the top 10 to 20 per cent of rural households. In the Central and Nyanza Provinces of Kenya, comprising 74 per cent of all smallholders in the country, the share in both income and land of the bottom 40 per cent declined between 1963 and 1974 and between 1970 and 1974 respectively.[7] In Ghana the share in cocoa earnings of the bottom 40 per cent of cocoa growers declined between 1963-64 and 1970, while that of the top 20 per cent rose sharply. In Nigeria the evidence from village studies in the food-producing areas suggests that over time the land share of the bottom 40 per cent has been declining, while there has been a corresponding increase in that of the top 10 per cent. There is also firm evidence of increasing concentration of cattle ownership in Botswana. While time series data are not available for Somalia, the pattern of commercialisation of livestock and the accumulation of surplus in the hands of breeders with large holdings suggest a similar trend. The rapid expansion in the number of "progressive" farmers in Malawi and Zambia, and the appropriation of major benefits from expansion of cash crops by farmers with 5 to 20 hectares in the Ivory Coast, would also suggest a rising share in rural incomes received by the top 10 to 20 per cent of smallholders.

Nothing has so far been said regarding the situation in Mozambique. Virtually no data on income distribution in Mozambique are available either for the colonial or the post-independence period. However, from everything that is known about the structure and operation of the economy prior to independence, it would seem highly likely that incomes were distributed extremely unequally. Furthermore, this inequality coincided largely with racial divisions. There can be little doubt that with the mass exodus of settlers and other expatriates, and the state take-over of all major modern-sector enterprises, there has been a sharp improvement in over-all income distribution. Subsequent developments, however, are likely to have adversely affected the incomes of many who were dependent on wages either within the country or in South Africa.

One important conclusion that emerges from the evidence considered above is that there appears to have been a worsening of rural-income distribution in countries experiencing high growth such as Botswana, the Ivory Coast, Kenya, Malawi and Nigeria, as well as in those experiencing low growth or stagnation such as Ghana, Somalia and Zambia. Furthermore, this tendency is observable in countries with a considerable range in incomes per head. The processes generating inequalities are discussed later; nevertheless, this finding suggests the key role played by patterns of growth, in contradistinction to other factors, in explaining differentiation.

Poverty

While few people would dispute the widespread prevalence of poverty in sub-Saharan Africa, especially in the rural areas, there is less agreement on its quantitative dimensions. Certainly, rigorous estimates of poverty incidence in African countries are few and far between. The difficulties caused by data deficiencies are compounded by the lack of a common methodology in measuring the incidence of, and trends in, poverty. At least three different methods have been used to estimate poverty: some estimates of over-all poverty are based on a given income per head and some assumptions about income distribution; a more common practice is to estimate the incidence of poverty on the basis of household income and expenditure surveys through calculation of a poverty datum line; and finally, indicators of some basic-needs satisfaction may be used to obtain estimates of poverty. Thus, the measured incidence of poverty would vary according to the methods used and the assumptions made. In this subsection we review briefly the results of some attempts to measure poverty in African countries.

Given the relatively low incomes per head in most African countries, it is not surprising that estimates using the first method indicate that the incidence of poverty in Africa is considerably higher than in Latin America but comparable to that in Asia.[8] Using 1969 data, and assuming that an annual income of lower than US$75 constituted poverty, a World Bank study concluded that 60 per cent of the population was poor. Using 1972 data and defining poverty as persons with incomes lower than US$117, an ILO study found that 65 per cent of the population on the continent may be considered poverty-stricken. More recently, a World Bank study, using mostly 1975 data, found that many African countries had a poverty incidence of at least 50 per cent.[9] Since the average incomes in rural areas are much lower than in urban areas, it may be assumed that the proportion of the rural population suffering from poverty would be even higher.

Some studies of household consumption in selected countries either tend to confirm the estimates of poverty incidence based on income levels or to suggest that these may, in fact, be underestimates. If we define poverty in terms of stipulated percentage of expenditure on food, the estimates of rural poverty range from 50 to 90 per cent for six countries or regions within countries.[10] Another set of studies defining poverty lines in terms of minimum "baskets" of goods and services have provided estimates of rural poverty which range from 50 per cent in Ghana to 70 per cent in Somalia.[11] The importance of methodology and definitions in estimating poverty is brought out by considerably lower estimates for rural poverty for Botswana (54 per cent), Kenya (36 per cent) and Somalia (30 per cent) reported in the studies included in this volume. In summing up evidence from a number of studies, Elliott concluded that

[it is] difficult to extract much analytical detail from these studies, but they point in a number of interesting directions. First, the most basic form of poverty, shortage of food, is widespread even in relatively well-endowed, "well-developed" countries with higher than average GNP/head. Second, even in richer subregions, with a long history of cash cropping (Eastern Region of Ghana, Nyanza in Kenya) a sizeable proportion of the population is in extreme poverty on our consumption criteria. Third, in general well over half the rural population, and in extreme cases like Tanzania all but a tiny fraction, have incomes that dictate consumption patterns that in any society would be regarded as unacceptable—a high proportion spent on starchy staples and inferior goods, an inability to switch to preferred (and nutritionally desirable) foods and exceedingly low expenditures on even basic household goods.[12]

With regard to some basic-needs indicators, the FAO has provided estimates of the proportion of undernourished people (table 4). These show a considerable range, from 40 per cent in Somalia to 8 per cent in the Ivory Coast. Information on life expectancy, infant mortality and literacy is pro-

Table 4. Percentage of undernourished people, 1969-71 and 1972-74

Country	1969-71	1972-74
Kenya	24	30
Malawi	19	14
Ivory Coast	9	8
Botswana	33	.
Zambia	35	36
Nigeria	.	34
Ghana	22	.
Mozambique	34	20
Somalia	42	36
		40

. = not available.
Source. FAO: *Fourth World Food Survey, 1977*, Appendix M, pp. 127-128.

Table 5. Physical Quality of Life Index

Country	Ranking income per head, 1976	PQLI, 1974-75	Ranking PQLI	Life expectancy at birth (years)			Infant mortality (per 1,000 live births)			Literacy (%)		
				1960	1975	Annual DRR % 1960-75	1960	1975	Annual DRR % 1960-75	1960	1974	Annual DRR % 1960-74
Somalia	9	19	9	35	41	1.0	.	.	.	.	50	.
Malawi	8	30	4	35	41	1.0	.	142	.	.	25	.
Mozambique	7	27	6	36	44	1.4	.	93	.	.	40	.
Kenya	6	39	1	43	50	1.5	.	51	.	.	40	.
Nigeria	5	27	6	34	41	1.2	207	163	1.6	25	.	.
Zambia	4	38	3	39	45	1.1	.	.	.	41	43	0.2
Botswana	3	27	6	.	.	.	.	.	.	.	.	.
Ghana	2	39	1	37	44	1.3	113	63	4.2	.	25	.
Ivory Coast	1	29	5	36	44	1.4	.	.	.	9	20	0.9
Low-income countries	150	33	.	36	44	1.4	142	122	1.1	10	23	1.1
Middle-income countries	750	67	.	49	58	2.6	72	46	3.4	61	63	0.4

. = not available.
DRR = Disparity Reduction Rate, which measures progress in these socio-economic indicators.
Source. James P. Grant: *Disparity reduction rates in social indicators* (Washington, DC, Overseas Development Council, 1978).

vided in table 5. It should be recalled that these data relate to the national level and represent averages, not distribution by sectors or income groups. The data presented indicate a considerable lag in basic-needs satisfaction

Agrarian change, differentiation and rural poverty

between these countries and the average for developing countries as a whole. However, even within the relevant income groupings, many of the African countries, particularly those in the higher income per head brackets, show below average PQL (Physical Quality of Life) indices.[13] There does not appear to be a correlation between the level of income per head and the extent of basic-needs satisfaction. While the relative rankings under the two indices of Mozambique, Nigeria, Somalia and Zambia are reasonably close, Botswana and the Ivory Coast which have ranks of 1 and 3 on the index per head fall to 5 and 6 in the PQL Index. On the other hand, Kenya and Malawi show a distinctly superior performance in terms of PQL Index in relation to their ranking on the income per head basis.

Finally, the issue of trends in poverty must be considered. It is not possible on the basis of the existing data to draw any firm conclusion on the subject. However, the earlier data on growth in agricultural production—especially of food—in conjunction with the fragmentary evidence on trends in income distribution, strongly suggest that at least in those countries where growth is slow or stagnant, such as Ghana, Somalia and Zambia, there must have been an absolute reduction in the consumption levels of significant groups of the rural population. There also appears to be a presumption of little improvement or decline in the living standards of a section of subsistence farmers and other marginal groups, even in countries experiencing more rapid growth. Agricultural workers appear to have experienced a decline in real wages in several countries: in Kenya these declined sharply in the 1970s, while in both the Ivory Coast and Malawi they fell by 35 and 20 per cent over 1960-61 and 1975-76 and 1968-69 and 1976-77 respectively.

The data on basic-needs satisfaction show an improvement in terms of life expectancy, infant mortality and literacy in the case of all countries for which relevant information is available. These are, however, not necessarily in conflict with the evidence presented on income distribution and poverty. Although health and education services are to some extent being provided for a larger proportion of the population, the data do not tell us the extent to which the rural poor are benefiting from these services. Nor is increasing inequality and even decline in material consumption standards in conflict with increasing provision of certain social services.

MECHANISMS GENERATING DIFFERENTIATION AND PERPETUATING POVERTY

The discussion in the preceding section has suggested a number of conclusions: economic growth has been accompanied by an uneven distribution of benefits; it appears to have intensified inequalities, both in low- and in high-growth countries; and significant groups of the rural population have either experienced little or no improvement in living standards or have suffered a decline in income and consumption. These conclusions are in line with the experience of the majority of developing countries in Asia and Latin America.[14] In this section the processes and mechanisms which have contributed to this outcome are analysed. This discussion is important because, in view of the differences in the level of development and economic structures between African and other developing countries, it cannot be automatically assumed that the explanations advanced for the latter are equally applicable to the former. Without overlooking the enormous diversity of individual countries within different regions, one may state that, on average, sub-Saharan African countries are at a lower stage of economic development with a greater preponderance of labour and output in the rural sector, different land tenure systems and a lesser degree of commercialisation of agriculture, with proportionately greater dependence on exports of a rather narrow range of primary products. Furthermore, African countries have been independent for a relatively short period and the structure and policies inherited from the colonial era have thus played a correspondingly greater part in shaping the evolution of their economies.

In pursuing this analysis it is important to bear in mind the distinction between differentiation and poverty. It is possible for a country simultaneously to experience growing inequalities in incomes and consumption, and a reduction in the numbers or proportion of households suffering from absolute poverty. Thus in this section we first look at the forces behind growing differentiation in the countries covered in this volume and then turn to the related analysis of rural poverty.

Processes behind differentiation

Economic differentiation in the pre-colonial period was restricted both in space and in extent. It was confined largely to areas with hierarchical and feudal social structures. The scope of differentiation in most areas was limited by land abundance, the traditional land tenure system, the state of technology and the availability of markets. Differentiation on a significant scale evolved during the colonial period. The main economic objectives of the colonial authorities, particularly in the early years of colonial rule, were to provide profitable opportunities for settlers, plantations and trading companies; to generate revenues to meet the costs of colonial administration; to secure cheap sources of foodstuffs and raw materials for their economies; and to find growing outlets for their manufactured products. These objectives were met in diverse ways in different parts of sub-Saharan Africa. In many countries in eastern, central and southern Africa, the main emphasis was on settlement by European farmers, the establishment of plantations and the exploitation of mineral resources. In western Africa stress was laid on mineral exploitation, plantations and, increasingly, the production by peasant farmers of export crops. The pattern of development in the first group of countries required alienation of land, and labour for European farms, plantations and mines. In the second group there was less need for land and labour in expatriate enterprises, policies being principally directed at encouraging peasant cash crop production, and the collection and marketing of these crops. In Kenya, Malawi, Mozambique and Zambia, these policies resulted, though in varying degrees, in the loss or reduction of land available to peasants in some areas, a strong discouragement of peasant cash crop cultivation and the use of a variety of more or less coercive methods to secure wage labour. This was accompanied by the concentration of public expenditure on social and economic infrastructure in areas—generally the most fertile parts of the country—where farms, plantations and mines were located.

In Ghana, the Ivory Coast and Nigeria ecological factors dictated the choice of the fertile southern forest zones for peasant cash crop production. There was likewise concentration of the necessary infrastructure in these areas. Originally production was organised on the basis of family labour; subsequently, however, further expansion was fuelled by migrants from the north and the neighbouring countries. The surplus from cash crop production was appropriated both by trading companies and, later, by the colonial authorities through taxes and deductions.

The alienation of land for settlement on the most fertile areas, the creation of a class of workers more or less dependent on wages, the concentration of public expenditure on social and economic infrastructure in certain favoured areas, and the promotion of export crops laid the foundation for differentiation and, in some areas, for impoverishment. These features of colonial policies were modified in various ways in the succeeding decades. The post-colonial period witnessed a rise in rural production and an increased emphasis on the commercialisation of agriculture. The latter in turn led to significant changes in the patterns of use and ownership of land, as in the utilisation of labour. In most countries these developments, described briefly below, were accompanied by further accentuation of differentiation.

The process of commercialisation of agriculture had already reached an advanced stage in Ghana and Nigeria in the 1950s, encompassing the great majority of farmers, particularly in the southern zones. In the 1960s this trend increased: whereas previously the overwhelming proportion of cash production had been for export markets, the growth of internal markets for food and for processed goods as well as changes in the relative profitability of food and export crop production, stimulated the commercialisation of food production.

Such changes proceeded at a faster rate in other countries where African commercial production had been held in check by the colonial authorities through a variety of policies. Growth in commercial crop production by African farmers was particularly noticeable in the Ivory Coast, Kenya, Malawi and Zambia. Although a high and growing proportion of farmers established some contact with the market, a characteristic of the pattern of rural development in all these countries was the concentration of the bulk of the non-estate commercial production in the hands of a relatively small proportion of farmers variously called "emergent", "commercial", "progressive" and "capitalist" farmers. In the southern region of the Ivory Coast, for instance, in 1973-74 only 15.6 per cent of all farms grew neither cocoa nor coffee, but a high share of cash crop production was accounted for by between 20 and 30 per cent of farmers with farms in excess of 5 or 7 hectares. In Zambia the penetration of the cash economy is less extensive: the agricultural census in 1970 indicated that 54 per cent of rural households sold no produce, although evidently the proportion is much lower in the favoured areas. In Malawi the number of farmers growing cash crops has risen rapidly, but the bulk of the cash crop production is controlled by a relatively small proportion of smallholders—probably around 10 per cent. As early as 1968-69 the distribution of cash income from all sources among smallholders was highly skewed, with a Gini coefficient of 0.58. For a number of reasons, including the big reduction in migrant workers and the concentration on selected farmers and areas under the four development projects, the extent of differentiation must almost certainly have further intensified in the subsequent period.

In Mozambique the great bulk of commercial production until independence continued to come from estates and expatriate farmers. This was in large part a reflection of the colonial policies which discouraged cash crop cultivation by Africans in order to ensure a regular supply of cheap labour for plantations, large farms and mines in Mozambique and South Africa. Nevertheless, the need for some crops which were not profitable for companies and large farmers led the Government to impose production quotas at fixed prices on peasants. This was the case, for instance, with cotton, which at the peak was grown by 700,000 registered cultivators.

In the livestock-dominated rural economies of Botswana and Somalia there was a major increase over the period in the sale of livestock products. However, even more than in other countries, a high share of such commercial production was accounted for by a tiny proportion of livestock owners.

The commercialisation of agriculture was thus associated in all these countries with the emergence of a new class of relatively affluent farmers and livestock owners. This process in turn was facilitated by changes in the traditional land tenure systems. Although there were many differences of detail, in essence the traditional land tenure system was similar in most parts of Africa. Its key features are that, while the individual or the family has a right to use of land, the latter belongs to the tribe, the clan or the community. It is allocated to members of the clan by the chief, the village headman or the council of elders. In addition to this process, land may also be acquired through inheritance. In general, the outsiders have to satisfy a number of conditions before acquiring the right to use of land on the same terms as the members of the clan or the community. The agricultural system was based on shifting cultivation, and production, based on a sexual division of family labour, was geared to subsistence needs. There was little accumulation or sustained increase in output per head. On the other hand, the system ensured access to land for all households, prevented emergence of inequalities in land and living standards, and maintained soil fertility.

With the onset of colonialism, the traditional land tenure system came

Agrarian change, differentiation and rural poverty

under pressure from two sources. First, as noted earlier, large tracts of land (especially in the more fertile areas) were alienated for use by European settlers and plantations. This resulted in some landlessness and the crowding of peasants in a restricted area. This in turn often led to partial abandonment of the shifting cultivation system, resulting in overgrazing, soil erosion and a decline in soil fertility. Among the countries studied here, land alienation on a significant scale occurred in Kenya, Malawi and Mozambique, and on a smaller scale in Botswana, the Ivory Coast and Zambia.

The second source of pressure on the traditional land tenure system came from the commercialisation of agriculture. The moment land was planted with cash crops, especially tree crops, it became vastly more valuable. Gradually the ownership of land became more individualised, and in areas of greatest commercial penetration, for instance in parts of Ghana and Nigeria, there developed a significant land market as well as the beginnings of a system of tenancy and share-cropping. Nevertheless the traditional system remained predominant in most countries, though with population growth and increasing scarcity of good-quality land, inheritance rather than direct allocation by clan authorities became the main method of acquiring land. However, although there was an increasing tendency towards *de facto* land ownership on a personal or family basis and a growing market for land transactions, it was only in Kenya that a systematic effort was made—through an ambitious policy of land settlement, consolidation and registration—to develop a land tenure system based on individual ownership. Land registration was initiated in 1954, and by the end of 1978 a total of 7.6 million hectares, or over half of the total registrable area, had been registered. In other countries the evolution in land tenure continued to be accommodated within the framework of the traditional system. Similarly in the pastoral economies of Botswana and Somalia, with the exception of some modern ranches, the grazing rights to the pasture land continued to be held on a communal basis.

While in most African countries the traditional land tenure system is evolving in the direction of private ownership of land, *de jure* or *de facto*, Mozambique illustrates an attempt, shared in varying degrees by a few other African countries such as Angola, Ethiopia and Tanzania, to build a system of agricultural production and land use along socialist lines. The ownership of land is vested in the State and agricultural production is organised under three different forms: many of the previous plantations and large farms have been converted into state farms; peasant agriculture is being organised within the framework of communal villages under which cultivation is undertaken both on a collective basis and on family plots; and production co-operatives are being developed throughout the country as a basis for socialist agriculture.

Ecological factors have sometimes interacted with government policies to produce distinct patterns of regional inequalities. In Kenya it has been stated that inequality has been built into the country's rural economy by the forces of nature. Only 17 per cent of the country's total land area is regarded as medium to good agricultural land, with more than four-fifths of it classified as having only low potential or as unsuitable for agricultural use. The average incomes in the more favoured areas are significantly higher than in low potential zones. In Botswana the rainfall is such that only about 6 per cent of the total area of the country is cultivable with dryland crops. The remote area dwellers, constituting about 50,000 of the total population of 800,000 and inhabiting ecologically poor zones, are considered the most disadvantaged group in Botswana. Regional inequalities arising in part from ecological factors are also a marked feature in Mozambique. In the Ivory Coast the ratio of average incomes between south and north has been estimated at seven to one. Though no firm figures are available for Ghana and Nigeria, the significant differences in average living standards between the rainy, forest zones in the south and the savannah areas in the north are well recognised.

Often the inequalities associated with ecological factors have been initiated or reinforced by the pattern of public expenditure, during both the colonial and the post-independence periods. An analysis of the public expenditure shows a bias in most countries in favour of the larger-scale and more prosperous farmers and relatively advantageous areas. In Kenya, for instance, a significant proportion of public expenditure on agriculture in the 1960s was devoted to land purchase, settlement and land registration; the great bulk of this expenditure benefited the relatively small number of the newly settled farmers concentrated in parts of the Central, Rift Valley and Eastern Provinces. The expenditure on research, extension and credit continues to show a pronounced bias in favour of cash crops and of relatively prosperous farmers and areas. In Malawi a high proportion of public expenditure on agriculture has been directed at progressive farmers through four large development projects. In Zambia, likewise, the extension services, research and credit facilities have largely served the needs of commercial farmers.

In Botswana considerable sums were spent on the drilling of public boreholes, and generous subsidies were also provided for the construction of private ones. The benefits from this investment accrued principally to large-scale herd owners.[15] Indeed, this pattern of public expenditure reinforced the trend towards concentration in livestock ownership. In contrast, relatively small sums have been spent on projects designed to improve smallholder arable farming where some of the poorest sections of the rural society are to be found. In Somalia, too, the bulk of the expenditure on crop farming and livestock has been concentrated on the development of ranches and costly irrigation and settlement schemes. In Nigeria a good deal of the expenditure on agriculture in the 1970s was devoted to state plantations, irrigation projects for rice and wheat and extension services promoting high-yield seeds, chemicals and fertilisers. The first two items did nothing to improve smallholders' productivity and income, while the third item benefited primarily the progressive farmers. A similar pattern with a bias in favour of large-scale farmers is discernible in Ghana.

Thus it is apparent that the State, through its pattern of expenditure on agriculture, has created or reinforced rural differentiation. This has been done through settlement schemes; through "land reform" resulting in greater inequalities in land distribution; through concentration of extension services, credit facilities and provision of seed to the more affluent commercial farmers; and finally, through the provision of subsidies on fertilisers, tools and machinery whose main beneficiaries have been the progressive farmers.

Persistence of rural poverty

In the preceding subsection the main processes contributing to a growing differentiation in the countryside in African countries have been outlined. While the differentiating mechanisms have regional and country specificities, the uneven pattern of growth is a normal feature of a private-enterprise economy, especially in the earlier stage of development. We now turn to an examination of the factors behind the persistence and in some cases intensification of poverty affecting large sections of the rural population. These factors are complex and interlocking, and deserve much fuller discussion than is possible here. We focus here on four issues: the growth of rural production; resource transfers to and from rural areas; the role of ecological factors; and migration.

Earlier it was noted that over the period 1960-78 four countries—Ghana, Mozambique, Somalia and Zambia—experienced either negative or low growth in income per head. If we consider the agricultural sector, Nigeria may be added to the list. Thus, unless there was a significant net inflow of resources into the rural areas in these countries, or an appreciable improvement in income distribution in favour of the low-income groups, there must have been no improvement and, in countries experiencing negative growth, a deterioration in the living standards of the majority of rural inhabitants. The earlier discussion on income distribution suggested that there may have been a worsening of income distribution in Ghana, Nigeria, Somalia and Zambia. Subsequently, it will be argued that in most countries there was also a net transfer of resources from the rural areas. Hence, in those countries slow or negative growth, in the economy as a whole or in agriculture, contributed to the maintenance or aggravation of rural poverty. However, even in rapidly growing economies such as Botswana, the Ivory Coast, Kenya and Malawi, the benefits of growth were concentrated among a relatively small proportion of the rural population and significant sections either experienced no improvement in living standards or suffered declining incomes. Thus, in the absence of active policies to combat poverty, growth *per se* cannot be expected to bring about widespread improvement in living standards. This is a widely accepted conclusion which receives tentative corroboration from the studies included in this volume.

Resource flows to and from rural areas can have an important impact on the level, growth and distribution for rural incomes and hence on poverty. A complete discussion of this phenomenon is complex and involves the analysis of exchange rates, trade restrictions, taxation, subsidies and public expenditure. There are few African countries for which such analysis has been carried out. Two aspects will be discussed here which throw some light on this subject—the terms of trade faced by farmers, and public expenditure. Both of these affect rural-urban as well as intra-rural income distribution; interest in this subsection will be focused on their impact on resource flows between urban and rural areas.

The limited evidence available on the farmers' internal terms of trade indicates that in practically all countries changes in relative prices have been an important instrument in extracting surplus from the rural areas, though the precise mechanisms used and the degree of surplus expropriation vary from country to country. In Ghana export taxes, so-called stabilisation funds, an overvalued currency, and inefficient industrialisation combined to produce a massive squeeze on the real incomes of farmers. Rocketing food prices led to an improvement in the terms of trade for surplus food producers but intensified the decline in real income for cocoa farmers and deficit food farmers. The terms of trade index for cocoa growers declined steadily from 1956, when it stood at 170, to 123 in 1960, 57 in 1970 and 47 in 1976. For food farmers, however, the terms of trade index improved from 100 in 1963 to 155 in 1976. As the great bulk of the marketed production consists of cocoa, the magnitude of the income transfers from cocoa growers is quite astonishing. Part of the explanation for these trends lies in the modifications in export prices on the world market, but a good deal has to do with domestic policies such as excessive overvaluation of the currency and the declining ratio of producer price to export price (which fell from 49 per cent in 1960 to 39 per cent in 1970 and 22 per cent in 1976).

In Nigeria, until the emergence of petroleum exports and revenues, primary product exports were subject to heavy taxes and deductions. These were phased out from 1969 onwards. However, a sharp rise in the real exchange rate (due to rapid domestic inflation and the rising value of the Naira) effectively reduced the real income of export growers by one-third between 1968 and 1979. As in Ghana, the terms of trade moved in favour of food growers, the index rising from 100 in 1968 to 272 in 1977.

Zambia is not a significant exporter of agricultural products: most of the marketed agricultural production is directed to the home market. The major consideration in fixing prices has been to keep the cost of food low for urban consumers; thus, for the majority of products for most years, the domestic prices have been below the import parity prices. The terms of trade faced by agricultural producers, comprising both food and industrial crop growers, declined by 20 per cent between 1964 and 1973, and by about 25 per cent again between 1973 and 1978.[16] Thus the food growers' terms of trade followed a radically different pattern from that in Ghana and Nigeria.

In contrast with the experience in these three countries, the terms of trade for farmers appear to have been relatively more favourable in the Ivory Coast, Kenya and Malawi. In Malawi the terms of trade for smallholders improved to some extent between 1968 and 1975, declining thereafter. In

Kenya in most years since 1964 there was a modest decline, though there were some years—such as 1970, 1972 and particularly 1975-77 (associated with coffee and tea boom prices)—when an improvement was registered. In the Ivory Coast the terms of trade of coffee and cocoa growers appear to have declined somewhat in most years. While the relatively favourable or, at any rate, less unfavourable terms of trade for farmers may provide a partial explanation for the generally better agricultural performance of these countries, it nevertheless remains true that, even in these countries, there were significant net resource transfers from agriculture to other sectors of the economy. In the Ivory Coast, for instance, the ratio of producer to export prices for coffee and cocoa averaged 54 per cent between 1960 and 1975, the ratio declining since the early 1960s. Between 1965-66 and 1974-75, export and stabilisation funds accounted for 38 and 31 per cent of total export proceeds for cocoa and coffee respectively. In Kenya it has been estimated that between 1964 and 1977 the surplus generated from the agricultural sector was equivalent to 75 per cent of the country's entire capital formation. In Malawi the marketing board generated large profits from the sale of crops grown by smallholders: in 1971-72, which was not an atypical year, the profits amounted to 63 per cent of the purchase value of these crops.

In Somalia the two main rural groups faced widely divergent terms of trade in the 1970s. Whereas the real producer price for maize declined by about 20 per cent, the price index for livestock rose from 100 to 222 between 1970 and 1978. In livestock export sales the share of producers rose from 52 to 62 per cent over the period.

A second factor affecting resource flows is the allocation of public expenditure between rural and urban areas. While a breakdown of all public expenditure between rural and urban areas is not available for any of the countries under consideration, the expenditure on agriculture (including forestry and fishing) may provide a partial indication of the extent to which the rural areas have benefited from current public and capital expenditure. An examination of the expenditure on agriculture since the early 1960s shows that in most countries the share of agriculture in total expenditure is less than 10 per cent, and often varies between 5 and 8 per cent. Furthermore, only in Malawi and Somalia is there evidence of a significant increase in the share of agriculture between the 1960s and the 1970s. A similar trend is noticeable in Ghana in the late 1970s. In Botswana and Kenya, on the other hand, although the expenditure on agriculture has been growing, its share in total expenditure has fallen since the 1960s.[17] On the basis of the available evidence, it would be reasonable to conclude that public expenditure shows a pronounced bias in favour of activities located in the urban areas. This conclusion is in line with the situation observed in most developing countries.[18]

Earlier, we discussed the role of the ecological configuration of a country in generating inequalities. By the same token, ecological factors, in combination with government policies, have been influential in perpetuating or intensifying poverty. In most of the countries under review the low agricultural potential of certain regions and areas has been reinforced by their relative neglect in terms of the development of social and economic infrastructure. The bias in government policy in favour of export products and against food crops, a characteristic shared by most countries for the period under consideration, accentuated the disadvantageous position of such regions which were typically unsuited for the cultivation of cash crops. Furthermore, in some cases—as, for example, in parts of Kenya, where demographic pressure has resulted in greater population density in semi-arid regions, or where there has been disproportionate increase in livestock, as in Botswana and Somalia—the ecological balance has been disturbed, with consequent deterioration in the quality of the soil through the disappearance of trees, shrubs, etc. As discussed bellow, one of the responses of the people in these areas has been to seek higher incomes through migration, either abroad of to other parts of the country. While this has mitigated population pressure and, in some cases, has resulted in increased household incomes through remittances, in other cases the migration of the young and the educated has exacerbated rural poverty through a reduction in agricultural productivity and a failure to produce adequate food for lack of manpower in the household.

Poverty is not confined to ecologically disadvantaged areas alone. In the more favourably endowed areas of most of the countries under review, the process of land differentiation, as indicated earlier, has advanced to a considerable extent. Thus in an increasing number of countries a significant and growing proportion of the rural households find themselves in a situation where adequate land is not available to produce sufficient food for the family and/or to generate the needed cash through cultivation of cash crops. Typically in situations such as these, some members of the household would seek to augment family income through migration or part-time and seasonal employment in the same area. In most of the African countries landlessness has not yet become a serious problem; however, in some, such as Kenya and Malawi, it is emerging as a significant factor in explaining rural poverty.

When confronted with scarcity of cash to pay school fees or meet other essential needs or with pressures arising from diminishing land fertility or availability, or simply when lured by the attractions of higher incomes, the response for decades has been to migrate in search of employment. These migratory movements have taken place either within the country as, for example, from the aride northern zones to the fertile southern regions in Ghana and the Ivory Coast; or from one country to another as, for instance, from Botswana, Malawi and Mozambique to South Africa and what was then Southern Rhodesia; from the Sahelian countries to Ghana and the Ivory Coast; or more recently from Somalia to the Gulf States. In addition, all the countries experienced large migration flows from the countryside to the main urban areas.

Before we analyse their implications for poverty, it may be useful to summarise the importance of migration flows in a number of countries. In

Agrarian change, differentiation and rural poverty

Ghana, as early as 1960, agricultural employees numbering 520,000 constituted 33 per cent of the labour force engaged in agriculture. A high proportion of these employees were migrants from the northern regions or from neighbouring countries. In the Ivory Coast, by 1974-75, there were over 180,000 permanent agricultural workers in a country with nearly 550,000 holdings. The former constituted around 10.5 per cent of the total labour force employed in agriculture. Data on temporary employees are not available but, if the pattern observed in Ghana holds, the total number of agricultural workers may have been in the region of 400,000. Reflecting the vast immigration of foreign unskilled workers in the postwar period, their share in the total population was estimated in 1975 at 30 per cent. It is thus likely that the great bulk of the agricultural workers were of foreign origin, most of the remainder being migrants from the northern region.

In Mozambique the existence during the colonial period of a significant large farm sector led to the creation of substantial wage employment in agriculture: in 1970, this was estimated at nearly 455,000, or over 20 per cent of total employment in agriculture, and compared with 1.6 million traditional holdings. Forced labour and other coercive methods were used freely to evoke this labour supply. In addition to wage employment in the domestic agricultural sector, large numbers from rural areas migrated to work on mines and plantations in South Africa and the former Southern Rhodesia. In 1966 nearly 110,000 workers were recruited from Mozambique to work exclusively in South African gold-mines. Allowing for unrecorded workers and those employed on the plantations and farms in South Africa and the former Southern Rhodesia, the total number of migrant workers in 1970 may well have been in the neighbourhood of 200,000-250,000. After independence, with the departure of expatriate farmers, there was a considerable decline in agricultural employment. A similar trend was observed in recruitment in South Africa, with the number recruited in gold-mines declining to 44,000 in 1976.

In Malawi for much of this period the dependence on foreign countries as a source of employment was even greater. In 1966, for instance, there were at least 200,000 male workers abroad, as compared with a total domestic wage employment of around 130,000. By 1974 these numbers had grown to 250,000 and 265,000 respectively, as compared with approximately 1 million holdings. The situation changed dramatically with the suspension of recruitment of labour to South Africa late in 1974. When recruitment was resumed in 1977 it was up to a ceiling of 20,000 workers annually. Apart from the extreme dependence on foreign countries for wage employment and the dramatic turnabout in 1974, the experience in Malawi is notable for the extraordinarily rapid expansion of wage employment on the estates, which had reached a figure of 130,000 in 1977. There was in addition a considerable expansion of casual and seasonal employment on smallholdings. Though the exact figures are not available, the 1968-69 agriculture survey indicated that nearly 29 per cent of smallholders used hired labour on their farms.

Despite rapid growth of the economy in Botswana, there continued to be dependence on South Africa for jobs. The numbers of migrant workers rose from 50,000 in the mid-1960s to 70,000 in the mid-1970s; this compares with a total of about 92,000 rural households in 1974-75. Nearly one-third of Botswana's male labour force typically work in South Africa. While in Botswana migration dates back several decades, it is a more recent phenomenon in Somalia. However, it has attained a sizeable dimension, recent estimates putting the number of migrants to the Gulf States at between 100,000 and 150,000 (compared with a total wage employment in the country of 120,000, and 620,000 rural households).

In addition to the rural/rural and international migration discussed above, the third pattern of migration was from rural to urban areas, especially the capital cities. All countries, whether land-abundant or not, experienced this flow which became especially important after independence. One characteristic of this migration was that a large proportion of the migrants were young and had received a formal education.

The various types of migratory movements described above touched a vast majority of rural households in these countries and involved them in the complex web of the market economy. There can be little doubt that over the years migration has enabled millions of rural dwellers to improve their incomes and acquire new skills. In recent years migrants to South Africa and to the Gulf States have been able to earn incomes which are several times higher than the wages available to them domestically. Likewise, employment in the estate sector, despite the relatively low level of wages in most countries, has enabled rural households to enhance their total incomes. The remittances sent by migrants often make a significant contribution to the total family incomes. Surveys carried out in Botswana and Malawi indicate that such remittances are proportionately more important for low-income than for other categories of rural households.

While migration often represents a means of acquiring cash or higher incomes for the individual and/or the family, it does not always enable the migrants and their families to escape poverty. In many countries they continue to constitute the poverty groups. For instance, plantation workers in several countries have extremely low, and often declining, real wages. The position of those who work as full-time or casual employees on medium- or small-scale farms may be even worse. A number of factors are responsible for this situation: perhaps the most important is the non-existence or weakness of the organisations of rural workers in many countries; in some countries such as Malawi the wages and incomes policy pursued by the Government has been an important factor in keeping the wages of plantation workers low; in the Ivory Coast, the unlimited availability of labour from the neighbouring countries has effectively set a ceiling to the wages of farm employees.

Perhaps the more important effects of migration are felt in the areas from which the migrants originate. It cannot always be assumed, as the Zambian example shows, that remittances will be sent back to the areas of origin.

Given the concentration of migrants among the educated and the young males, the unbalanced household composition in the emigrating areas can greatly aggravate poverty. Paradoxically, labour shortage may emerge as the major constraint to increased production and food self-sufficiency. In several countries such as Botswana, Somalia and Zambia, one of the characteristics of the poorer households is their unfavourable labour position and high rates of dependency. A particular and pervasive instance of this is to be found in the households headed by women. Botswana illustrates this situation well, with 40 per cent of rural households temporarily or permanently without adult men. Three-quarters of such households own no cattle, as compared with one-quarter for male-headed households.[19] The average incomes of the former are less than half those of the latter. Furthermore, migration has imposed heavy physical and psychological burdens on the women, old men and children left behind. The absence of male members of the household has in some cases contributed to a decline in soil fertility through more prolonged cultivation of the cleared plots. It may be surmised that these factors have played a role in the agricultural crisis—particularly with regard to foodstuffs—now confronting a large number of African countries.

CONCLUSION

The preceding analysis of growth, differentiation and poverty in rural areas suggests a number of policy conclusions. If one of the main objectives of development planning is to achieve rapid and broad-based growth in rural areas, it is essential to reorient policies in a number of critical areas, some of which are touched upon here. In the first place, as noted earlier, the land tenure systems in most African countries are in a state of transition from the traditional communal systems to various types of modern systems based on individual ownership, or sometimes state or collective ownership. With population continuing to grow at a high rate, there is a real danger that the situation in African countries might evolve into something resembling that in most Latin American and Asian countries with high rates of landlessness, a significant incidence of tenancies, share-cropping, peasant indebtedness and other related institutions. Such an outcome could lead to a durable and socially disruptive polarisation of rural wealth and income with negative effects on growth potential. Many African countries are still in a position where it is probably not too late to evolve policies on land use and ownership designed to promote simultaneously growth and widespread improvement in living standards, on the one hand, and, on the other effective participation of the peasantry in the process of development.

Second, slow growth or stagnation in agriculture is one of the major reasons for the poor over-all performance of the economy and the persistence and accentuation of rural poverty in most African countries. Thus, both in order to speed up growth and to attack rural poverty, it would be essential for government policies to be reoriented so as to give incentives to agriculture and other rural activities and to devote a growing proportion of government expenditure to the improvement of the rural economic and social infrastructure, as well as to research on technologies designed to improve yields and productivity. This proposal has a number of implications: the overvaluation of exchange rates, the fostering of inefficient industrialisation through poor protective measures, high taxation of export crops directly or through marketing boards, inefficient and costly marketing and transportation systems—all these have served in the past to reduce the level and growth of rural incomes. A reversal of these policies (of which there is evidence in some countries) can go a long way towards unleashing productive forces in the countryside. But a sustained improvement in production, yields and productivity is dependent upon a steady stream of biological, chemical and mechanical innovations to raise the agricultural production system to progressively higher levels of efficiency.

An issue of particular complexity and urgent importance is the striking of an appropriate balance between food crops for export and for domestic consumption. Neglect of the latter over a number of years has resulted in the paradoxical situation that a growing number of African countries with an overwhelming proportion of their population in rural areas are increasing their dependence on food imports. It is important, both for rapid growth and for poverty relief, that the imbalances in resource allocation and other policy biases against the food sector be eliminated rapidly.

The removal of disincentives to agriculture and the allocation of a larger share of public expenditure to rural infrastructure and productive activities are unlikely, of themselves, to result in the swift reduction of rural poverty. It was noted earlier that, even in countries where rural production increased rapidly, important groups of the rural population failed to derive any significant benefits. Thus, the above measures need to be adopted in tandem with efficient and equitable systems of land use and ownership and, most important, of mechanisms for credit, seeds, fertilisers and extension services to reach the broad mass of the peasantry and not simply the progressive farmers. This may be possible only through group action, and thus the organisation of peasants in associations, groups, co-operatives and in other related ways would become the key element in a broad-based rural development strategy.

Notes

[1] Unless otherwise stated, the data used in this chapter are drawn from the nine country chapters included in this volume. The term "differentiation" is used to refer to economic (mainly income and wealth) inequalities.

[2] In interpreting this table and other tables, one should remember that growth rates are often based on observations in the initial and the terminal year. This introduces some well-known biases in the growth rates calculated for various countries.

Agrarian change, differentiation and rural poverty

³ The data on annual terms of trade are contained in UNCTAD: *Handbook of international trade and development statistics* (New York, United Nations), various issues.

⁴ D. C. Davies: "Human development in sub-Saharan Africa", in W. Bussink, D. Davies, R. Grawe, B. Kawalsky and G. P. Pfefferman: *Poverty and the development of human resources: Regional perspectives,* World Bank Staff Working Paper No. 406 (Washington, DC, World Bank, 1980), pp. 55-95.

⁵ This is suggested by a few crude estimates of the Gini coefficient of over-all income distribution in some countries for different years, e.g. 0.63 for Kenya, 0.59 for Malawi, 0.57 for Botswana and 0.53 for the Ivory Coast; see S. Jain: *Size distribution of income: A compilation of data* (Washington, DC, World Bank, 1975).

⁶ E. Crawford and E. Thorbecke: *Employment, income distribution, poverty alleviation and basic needs in Kenya,* Report of an ILO consulting mission (Ithaca, NY, Cornell University, 1978; mimeographed).

⁷ P. Collier and D. Lal: *Poverty and growth in Kenya,* Studies in Employment and Rural Development, No. 55 (Washington, DC, World Bank, 1979).

⁸ D. Ghai, E. Lee and S. Radwan: *Rural poverty in the Third World: Trends, causes and policy reorientations* (Geneva, ILO, 1979; mimeographed World Employment Programme research working paper; restricted).

⁹ Davies, op. cit.

¹⁰ Ghai et al., op. cit.

¹¹ Reported in A. Bequele and R. van der Hoeven: "Poverty and inequality in sub-Saharan Africa", in *International Labour Review* (Geneva, ILO), May-June 1980, pp. 381-392. See also ILO/Jobs and Skills Programme for Africa (JASPA): *Economic transformation in a socialist framework: An employment and basic needs oriented development strategy for Somalia* (Addis Ababa, 1977); idem: *Options for a dependent economy: Development, employment and equity problems in Lesotho* (Addis Ababa, 1979); ILO: *Employment, incomes and equality: A strategy for increasing productive employment in Kenya* (Geneva, 1972); idem: *Growth, employment and equity: A comprehensive strategy for the Sudan* (Geneva, 1976); D. Ghai, M. Godfrey and F. Lisk: *Planning for basic needs in Kenya: Performance, policies and prospects* (Geneva, ILO, 1979).

¹² C. Elliott: *Rural poverty in Africa,* Occasional Paper No. 12 (Swansea, Centre for Development Studies, University College of Swansea, 1980).

¹³ This is a composite measure based on life expectancy, infant mortality and literacy. There are some problems with this measure, the components of which tend to be colinear.

¹⁴ ILO: *Poverty and landlessness in rural Asia* (Geneva, 1977).

¹⁵ I. Livingstone and R. K. Srivastava: *Poverty in the midst of plenty: Problems of creating incomes and employment in Botswana,* Development Studies, Occasional Paper No. 4 (Norwich, University of East Anglia, 1980).

¹⁶ ILO/Jobs and Skills Programme for Africa (JASPA): *Basic needs in an economy under pressure,* Report to the Government of Zambia by a JASPA basic-needs mission, Sep. 1980 (Addis Ababa, 1980; restricted).

¹⁷ Data from United Nations: *Statistical Yearbook*; International Monetary Fund: *International Financial Statistics Yearbook*; and country statistical abstracts, selected years.

¹⁸ M. Lipton: *Why poor people stay poor: A study of urban bias in world development* (London, Maurice Temple Smith, 1977).

¹⁹ Livingstone and Srivastava, op. cit.

SOCIAL JUSTICE AND DEVELOPMENT POLICY IN KENYA'S RURAL ECONOMY

2

William J. House and Tony Killick [1]

The purpose of this chapter is to explore the problems of poverty and inequality in Kenya's rural economy, and the influence on them of past development strategies. The following sections survey the extent, nature and causes of poverty and inequality, and examine the impact of past policies. By way of introduction, however, we commence with a sketch of the nature of the country's rural sector and its contribution to the over-all national economy.

AN INTRODUCTION TO THE RURAL ECONOMY [2]

Nature of the rural economy

Inequality has been built into the country's rural economy by the forces of nature. It encompasses a remarkable variety of topographical and climatic conditions, of soil types and, therefore, of vegetation. In the highlands tea and coffee are the main crops; at lower altitudes maize, wheat and livestock are the major products.

As can be seen from table 6, only a small part of the country's total land surface is regarded as good agricultural land, with more than four-fifths classified as having only low potential or as unsuitable for agricultural use. Only 17 per cent of the area is regarded, in the light of present knowledge, as being of high or medium potential. Moreover, most of the good land occurs in the south-western part of the country, with virtually all the northern and most of the eastern parts being little cultivated and maintaining only sparse numbers of livestock.[3]

The geographical spread of the country's population, 85 per cent of whom still live in rural areas, naturally reflects the influence of these varying agricultural conditions. While the over-all density of population is low, there is a

Table 6. A classification of land area by agricultural potential

Classification	Area ('000 hectares)	% of total
High potential	6 785	11.9
Medium potential	3 157	5.5
Low potential	42 105	74.0
Unsuitable for agriculture	4 867	8.6
Total	56 914	100.0

Source. Kenya: *Statistical Abstract 1978*, table 81.

strong positive correlation between population densities and the quality of land.

It is not too much of a simplification to classify Kenya's agricultural products as either domestic foodstuffs or exports. With the exception of meat and occasionally maize, few products are both consumed locally and exported on a major scale. Coffee and tea are the main export crops; maize, livestock and dairy products are the most important of the foodstuffs. This composition of agricultural output has changed little during the past 20 years or more, although cane sugar was first grown commercially in the early 1960s and is becoming an important crop; poultry farming has been another growth industry.

More significant changes have been occurring in the types of production units on which several of the crops are grown. During much of the colonial period African farmers were prohibited from growing coffee and tea, but in recent years these and other cash crops have been widely adopted. This increased contribution of smallholder cultivation is reflected in the following estimates of the share of smallholder cultivation in the total *marketed* agricultural output (averages for the periods shown):[4]

1954-59	29.2 per cent	1970-74	51.6 per cent
1960-64	37.7 per cent	1975-78	53.3 per cent
1965-69	48.3 per cent		

The above figures understate the full contribution of smallholders because *(a)* they exclude subsistence production, and *(b)*, as is shown later, some of the areas still officially classified as "large farm" have, in fact, been subdivided and are cultivated on a smallholder basis. Genuinely large farms are still very important, however, especially in the cultivation of coffee, tea, sisal and wheat, and in ranching.

The rising importance of small farms in the total marketed output is partly due to the increasing monetisation that is occurring as the economy develops. Apart from a limited number of nomadic pastoralists in the remoter regions of the country, there are a few subsistence farmers in the sense of farmers who do not normally produce any marketable surplus. If only to a limited extent, the vast majority of crop farmers are integrated into the monetary system, and it now seems likely that the value of monetised agriculture exceeds that of subsistence production.

Although agriculture obviously dominates the rural economy, it would be a mistake to treat the two as synonymous. Marketing, distribution, education, government services, food processing, furniture making, tailoring and other trades also contribute significantly. Nearly a quarter of total rural employment is accounted for by non-farm activities, and probably a similar proportion of total rural incomes.[5]

Agriculture in the national economy

Table 7 sets out some key indicators about the place of agriculture in the national economy. It can be seen that total agricultural value added (monetary and subsistence) made up one-third of total GDP in 1976, with monetary agriculture contributing about a fifth of total monetary GDP. Agriculture is, by a large margin, the largest single sector of production. The contribution of agriculture and other rural work to total employment is even greater. Non-rural employment makes up only 17 per cent of the national total; the share of agriculture alone is almost two-thirds. Although labour productivity is low, it has at least grown over the years. One of the features of Kenya's successful growth record since independence was that agriculture contributed rather fully to this expansion. As is normal in the process of development, the share of subsistence production has been falling. Monetary agriculture, however, has grown in real terms nearly as quickly as total monetary GDP, especially in the 1970s. For the whole period from 1966 to 1977 monetary agriculture grew at over 4.8 per cent per annum in real terms, against an average for the whole monetary GDP of 6.1 per cent per annum. This record is a good deal better than that which has been achieved in many developing countries.

Another feature brought out by table 7 is that the savings rate is apparently high in agriculture. As item 5 of the table shows, Sharpley estimates for 1976 that the net outflow from agriculture was equivalent to almost nine-tenths of total capital formation in the entire economy. Although this surplus was only a *potential* source of investment, it indicates that agriculture has been a potent source of surplus that could be transferred to finance development in the rest of the economy.

Agriculture has also contributed significantly to the country's success during the first 15 years of independence in achieving rapid economic growth without running into major balance-of-payments crises. Especially as a source of export earnings, but also as a means of import substitution, the sector has been a large net contributor of foreign exchange. This is illustrated for 1976 (admittedly a somewhat favourable year) in item 6 of table 7, which shows net earnings/savings attributable to agriculture as equivalent to 57 per cent of the country's total import bill.

Table 7. Contribution of rural activities to the national economy, 1976

Item	Percentages
1. Total agriculture as % of total GDP at factor cost, in current prices	33
2. Monetary agriculture as % of monetary GDP	22
3. Rural employment as % of total employment	83
4. Agricultural employment as % of total employment	64
5. Agricultural surplus as % of total gross national capital formation	89
6. Net contribution of agriculture to import capacity (% of total imports)	57

Sources. Items 1 to 4: Kenya: *Economic Survey 1978* (Nairobi), table 2.1. Item 5: J. Sharpley: "Resources transfers between agricultural and non-agricultural sectors, 1964-77", in Tony Killick (ed.): *Papers on the Kenyan economy: Performance, problems and policies* (Nairobi and London, Heinemann Educational Books, 1982), p. 315. Item 6: Kenya: *Economic Survey 1979* (Nairobi), table 6.6 and idem: *Economic Survey 1978*, table 4.9.

Against the background of this description of the nature and role of the rural economy, the problems of poverty and inequality within that economy will now be examined.

EXTENT, NATURE AND CAUSES OF POVERTY IN RURAL KENYA

A national survey

Kenya has the reputation of having a highly skewed distribution of income, with many of the benefits of post-independence economic growth having been siphoned off by a small, but politically powerful, élite.[6] For example, a World Bank staff member estimated for 1969 that the poorest 40 per cent of Kenya's population received only 10 per cent of the total income; the richest 10 per cent received 56 per cent; and the top 5 per cent received 44 per cent (reported in Killick, 1976). The estimated Gini coefficient for this distribution was 0.6, which led Chenery and his associates to classify the degree of Kenya's income inequality as "high".[7]

More recently, Crawford and Thorbecke have attempted to assess the distribution of income over-all and by sector.[8] Their data (relating to 1974-75) reveal marked inequalities in the distribution of income.[9] Rural and urban modern sector recipients compose 17 per cent of the population, yet they receive 48 per cent of the total income. The 80 per cent of the population in traditional agriculture share 47 per cent of the income. Crawford and Thorbecke speculate that the true national Gini coefficient is likely to be of the order of 0.5 to 0.55.

Moreover, a number of researchers have found large urban-rural and inter-regional inequalities in Kenya. In 1972, for example, 45 per cent of modern sector employment was concentrated in the 11 towns with a population which (in 1969) exceeded 10,000, yet their joint share of the labour force was only 16 per cent.[10] Their average level of money wages was 2.5 times the rural average wage and the absolute differential was estimated to have increased from K.sh. 320 per month in 1964 to K.sh. 406 in 1972.

From Crawford and Thorbecke's data we estimate non-agricultural average household income to be K£561, which is three times larger than average income in agriculture. Allowing for larger household sizes in agriculture, incomes per head are probably about four times larger outside agriculture.

At the regional level there are marked disparities between incomes and access to modern sector employment. For example, Bigsten has estimated regional value added per employee and regional income per head for 1971.[11] Value added varies from K£918 in Nairobi and K£551 in Coast Province to K£236 in Central and K£102 in Western Province; income per head is highest again in Nairobi and Coast Province, averaging K£338 and K£76 respectively, and lowest in Western Province at K£16. Such income differentials are obviously related to the extent to which the labour force has access to the better-paying jobs in the modern sector. Rempel estimated the proportion of the labour force of each district employed in the modern sector and, as might be expected, this correlates strongly with regional productivity and income per head.[12]

Access to education and health facilities tends to follow the differences in income. As can be seen in table 8, Nairobi, Central and Coast Provinces appear very favoured, although the position of the latter is strongly influenced by the inclusion of Mombasa town.

We turn now to examine inequalities within and between the rural areas of Kenya.

Inequalities within the rural economy

Crawford and Thorbecke have estimated the distribution of income in rural Kenya for 1976.[13] They rely mainly on the Integrated Rural Survey (IRS) for 1974-75, and certain limitations are apparent. The IRS estimates suggest that nearly 50 per cent of the 2.1 million households in rural Kenya receive average household incomes of K£100 per annum or less. Of the 1.42 million smallholder households (which comprise 80 per cent of the total Kenya population), 44 per cent earn average incomes of less than K£100, and 51 per cent average K£250 per annum. The former group are very likely to be living in poverty, and we return to this subject below. Even the significant number of smallholders averaging K£250 per annum can hardly be described as rich, since their average incomes lie below the average for Kenya as a whole of K£291. Only 5 per cent of smallholder households average K£600, while the richest group in the agricultural economy are the 3,400 large-scale farmers, whose farms exceed 20 hectares, and whose incomes average K£2,500 per year.

Our prime concern in this section is with the smallholder community, for two reasons: first, in concentrating on this community we are dealing with 80 per cent of Kenya's population; and, second, the sector has in recent years received major attention from the Government's Central Bureau of Statistics, which has been responsible for the IRS.

One difficulty in trying to determine the factors associated with the

Table 8. Distribution of public services by province, various years

Province	A	B	C	D	E	F	G	H	I	J	K	L
Nairobi	4.4	61	46	98	31	21	39	.	6.59	4.4	70.8	338
Central	15.3	64	18	82	53	15	18	0.97	0.50	9.7	9.7	42
Coast	8.6	32	11	53	25	6	19	0.90	0.97	6.3	13.1	76
Eastern	17.4	47	7	59	20	9	14	0.53	0.64	4.9	6.4	24
North-Eastern	2.2	4	1	61	20	0	11	0.71	0.04	3.8	3.5	.
Nyanza	19.4	31	7	55	15	10	10	0.33	0.58	1.9	3.3	25
Rift Valley	20.4	29	6	59	19	23	19	.	0.34	5.5	8.8	40
Western	12.3	40	8	57	23	19	11	0.51	0.18	4.7	4.1	16
Kenya	100.0	39	13	.	.	14	16	.	.	5.2	.	.

. = not available.

Key. A: % of Kenyan population (1969).
B: % enrolment of 5-14 year age-group in primary school (1969).
C: % enrolment of 15-19 year age-group in secondary school (1979).
D: % of teachers who are qualified (1976).
E: Development expenditure per head on secondary education (K£) (1974-78).
F and G: Health centres and hospitals respectively, per million people (1975).
H: Recurrent health expenditure per head (K£) (1973-78).
I: Development expenditure per head on curative health (K£) (1974-78).
J: Development expenditure per head on roads (K£) (1974-78).
K: Recurrent expenditure per head (K£) (1974-78).
L: Income per head (K£) (1971).

Source. A. Bigsten: *Regional inequality and development: A case study of Kenya* (University of Gothenburg, 1978).

relative incomes of different groups of smallholders is that the data were collected by holding, and many of the variables influencing household income are themselves partially correlated. For example, while size of holding and household income are positively related, income does not rise in direct proportion with land area: thus, the average income associated with a 3.0 to 3.9 hectare holding is only one-quarter higher than the average income associated with one of 0.5 to 0.9 hectare. Furthermore, household income itself is a poor indicator of relative welfare because of the positive association between income and household size.[14]

For these reasons Smith preferred to conduct his analysis on the basis of "per adult equivalent income groups", where it was assumed that a child less than 15 years old was equivalent to half an adult. This weighting attempted to reflect the relative earning power and consumption requirements of the average child and adult.[15]

Table 9 presents a picture of income distribution amongst smallholders and illustrates the degree of income inequality, both between and within the agro-ecological zones.[16] In general, the dispersion of smallholder incomes amongst the agro-ecological zones follows the pattern established earlier for the provincial distribution of incomes. The coffee and cotton zones to the

Table 9. Proportion of adult equivalents by agro-ecological zone and by per-adult equivalent income group, 1974-75

Zone	Average income (K.sh.)	Gini coefficient	Per adult equivalent income group in K.sh.						Total, %	No. of adult equivalents ('000s)
			0-249	250-499	500-999	1 000-1 499	1 500-2 499	2 500+		
Tea, east of Rift	847	0.08	9.8	36.8	33.5	7.8	7.2	4.9	100	750.2
Coffee, west of Rift	732	0.15	25.0	26.0	22.9	12.5	10.9	2.6	100	1 511.5
Upper cotton, west of Rift	597	0.10	20.1	33.9	31.8	11.9	1.9	0.3	100	1 503.1
Tea, east of Rift	848	0.18	9.1	20.9	45.6	10.9	9.2	4.4	100	931.1
Coffee, east of Rift	883	0.11	8.9	24.7	35.6	15.9	12.0	2.8	100	1 680.7
Lower cotton, east of Rift	602	0.16	15.2	38.4	31.7	10.8	3.0	1.0	100	612.3
High altitude grass	.	.	0.0	4.5	14.4	38.8	37.5	4.6	100	62.4
Coast composite	598	0.12	18.8	33.2	34.7	8.4	4.0	1.0	100	422.3
% total adult equivalents	–	–	15.54	29.0	32.76	12.32	7.95	2.43	100	–
Total (K.sh.)	747	–	141	379	716	1 204	1 943	3 471	–	–

. = not available.   – = not applicable.
Source. Derived from L. D. Smith: *Low income smallholder marketing and consumption patterns: Analysis and improvement, policies and programmes* (Rome, FAO, Marketing Development Project, Sep. 1978), table 22, p. 31; and M. J. Dorling: *Income distribution in the small-farm sector of Kenya: Background to critical choices*, Paper presented to the Kenya Chapter of the Society for International Development (Nairobi, July 1979), table 1, p. 12.

west of the Rift Valley incorporate Nyanza and Western Provinces, while the western tea zone is largely in Rift Valley Province. The tea and coffee zones to the east of the Rift are mainly encompassed by Central Province, with some parts in Eastern Province. The cotton zone to the east of the Rift lies mainly in Eastern Province.

The low-income zones of coffee and cotton west of the Rift, of cotton east of the Rift and the coast exhibit below-average incomes and large proportions of their population with incomes less than K.sh. 500 per adult equivalent per annum. As expected, the high-income zones show a below-average proportion of their population in the bottom-income classes, and an above-average proportion in the higher-income groups. Even so, in an above-average income zone, such as the tea-zone, east of Rift, 30 per cent of the adult equivalent proportion of the population still have incomes lower than K.sh. 500.

Smith went on to analyse sources of incomes by income groups and states that "it seems plausible to suggest that it is the level of non-farm income which is a key element in determining the productivity and output of the farming enterprise and the over-all level of household income". He gives as an example the relatively high level of income from regular employment of those households with a per adult equivalent income of over K£50 per annum, which provides a regular source of cash and security, and a source of collateral for borrowing funds. But while such factors may have allowed a proportion of smallholders to achieve relative affluence, stringency is the general rule. As Smith concludes—

Different groups of smallholders obtain varying degrees of success in this environment and the survey results for 1974-75 suggest that, whilst a very small minority of smallholders seem to prosper, the majority were earning a living which only just enabled them to enjoy the basic necessities of life—a simple diet, some clothing and shelter—but gave little basis for accumulating resources or investing for increased prosperity in the future. [17]

There is also other evidence which points to perhaps greater disparities within the smallholder population. Smith establishes a very close and significant relationship between income and the level of assets per adult equivalent. Each value increases with income, rising from under K.sh. 600 to over K.sh. 4,500 per adult equivalent, and the amount of land rises from 0.32 to 1.02 hectare per adult equivalent over the income classes. If the value of assets is an indicator of past income and economic performance then the snapshot picture of income levels in 1974-75 reveals not random events but fairly typical relative disparities for previous years as well.

As Smith observed, the richest smallholder families appeared to have, on average, a relatively high level of non-farm income, especially from regular employment, which provided security: they also used more farm inputs and hired labour and had higher levels of farm output. In addition, they had a higher level of assets which gave them not only more security but also more collateral with which to borrow additional funds. This general pattern established at the national level also holds for regional or agro-ecological zones. Total farm income, non-farm income (particularly regular employment income) and the value of assets are in general all positively related to income per adult equivalent at the regional levels.[18]

These income differentials between smallholder areas reflect developments over the past 25 years. Using the value of marketed output as an index of development, Heyer has shown that from 1957 Central Province (which largely comprises the tea and coffee zones to the east of the Rift) drew rapidly ahead of other areas. While coffee formed the basis of this growth, other products, such as maize, hides and skins, pulses and fruit and vegetables also figured significantly. Tea, dairy products and pyrethrum only became important in the 1960s. The advance of Central Province is largely attributable to its natural advantages in cash crops. During much of the colonial period Africans were not allowed to engage in the cultivation of cash crops; however, once these restrictions had been lifted the natural potential was quickly exploited.[19] Central Province was the focal point of the struggle for political independence and it therefore received substantial development and infrastructural resources with a view to containing the political situation. At the same time, land consolidation and registration of title took place, with the result that, from the 1950s, Central Province was set on its forward course towards its present dominant position.

Some of these past developments are reflected in the regional distribution of household assets among smallholder families, which are accumulations out of past levels of income. From data collected in the IRS,[20] Central Province invariably has a greater than average mean holding of assets, while households in Eastern, Nyanza and Western Provinces are often below average in their possessions.

To sum up on the question of inequalities within the rural economy, substantial differences have been shown to exist between the various provinces and zones of the country. Within the smallholder community inequalities, not surprisingly, are limited, with a large proportion of this section of the population living on small incomes. Were the data sufficiently comprehensive to provide a size distribution of income for the entire rural population, much larger inequalities could confidently be predicted. In fact, an estimate for Central Province does give a Gini coefficient much larger than those reported in table 9.[21] The next step, however, is to examine the extent of poverty in rural Kenya.

Poverty in rural areas

Crawford and Thorbecke constructed a "poverty level" diet composed of maize and beans, designed to provide an adult with 2,250 calories per day. Using the average prices reported in the 1974-75 IRS, they placed the rural poverty household income level at K.sh. 1,700 per annum for food purchases alone, and at K.sh. 2,200 per annum for all purchases.[22] Smith calculates that

Table 10. Smallholder expenditure patterns per adult equivalent by income classes, 1974-75

Item	Per adult equivalent income group, K.sh.						
	0-249	250-499	500-999	1 000-1 499	1 500-2 499	>2 500	Average
% of adult equivalent	15.54	29.00	32.76	12.32	7.95	2.43	100.00
Total value of food	285	334	491	641	843	1 118	475
Food and non-food expenditure	349	412	595	817	1 050	1 353	586
Average income	141	379	716	1 204	1 943	3 471	747
Food as % of total income	202	88	69	53	43	32	64
Total expenditure as % of total income	248	109	83	68	54	39	78

Source. Smith, op. cit., table 32, p. 43.

such a poverty diet would have cost K.sh. 309 per adult equivalent per annum. The IRS reveals expenditure patterns, at the national level, as shown in table 10.

According to the table, the poorest 15 per cent of smallholders fall below the poverty line for expenditures on food and their diets are clearly inadequate. Furthermore, since the poverty line falls within the second income group, an unknown proportion of the 29 per cent of adult equivalents here are poor by this standard.

These estimates are confirmed by Shah, who estimates that the average calorific intake of the lowest income group,[23] from the nutritional status of the combination and value of foodstuffs consumed, is only 1,960 calories per adult equivalent. This meets only 80 per cent of the Crawford-Thorbecke standard. Indeed, the estimated average calorific intake of the next 29 per cent of the population is only 2,205 calories from a value of food consumed of K.sh. 334, which is less than the required 2,250 calories. Therefore, at the national level, fully 15 per cent of the population clearly receive an inadequate diet, while some of those in the following 29 per cent fall marginally below the poverty line.

As table 10 reveals, the two lowest income groups must be running down their already small stocks of assets, including perhaps selling off their land, since total expenditure exceeds income. However, this was partly a result of poor weather during the years under review, and some of these families could have been expected to recover in the subsequent periods. A further important aspect of poverty brought out by the IRS and by the in-depth case studies carried out by Hunt and Rukandema is the extremely low holdings of household assets by those at the bottom end of the income scale.[24]

Alternative estimates of poverty are presented by province in table 11. The table shows that 571,000 households fall below the K.sh. 2,200 total consumption level, which represents 38.5 per cent of all smallholder house-

Table 11. Distribution of smallholder households falling below poverty line, by province

Item	Province						
	Central	Coast	Eastern	Nyanza	Rift Valley	Western	Total
% of household with food consumption < K.sh. 1,700	17.8	37.7	27.1	50.7	36.5	47.7	35.8
No. of households ('000s)	58.5	26.3	95.8	195.8	32.8	121.6	530.9
% of total < K.sh. 1,700	11.0	5.0	18.0	36.9	6.2	22.9	100.0
% of households with total consumption < K.sh. 2,200	18.2	43.5	28.5	55.5	38.8	51.6	38.5
No. of households ('000s)	60.1	30.4	100.6	214.2	34.9	131.3	571.5
% of total < K.sh. 2,200	10.5	5.3	17.6	37.5	6.1	23.0	100.0

Source. Adapted from E. Crawford and E. Thorbecke: *Employment, income distribution, poverty alleviation and basic needs in Kenya*, Report of an ILO consulting mission, Apr. 1978 (Ithaca, NY, Cornell University, 1978; mimeographed), table 3, p. II-9.

holds; and 539,000 households, or 35.8 per cent of the total, fail to meet the food consumption poverty line of K.sh. 1,700.

At the provincial level, the incidence of poverty is above average in Coast, Nyanza and Western Provinces. Nyanza alone has 37.5 per cent of all those with inadequate total household consumption and, together with Western Province, accounts for 60.5 per cent of the total poor. When similar calculations are made by agro-ecological zones, the tea, coffee and upper cotton zones, all west of the Rift, account for 65 per cent of the total poor.[25] A close corresponding pattern appears for the food consumption poverty groups and, as we would have expected, between the provincial and zonal patterns of poverty. The within-province incidence of poverty reveals that in Nyanza and Western Provinces over one-half of smallholder households are classified as poor. One the other hand in Central Province less than one-third of the population are living below the poverty line. The current development plan argues that "poverty is not easily identified on a regional basis",[26] and a World Bank researcher has claimed that within rural Kenya poverty is not regionally specific.[27] Yet our analysis has shown that Nyanza, Western and Eastern Provinces contain 78 per cent of all poor Kenyan households. While no area is free of poverty, its incidence is demonstrably concentrated in these regions.

In drawing the profile of poverty in rural Kenya, mention must be made of the special problems faced by three relatively disadvantaged groups in rural Kenya. It will also be useful to survey the results of region-specific studies of poverty and inequality.

Smock shows that *women* are the backbone of the rural economy but enjoy few of the advantages that such a role might be expected to convey.[28] While the menfolk seek wage employment in rural and urban centres, women play the dominant role in small-scale farming, often on smaller plots and for smaller returns than in the male-headed households. Certain factors have operated to increase women's workloads in addition to their traditional tasks of water carrying and wood collecting: expansion of educational opportunities has reduced the availability of child labour on the farm; land pressure has made the collection of firewood more difficult; and new varieties of crops have necessitated increased weeding. At the same time, of course, they are the child-bearers, with the average rural family having eight children. So women perform their arduous economic functions during a continuing cycle of pregnancy, childbirth and child rearing, in addition to doing housework and cooking. These burdens no doubt help to explain why women have low rates of participation in modern sector employment and compare unfavourably with men in literacy and educational attainment.

The *landless* who have become squatters have been identified by the Government as one of five target poverty groups. They occupy state or private land to which they have no title and from which they are estimated to raise meagre household incomes of K£100-125. Alternatively, the landless migrate to an urban centre or take up low-paid wage employment on large farms or in settlement schemes. Significant numbers appear to fall below the Crawford-Thorbecke poverty line.[29]

The third special group in rural Kenya is the traditional *pastoralists,* such as the Masai and Samburu, who occupy large areas of rangeland. Their stock is individually owned and kept primarily for meat and milk for subsistence, and their husbandry patterns are concerned almost exclusively with the basic day-to-day needs of their livestock.

Such a lifestyle cannot continue for much longer, however. As one study in the Masai area of Kajiado District has shown, the subsistence demands of a rapidly growing population will outstrip the capacity of the resources of the District to provide for them by the turn of the century.[30]

Two area-specific studies of poverty and inequality undertaken in recent years should be mentioned. Both Hunt's research in Mbere in Eastern Province and Rukandema's results from Kakamega District in Western Province confirm the general nature of the results of the IRS. Significant numbers of families are shown to be very poor; their poverty takes the form of a lack of disposable income with which to purchase non-subsistence goods and services, rather than an inability of families to feed themselves.

Deficiencies in other basic needs

The fourth development plan identifies three major nutritionally deficient groups of Kenyans. Amongst the smallholder population 3.29 million persons, or 32 per cent of the total, are thought to suffer from protein energy malnutrition (PEM). A further 250,000 unemployed and underemployed urban dwellers and 670,000 pastoralists are believed to suffer from a similar condition.[31]

In addition, a survey was conducted in early 1977 which investigated the nutritional status of children aged 1 to 4 years.[32] The results showed that the incidence of severe PEM is rather low in rural Kenya, although they may have been influenced by the good harvests experienced in that year. Children in Central and Eastern Provinces appear to suffer most from moderate PEM, while those in Nyanza, Western and Coast Provinces suffer the least. The results for Central Province are confirmed by Wanjohi, whose survey in the Mwea-Tabere rice scheme location indicates that malnutrition is a problem even though the average annual cash income, of K.sh. 800, is relatively high.[33]

These results are somewhat paradoxical since we have seen Central Province and the coffee-growing areas east of the Rift to have relatively high smallholder incomes and consumption levels, as well as having a below-average proportion of households below the poverty line. At the same time, the western provinces appear to have relatively low incomes and consumption, with the largest proportions below the poverty line, yet they suffer the lowest incidence of malnutrition.

One possible explanation is that the smallholders of Central Province have substituted cash crops for food crops and use their relatively high cash incomes to buy such non-food items as clothing and durable goods, as well as for reinvestment. As both parents become engaged in the market economy and incomes rise, nutrition is neglected and the possession of tangible assets appears more desirable than a staple diet. Data in the IRS support this suggestion, showing that the mean number of assets in Central Province households is almost invariably above the national average. This is true for items such as radios, clocks, watches, irons, lamps, torches and basic household furniture. We would venture the hypothesis, therefore, that malnutrition is inversely related to the share of household food production in total household income and positively related to the total value of income. The level of income remains an important determinant of malnutrition since, when the nutritional status of children is cross-tabulated against the occupation of the household head, the greatest incidence of stunted children occurs amongst agricultural labourers, some rural smallholders and those with no occupation. Furthermore, those occupations with the potential for generating reasonable cash incomes but with a low potential for domestic production of food, such as rural transport workers and village craftsmen, also showed high incidence of malnutrition in children. Blankhart has provided specific examples of poverty-related malnutrition.[34]

In comparison with other low-income countries, certain demographic indicators in Kenya appear favourable, perhaps confirming that severe malnutrition is largely absent. For example, the crude death rate per thousand of the population in 1977 was 14 in Kenya, compared with an average of 20 for all low-income countries, and this had fallen by 25 per cent during the previous 15 years in Kenya, compared with a fall of 21 per cent in other poor countries. Life expectancy at birth is estimated at 51 for males and 56 for females in Kenya compared with 41.7 years in all low-income African countries, and 44 years in all low-income areas.[35] Meanwhile, the annual growth rate of population of 3.9 per cent in Kenya, compared with an average for all low-income African countries of 2.5 per cent, is hardly consistent with major problems of severe malnutrition.

IMPACT OF GOVERNMENT POLICIES

In order to achieve some depth of treatment within a limited space, it has been necessary to concentrate in most of this section on the impact of some selected aspects of government policy. However, neither the incidence of rural inequality and poverty nor current policies towards rural development in Kenya can be understood without an appreciation of the influence of history: for example, current policies on land tenure and on agricultural marketing are a linear continuation of approaches adopted during the colonial period. This section therefore commences with a brief account of the evolution of rural development policy from colonial times.

Evolution of rural development policy

The treatment of agriculture during most of the period of British colonial rule offers a textbook study of racial exploitation.[36] From the beginning of the twentieth century, Europeans (mainly British) were encouraged to settle in Kenya. Agriculturally, the country was bifurcated into the "scheduled areas" (containing much of the best land), which was set aside for the settlers, and the "reserves" where most Africans were expected to live.

Policies were systematically designed for the benefit of the settler farmers: taxes were levied on the native population in order to secure supplies of African labour for European farms; Africans were prohibited from growing coffee, tea and other crops which could compete with the settlers' products; the settlers were protected against competition from imports and were offered subsidised freight rates for railing their exports to the coast; the construction of the infrastructure—the railway, the roads, Mombasa harbour—was sited to meet settler needs; finally, marketing boards were designed to serve their interests.

In the long run, especially in the international climate which grew up after the end of the Second World War, this was not a viable policy framework. African agriculture had been largely neglected, with such government assistance as was given being mainly restricted to the improvement of subsistence food production and the reduction of famine. The inequities, distortions and demographic forces, as well as the mounting nationalist campaign for political independence, created pressures for change that could not forever be resisted: in 1949 the authorities relented sufficiently to allow coffee to be grown by Africans; in 1952 smallholder tea was planted in Nyeri; and in 1954 the "Swynnerton Plan" was published, which stands as a watershed in Kenya's colonial policies towards agriculture.

The main components of the Swynnerton Plan were *(a)* a major programme of land reform; *(b)* increased availability of credit; and *(c)* a reorientation of research, extension and marketing bodies. All these changes were to be explicitly designed to promote the development of African smallholder farming. This strategy, as will be shown later, was broadly implemented. It remained the basis for colonial policy until independence in 1963, and for some years thereafter. It was only in the late 1960s that the Government began to move towards a more broadly based strategy.[37]

There were, of course, some changes after independence. Many white-owned farms were transferred to African ownership, often for smallholder cultivation. More was done for the development of previously neglected African-populated rangeland areas, including a large livestock development project which by 1974 had developed 800,000 hectares.[38] It was also during the 1960s that the Government began a vigorous rural water development policy.

There were other changes after independence which, however, many viewed with unease. The share of government spending on agriculture declined, symptomatic of a national development strategy with a pronounced urban orientation. It was clear that, within agriculture, policies were creating an indigenous capitalist rural economy. Swynnerton, the originator of much of this policy, could scarcely have been more explicit about the nature of the economy which he favoured—

Former government policy will be reversed and able, energetic or rich Africans will be able to acquire more land and bad or poor farmers less, creating a landed and landless class. This is a normal step in the evolution of a country.[39]

The strategy thus had built into it numerous biases in favour of the "energetic or rich African": research and extension services were strongly biased in this direction; the marketing system was more efficient for export crops and served large-scale farmers better than small; credit policy contained similar distortions.[40]

Implicit in this was acceptance of a "trickle-down" view of development—an assumption that, in reasonable measure, all would share the benefits of increased production without any special need for state intervention. But both globally and within Kenya this assumption has met with growing scepticism. Rural capitalism began to show an unacceptable face. Mounting concern about unemployment and poverty led the Government to invite the ILO to send a mission to Kenya to investigate these problems; its well-known report was published in 1972 and has since come to exert an influence on policies comparable with the earlier impact of the Swynnerton Plan.[41]

The ILO urged the adoption of an integrated approach to the development of the rural economy. This was to include the intensification of land use, with a concentration of effort directed to the poorer households; a redistribution of land towards more labour-intensive units; the settlement of unutilised or underutilised land; greater provision for famine relief; rural works programmes; improved social services and amenities. More generally, the report urged a major shift in national development strategy towards the alleviation of poverty through the expansion of productive, income-earning employment opportunities and what has since become known as "redistribution with growth".[42]

On paper, if not always in deed, the Government has gone far towards accepting the main thrust of the ILO proposals.[43] The Third (1974-78) Plan marked a break with the heavy emphasis on maximising GDP growth that had characterised earlier efforts, giving improved income distribution and greater employment top priority. The influence of the ILO is even more apparent in the Fourth (1979-83) Plan. Growth is scarcely discussed at all, and its authors claim with justice that "in this Plan, the efforts of the Government to deal with emerging problems and to take advantage of new opportunities will be organised around the theme of the alleviation of poverty throughout the nation".[44] While reservations are later expressed concerning the extent to which the Plan's proposals will be carried out, they will surely have some effect and do represent a genuine shift away from the rural development policies of the 1960s. There is a real sense in which Kenya's stated rural development strategy has shaken off the colonial heritage.

Having sketched a historical background, we turn now to examine in greater depth some key aspects of policy, as they relate to the circumstances and prospects of the rural poor. We begin with the all-important topic of land policy.

Distributive effects of land policies

Background

The point was made earlier that nature has imposed a degree of inequality on Kenya's rural economy. Only a very limited percentage of the country's surface combines good soils, a workable terrain and adequate rainfall. But superimposed on these natural inequalities is a more man-made condition, in the form of differential access to land within any given ecological zone.

We showed earlier that, while there are several other important determinants, both income and consumption among smallholders is correlated with farm size. But, even within the narrowly defined smallholder class, there is an appreciable skew in the distribution of land ownership. It can be estimated, for example, that, even if we exclude holdings of 8 hectares and over, the smallest-scale farmers, accounting for three-fifths of the smallholder class, occupy only 28 per cent of the land, while the largest-scale farmers, making up 14 per cent of this class, occupy 37 per cent of the total area.[45] The measured degree of skewness would be vastly greater if the data allowed us to include the larger holdings.

Collier similarly shows that the distribution of land in Africanised large-farm areas is still highly concentrated and that co-operative settlements have made only a small contribution to the goal of equalisation.[46] In Nakuru, for instance, 16,500 farmers held plots of slightly more than 1 hectare on average, while just 38 farmers each held farms in excess of 400 hectares. Collier also produces evidence for Central Province showing the degree of skewness to be increasing over time: between 1963 and 1974 the share of the two-fifths of the farmers with the smallest plots went down from 26 per cent to 18 per cent of the total land area, with a corresponding gain by those with the largest plots.

Quite apart from the existence of large numbers of farmers whose holdings are too small to yield more than very low incomes, there is also a substantial incidence of outright landlessness, with the related problem of illegal squatting. The origins of this problem go back to policies pursued by the British colonial authorities, as has been well described by Mbithi and Barnes.[47] The official estimate is that there are 410,000 landless and squatters in Kenya, and Mbithi and Barnes suggest that this number has been growing at about 5 per cent annually.

What alternatives are open to those with only tiny plots or none at all? On paper, one possibility would be for large-scale migration of the disadvantaged into the still substantial parts of the Rift Valley and elsewhere which could efficiently supports considerably larger populations.[48] Tribal hostilities are, however, likely to rule out this solution which, in any case, would do no more than buy time in the face of a national population that is likely to double in size during the next 15 years.

Farm employment might seem to offer a more promising alternative, but this apparently offers exceedingly low wages. A recent study by Leitner described Kenya's agricultural labourers as "the most numerous and most degraded workers in Kenya".[49] He found contract wages of K.sh. 86-117 per month for men and K.sh. 52-100 for women, with even lower money wages for non-contract work. While money wages are much higher in most urban jobs, such jobs are becoming increasingly difficult to obtain. The modern sector is absorbing a diminishing share of the total labour force. The urban informal sector does augment the jobs available in modern activities, and House found that 43 per cent of the heads of informal sector businesses in Nairobi were landless. A good many of these were below the poverty line, however, as well as an even higher proportion of their employees and apprentices.[50]

By and large, then, the alternatives to farming do not offer the landless and poor a secure escape from their condition.

The resettlement of European farms

Being the country's most valuable economic asset, land and its allocation and use excite powerful passions. The remorseless push exerted by the land hunger just described intensifies these emotions. The past association of landownership patterns with racial exploitation adds a further explosive ingredient. On the other hand, the existence of large areas of the best land under settler ownership also gave the rulers of newly independent Kenya an unusually wide range of policy options concerning land use and ownership. That land would be transferred from settlers to Kenyans was understood by all; but in what ways would this transfer be achieved?

The Government's initial answer was in the form of a "million acre" settlement scheme. This was to end the bifurcation of land between African peasant farmers cultivating small plots in the "reserves" and European farmers on large, protected and prosperous holdings. It was also to provide land to African tenant farmers who has been dispossessed during the consolidation of African lands in the later colonial years. More generally, it was to relieve the pressure of population and the hunger for land in the traditional smallholder areas. A million acres (roughly 400,000 hectares) was thus to be purchased—from Europeans—mainly through "aid" provided by the British Government and resettled under a mixture of high- and low-density settlements, as well as large-scale farms owned co-operatively or individually.

The "million-acre" scheme was subsequently supplemented by others, chief among which were the *Haraka* and *Shirika* programmes. Under the latter, each farmer was to be allotted a plot of about 1 hectare for food production, the rest of the farm being run as a large-scale unit, with a manager provided by the Government.

In trying to present data on the extent of land transfer and resettlement, the researcher is faced with a near-conspiracy of silence from official statistics. Land has become such a sensitive issue that few figures have been published. An attempt has been made to piece together such information as can be obtained, although it is impossible to vouch for its reliability.

First, table 12 provides summary statistics on the area officially resettled for smallholder cultivation. Table 13 provides further details of the smallholder settlement as it stood in 1968, together with additional information on the area resettled as large farms. It shows that by 1968 total area transferred as large farms was a little larger than the area officially subdivided into smallholdings, and that the average large farm (411 hectares) was over 40 times as large as the average smallholdings created under the settlement programme (9.4 hectares).

Since the 1968 data, virtually no official figures have been published on areas resettled under large farms. However, as far as can be judged, the ownership situation of former settler farms was roughly as follows, circa 1978:[51]

Mixed farms	'000 hectares
Resettled to smallholdings	800
Transferred as large farms	860
Remaining in non-Kenyan ownership	100
Other large farms (plantations and ranches)	
African-owned	440
Non-African owned	1 200

As regards the mixed farms, the high proportion recorded under "large farms" is rather misleading. A good number of these were purchased by co-operatives, "partnerships" and companies for group farming, and in many cases have been informally subdivided for smallholder cultivation.[52] From

Table 12. Summary statistics of smallholder resettlement on former European farms

Approximate date	Area resettled (hectares)	No. of farms or households[1]	Average farm area (hectares)
1968	430 000	45 900	9.4
1973	614 000	51 300	12.0
1978	800 000+[2]	71 000	11.3+

[1] The 1968 figure refers to the number of small farms established; the 1973 and 1978 figures relate to numbers of families settled. [2] As explained in the text, this figure would probably rise to about 1.1 million hectares if we add the large-farm area informally subdivided into smallholdings.

Sources. 1968: Kenya: *Economic Survey 1969* (Nairobi), p. 24 (large farms excluded). 1973: idem: *Economic Survey 1974* (Nairobi), pp. 227-230. 1978: idem: *Economic Survey 1979* (Nairobi), p. 290.

Table 13. A classification of land transferred to Africans as at 1968

Farm type	Area transferred ('000 hectares)	No. of farms	Average area (hectares)
Smallholder settlements			
High density	319	26 700	11.9
Low density	76	5 200	14.6
Squatters	35	14 000	2.5
Subtotal	430	45 900	9.4
Large farms			
Individually owned	386	1 192	324
Operated by co-operatives or the Department of Settlements	118	34	3 464
Subtotal	504	1 226	411
Total, all types	934	47 126	19.8

Source. Kenya: *Economic Survey 1969* (Nairobi), table 2.2.

the Fourth (1979-84) Plan it appears that of the 860,000 hectares of large mixed farms shown above about half have been subdivided, the other half remaining individually owned.

Even after this adjustment, however, the fact remains that much good land has been transferred intact as large farms to individual Africans, which explains why this is such a sensitive issue. There is evidence to suggest that prominent politicians, public servants and urban businessmen have been able to acquire for themselves large areas of good farming land.

When we turn to the effects of the settlement programme on rural poverty and income inequality, it is important to avoid exaggerating the importance of the programme in the national context. About 1.7 million hectares of mixed farm land have been transferred, compared with a total area of all types of agricultural land of over 50 million hectares, or with 6.8 million hectares of high-potential land. The land transfers thus affected about a quarter of the best areas, but only about 3 per cent of the nation's total stock of agricultural land. The number of settlers relative to the total rural population was even smaller, as can be seen by comparing the 71,000 resettled families shown in table 12 with Kenya's total of 1.7 million smallholdings, and with her 1.2 million rural families officially recorded as living below the poverty line. Even if all the remainder of the large mixed farms were converted to smallholder cultivation, this could make only a minor dent in the task of alleviating rural poverty.[53]

Bearing in mind, then, that we are writing of a programme that is of limited magnitude, what were the effects on poverty and inequality? It is surprisingly easy to provide a qualitative answer to this question, although impossible to quantify the effects. First, resettlement clearly reduced rural poverty: it provided good land to numbers of formerly landless or poor families and, in so doing, it alleviated land hunger in the reserves; it also permitted greater smallholder production and higher incomes per head in the highland areas, although the impact must have been modest and highly selected in relation to the total incidence of rural poverty.

Second, the transfer of valuable income-earning assets from the European settlers to Africans probably reduced racial inequality in the country but was associated with increased inequalities within the African population. This process occurred at a number of levels. One effect was to permit a small group of wealthy, or powerful, Africans to become large landowners, where formerly such a class had scarcely existed. Inequalities were also generated within the resettled smallholder population. As is shown in table 13, there were substantial differences in the average plot sizes of the low-density, high-density and squatter resettlements. Work by Scott et al. has shown that the private profitability of low-density farms was greater than on high-density farms, although social profitability was greater in a high-density situation. Berg-Schlosser estimated that low-density farmers obtained at least double the incomes of farmers in the high-density schemes.[54]

Of greater seriousness, however, were the inequalities created between the resettled farmers and those remaining in traditional smallholder areas. This is implicit in the statement already made to the effect that the programme affected about a quarter of the country's best land, but only about a twentieth of the country's total rural population. The average size of the settlement smallholdings of 9.4 hectares (table 13) can be compared with the fact that in 1976-77 three-quarters of the country's total number of smallholdings were of less than 2 hectares and that only 3 per cent of the total had 8 or more hectares. To make matters worse, some of those resettled already owned plots in the traditional areas; and the resettled smallholders were given disproportionate access to state extension and credit facilities, even though they were farming much better land than many farmers elsewhere. As Holtham and

Hazlewood put it, "The benefits of this programme accrued to a minority... and not a minority composed of the poorest."[55] The area of White mixed farming was geographically concentrated and so, therefore, was resettlement: 70 per cent of the area transferred under the original resettlement schemes was located in the Central and Rift Valley Provinces, and this inevitably aggravated regional and ethnic income differentials.[56]

Land tenure reform

The land tenure reforms initiated by the 1954 Swynnerton Plan have constituted a second major aspect of land policy. Concerned with tenure in traditional smallholder areas, this reform has three aspects: first, the land is *adjudicated* to ascertain the ownership rights of the individuals or groups in question; second, where necessary, there is then a *consolidation* of separate fragments of land into single units; finally, these units are *registered,* providing owners with title deeds to their land.

Although it was introduced long before independence, this reform has remained a major aspect of government policy. At the end of 1978 a cumulative total of 7.6 million hectares had been wholly or partly covered by the reform, or over half of the total registrable area. By any standard, this has been an ambitious programme and, over the years, it has claimed a sizeable share of total government spending on agriculture.

Swynnerton envisaged that a number of benefits would flow from the reform: provision of legal titles, and the accompanying security, was expected to encourage long-term investment in holdings and to provide collateral for farmers to obtain credit to support such investments; reduced expenditure of money and time in land litigation was also expected to have favourable effects on productivity; consolidation would facilitate crop rotation, waste less walking time, make it easier for cultivators to protect their crops against theft and pests, and make it more economical for the State to provide services to individual farmers; and finally, more wage labour was expected to be employed. To the extent that it led to such results, the reform could be expected to contribute directly to the alleviation of poverty in the rural economy, through its beneficial effects on output and employment.

The Government of Kenya has never had many doubts that the reform has been desirable, as indicated by statements in successive development plans.[57] Moreover, a far better record of implementation has been achieved in this programme than is true of other aspects of agricultural policy. Fleming[58] is among independent observers who support the Government's confidence in its land-reform policies. While recognising the difficulties of demonstrating the point conclusively, he argues that tenure reform has increased farm productivity, and cites a 1968 study which found substantial development in half of the farms after registration, mainly in response to increased security of tenure. Barber provides data on revenue from cash crops in four districts of Central Province, showing that the largest proportionate increases in revenue occurred in those districts in which tenure reform was furthest advanced.[59]

Other researchers have been more sceptical. Heyer speaks strongly for the critics (see also Okoth-Ogendo[60])—

The reform has produced little real change. Only in some areas was fragmentation really severe, and in these few areas there may have been some improvement through consolidation, but a good deal of fragmentation persists off the register. As far as registration is concerned, there is some doubt as to whether it has even reduced the amount of litigation that takes place. Relatively few farmers have obtained credit, and this could have been extended without registration anyway. The market in land already existed, and the controls over the purchase and sale of land that were introduced as part of the programme of land reform have severely restricted the market in practice. The incentive to invest was never really a problem. It is argued that until complementary action is taken to attack the underlying causes of the problem, land reform cannot be a success. Once the complementary measures are taken, land reform may be superfluous (pp. 17-18).
... there is little to show that the production increases of the 1950s and the 1960s were attributable in any way to the land reform programme. They were much more clearly the result of the relaxation of restrictions on the growth of cash crops and the keeping of dairy cattle. Furthermore, the land reform programme had the major disadvantage of leaving large numbers of people landless and unemployed (p. 11).

While it is probably going too far to suggest that increased output was not related "in any way" to the reforms, there is no doubt that the programme has worsened landlessness and, probably, unemployment. Land registration and the writing of title deeds creates the legal basis for a market in agricultural land; indeed, this is one of the objectives. However, while the creation of such a market is desirable for economic efficiency, it brings with it the danger that the well-to-do will buy out poor farmers, resulting in widening disparities in the ownership of land. Swynnerton explicitly recognised that this would happen and there is clear evidence that this is occurring to some extent. For example, detailed investigations of some of the districts covered by the Special Rural Development Programme found an increasing skewness in the distribution of land ownership and, therefore, growing landlessness in several of the districts.[61]

The adverse social consequences of such a development would at least be ameliorated if there were a parallel increase in opportunities for rural employment at a reasonable wage, as Swynnerton envisaged. It is doubtful whether this has occurred on a sufficient scale, however. The limited availability of regular employment and the low wages prevailing in the rural economy have already been indicated. As Mbithi and Barnes point out, the main effects may have been to increase per-worker productivity rather than the total numbers employed—

The criticism that consolidation initially creates unemployment and landlessness is a valid one. In Central Province the landowner possessing many fragments was frequently unable to cultivate all of them and so allowed tenants *(ahoi)* to cultivate the undesirable and more remote of these fragments. At the time of consolidation the *ahois'* cultivation rights were extinguished and they found themselves landless and with little hope of immediate employment, for the act of bringing together into one several dispersed fragments lends itself to more efficient use of labour and the owner has less need to seek outside help to manage the new holding.

In fact, the Government's fourth plan has admitted that all is not well: "With the implementation of land adjudication and registration, the volume of land transactions among smallholders has increased. Consequently, the incidence of concentration in land ownership among the better-off small-scale farmers has increased." It proposes controls to guard against this danger but also admits with dangerous candour, in a country where virtually everyone aspires to own a *shamba,* that "access to land is ... no longer a possibility for every family".[62]

To sum up the impact of land tenure reform on rural poverty and inequality, there have been two opposing influences at work: first, it is likely that the reform has been one of the factors contributing to increased smallholder productivity and incomes, and in this respect, therefore, it has helped to reduce rural poverty; second, however, it has worsened the incidence of landlessness and increased the concentration of landownership, thereby tending to exacerbate both poverty and inequality. Exactly where the net balance between these opposing trends lies is impossible to determine on the basis of present knowledge.

Government economic services to agriculture

The State provides a range of economic services intended to contribute to the development of agriculture. These include research and extension; the provision of credit; the distribution of inputs; and the provision by statutory corporations of output marketing facilities. All of these are important and are liable to influence the distribution of incomes and poverty in rural areas, but space limitations require us to confine attention to the impact of the research and extension services.

Impact of agricultural research

In the past, major resources have been devoted to agricultural research, both absolutely and by comparison with other African countries, although it seems that for some time it has been obtaining a diminishing share of the resources of the Government and the Ministry of Agriculture.[63] Nevertheless, in absolute amounts the research programme remains large, and it is the orientation of the programme which has attracted most criticism. To quote Heyer and Waweru[64] again: "Kenya compares well with other developing countries in the level of expenditure on research. Kenya does less well when the content of the research programme is examined. Although many developing countries share the problem of inefficient research programmes, few face the additional problem of bias away from small farms to such a degree."

As was mentioned earlier, the impact of the colonial experience on the content of research was strong and continued to be felt well after independence. More generally, it can be said that the research undertaken, and the way it has evolved, has reflected the political, economic and demographic conditions of the country. As they existed before and for some years after independence, these conditions created what are now perceived as biases in the research effort:

(1) There was a bias towards investigating the problems of *large farms and cash crops*. With exceptions to be noted shortly most past crop research was on coffee, tea, pyrethrum, sisal and wheat. It was especially concerned with the production problems of large-scale farmers and this limited the relevance of the results for application by smallholders, even those growing the cash crops in question. Research into farm mechanisation, for instance, has mainly been concerned with machinery suitable for use over large areas; little has been done on the development or adaptation of equipment which could be economically employed on smallholdings.[65] Before 1964 this bias reflected the domination of the colonial Government by settler interests; subsequently it has reflected inertia and the continuing strong influence of large-scale farmers (Black and White) on the allocation of public resources.

(2) There has been a bias towards the problems of *arable farming on high-potential land*. Most past research has related to crops rather than livestock, and has for the most part been concerned with ecological conditions in high-potential areas. The Third (1974-78) Plan admitted that limited progress had been made in developing crops and improved techniques for the medium-potential and more arid parts,[66] even though they occupy three-quarters of Kenya's agricultural area.[67] However, Wyeth points out that the concentration on high-potential areas could be defended on the grounds that most of the rural population live in those areas.[68] Conditions are changing, yet even today less than a quarter of the rural population lives in the medium-to-low-potential zones. But while the bias in favour of the high-potential areas was a bias in favour of the majority, both this, and the related bias towards arable farming, clearly operated to widen the naturally large inequalities between incomes in the high-potential areas and in the remainder of the rural economy.

As will be shown shortly, there is evidence of a similar set of biases in the extension service. There has been a mutually reinforcing process at work: research has favoured large-scale and "progressive" farmers; therefore extension workers have had more to "extend" to them; this greater contact has, in turn, led to a greater feedback of information on problems needing further research; and so the research bias has been reinforced. There is, however, one research result to which the accusations of bias do not apply: all agree that the development of higher-yielding and drought-resistant strains of maize has been a major outcome of past research, which has been widely applied on smallholdings and, in the drought-resistant varieties, has been applicable to lower-potential areas.[69] The political, demographic and economic forces which shaped past research efforts have changed markedly over the past decade or more: large-scale farmers are no longer dominant; population pressures are pushing people into areas with few reliable sources of water; landlessness is growing; and the stated objectives of government policy have shifted towards a greater concern with rural poverty. Nevertheless, research efforts have not adapted quickly to this changing environment, as is tacitly admitted by the Fourth (1979-84) Plan.[70]

It would be unfair to imply that the Government has been entirely unresponsive to these emerging needs. The Third Plan announced a new research project aimed at finding better ways of exploiting the drier, more marginal areas, and another for testing mechanical equipment to be used on small farms.[71] Some genuine progress has been made. Jamieson noted a shift in research resources towards previously neglected areas:[72] for example, a dryland farming research station at Katumani is being built up, whereas it was previously described as being "virtually abandoned". Formerly neglected stations on the coast and at Thika are also being developed, while more traditionally oriented stations at Kitali and Njoro are being held at their past strengths.[73] The Fourth Plan seeks to give this change of direction further impetus, although it should be noted that the implementation of research projects presented in the Third Plan was poor.[74]

Impact of the extension service

As with its research programmes, Kenya has devoted an above-average volume of resources to its agricultural extension service. Although estimates vary between 300 and 700 farmers per extension worker, even at the upper end of that range the ratio is much better than in the typical developing country where a figure of the order of 1,500 farmers could be expected.[75] The favourable nature of this comparison needs, however, to be qualified in that many of Kenya's extension workers have had little or no training.

It is not difficult to characterise the impact of the extension effort on rural poverty and inequality. In a fairly explicit way, Kenya's service has pursued a "progressive farmer" strategy. This singles out those farmers regarded as most likely to respond to advice for special attention on an individual farmer basis. The expectation is that these farmers will be natural leaders of their local farming communities, will be regarded as exemplars and will be copied by most other farmers. In this manner innovations are expected to be diffused throughout the community.

The diffusion theory on which the approach is based has been questioned in recent years but, even if the theory were valid, its application in practice would rather obviously have an inegalitarian bias.[76] This has been the case in Kenya. Ascroft et al.[77] conducted detailed investigations in Tetu, in Central Province. They classified farmers according to their "progressiveness" (i.e. their willingness to adopt agricultural improvements such as hybrid maize, fertilisers and grade cows) and then investigated the extent to which progressiveness was correlated with access to extension advice. Not surprisingly, they found that the most progressive farmers were three to four times more likely to have contact with extension agents than the least progressive farmers. More significantly, farm size was an increasing function of progressivity, so that it was the wealthiest farmers who received most help.

Other work has yielded similar results. Leonard's study of the extension service in Western Province also found farmer progressivity to be strongly correlated with wealth and, as might have been expected, that the most progressive farmers obtained disproportionate attention from extension officers. Hunt found the same result in her work on Mbere. Leonard summarised his findings by pointing out that if extension were distributed perfectly equally, one visit would be made to every other farm each year.[78] But in Western Province the progressive farmer received 2.91 visits per year, the "middle" farmer received an average of 0.44 visits and the non-innovator only 0.07. Since a large part of extension workers' time was spent on individual farm visits, the concentration on the progressive farmer was at the expense of the non-progressives. Leonard obtained similar, if less extreme, results for veterinary extension visits, even though, as he points out, the Rogers diffusion theory cannot provide a rationale for a progressive-farmer bias in the veterinary service.

This has not been the only type of inegalitarian bias. As mentioned earlier, farmers on government settlement schemes, who were already favoured in the area and quality of their land, have also been favoured by a disproportionate access to extension advice. Finally, extension policies have discriminated against women farmers. Indeed, according to the ILO it was deliberate policy in the 1960s not to give advice to women farmers. Since female-headed households in the rural areas have lower incomes than male-headed households, they are doubly discriminated against—because they are poor and because they are women.[79]

During the later 1960s the Government of Kenya began casting around for practical methods for overcoming the biases just described. Its commissioning of the 1972 ILO comprehensive employment strategy mission report was one move in this direction; the earlier creation of the Special Rural Development Programme (SRDP) had been another. The SRDP, which was initiated in 1970, introduced a variety of experiments in integrated rural development in 14 localities from all regions of the country.[80] Its special focus was on reaching the rural poor, and those of its experiments which proved successful were intended to result in methods which could be easily applied to other parts of the country.

The programme included a number of experiments in extension work, especially at Migori, Tetu and Mbere.[81] New procedures were devised for training extension agents, for involving field personnel in the programming of their activities and for monitoring the achievements of extension agents. In addition to experiments directed at the extension agents, a variety of new approaches were devised for involving farmers. Extension was focused on average farmers, delivered via village groups and organised according to schedules and sites selected by these groups. Work was concentrated on a limited number of innovations rather than a broad programme involving complex options. The experiment entailed extension conducted with clusters of farmers, each one organised as a recognisable group with an elected leader. The leaders, in turn, were organised into village committees which took the initiative in requesting extension. The inegalitarian bias in traditional extension work at Tetu was summarised earlier, but the SRDP also put forward an apparently superior alternative. It seems that the progressive-farmer strategy is not only inequitable but inefficient as well. Wyeth puts the point clearly—

... it has been discovered that farmers are less likely to follow the example set by progressives than that set by any other group. The reason is that the best farmers are effectively separated from the rest of their communities by the very fact of their success. Unlike others, they often read, write and calculate; they have more contacts than usual with businessmen and extension agents; and they have more funds... Other farmers see little reason to believe that what is good for a group of people so different from themselves can also apply to them. The situation worsens as innovations take the progressives even further ahead, for then the innovations seem even less relevant.

The Tetu experiment, and work elsewhere, has indicated strongly that focusing on "average" farmers through group extension methods is likely to be more effective.[82] The group work in Tetu made a conscious effort to avoid an over-representation of progressive farmers among those attending, and concentrated on encouraging farmers to adopt hybrid maize varieties. In the event, 97 per cent of those attending subsequently adopted hybrid maize and it was estimated that each of those who received the training passed their new knowledge on to an average of three other farmers. The group method was thus found a highly effective way of diffusing this improvement among the farming population.

In endeavouring to assess the equity impact of the extension services, we have to consider the ability of these to raise agricultural incomes. There are some observers who suggest that extension obtains few results. If that were the case, the service would be largely irrelevant to income distribution and poverty, except to the extent that it could be made more effective in the future. However, many would dispute such a view. The most important agricultural innovation since independence has been the widespread adoption of hybrid maize, and there is a good deal of evidence that this was disseminated among smallholders largely through the efforts of extension agents.[83] The fact that farmers themselves behave as if the service is valuable, competing for the time of the agents, in itself creates a presumption that the service does help to raise incomes. If this is accepted, it is nevertheless clear from the foregoing that in the past the service has tended to widen income disparities, by offering superior access to the wealthier progressive farmers. That this is the case is now generally accepted, and the Fourth Plan aspires to provide the remedy.

Ascroft et al. argue that extension creates poverty in Kenya. However, their concern is with relative, rather than absolute, poverty. They show that extension has widened the gap between the most prosperous and the poorest farmers, but that this has been achieved by enriching the most prosperous rather than by impoverishing the poorest. In other words, their conclusion merely serves to confirm the familiar one that the service has worsened the distribution of rural income. Mbithi, however, has suggested that some actual impoverishment may also have been caused, although this is more a hypothesis than a research result.[84] He argues that the traditional progressive farmer approach gave the ordinary, less educated farmer a feeling of helpless ignorance and inferiority: "... the farmer came to believe that he knew nothing, and should listen, initiate and obey". Farmers lost confidence in their own ideas and came to rely increasingly on government agencies; these, however, had only limited help to offer, especially to the poorer, less educated farmers. This would be an interesting hypothesis for further exploration. On the basis of existing evidence, however, it seems unlikely that the service could have worsened absolute poverty in more than a minor way. Its main thrust has surely been to reduce it, even if not to an optimum extent.

Questions of rural-urban balance

Evidence of a pro-urban bias

It has been shown above that average real income in the non-agricultural sector is a great deal higher than average income in agriculture. Large urban-rural income disparities are the outstanding fact, however, and these are also reflected in big differences in access to wage employment. It has been estimated that in 1976 some 46 per cent of the urban labour force was in modern sector wage employment, against only 20 per cent in the rural areas.[85] We have also shown that great inequalities exist between the provinces, much to the advantage of the capital city.

That the enjoyment of real incomes, modern employment, amenities and services so strongly favours urban communities is itself partly a result of a persistent tendency for the internal terms of trade to be to the disadvantage of agriculture. Estimates of the domestic terms of trade show a fairly clear downward trend between 1964 and 1975, despite temporary recoveries in 1969-70 and 1972.[86] There was a dramatic reversal in 1976-77 as a consequence of abnormally high world prices for coffee and tea, almost all of which were passed on to the farmers; this, however, was an essentially temporary phenomenon and the index began to fall again in 1978. For almost all the post-independence period the farmers have had to contend with relative prices moving to their disavantage, with a corresponding gain to those prevailing in other sectors of the economy. The outcome of the conditions described above has been a large net outflow of resources from agriculture and the rural economy. In results already mentioned earlier, Sharpley has estimated the capital outflow from agriculture relative to the total value of capital formation in the entire national economy: the agricultural surplus over the period 1964-77 is shown to have been equivalent to 75 per cent of the country's entire capital formation (or 71 per cent excluding the abnormal years 1966-77). In addition to this financial outflow, there has also been a large outflow of labour and human capital, in the form of rural-to-urban worker migration. This is most strikingly illustrated by the fact that 85 per cent of males aged 20-34 living in Nairobi in 1969 were immigrants from other parts of the country.[87]

Given these facts, it is fair to say that the development of Kenya's economy has displayed a pro-urban bias. Since, both absolutely and relatively, the incidence of poverty in Kenya is highest in the rural areas, this urban bias has been instrumental in aggravating poverty, as well as urban-rural inequality. The question that concerns us next is the extent to which this bias has been the result of government policies. This will now be discussed, as well as the influence of policy on the concentration of income within the rural economy. Here too we are forced to be selective in our coverage and concentrate on the impact of the state provision of social and infrastructural facilities and of industrial policies.

Government provision of social and infrastructural facilities

Our discussion of the impact of social services and infrastructural investments by the State can usefully commence with a reference back to table 8. This sets out ten different indicators of access to public services, province by province. In five out of the nine cases for which complete data are available, Nairobi is recorded as having the best access, and it comes second in three out of the remaining four cases. Column K of the table, which records total recurrent expenditure per head on the part of the Government and thus incorporates all government services, is probably the most significant of all the indicators. The statistics on this are enormously to the advantage of Nairobi, although a strong bias is inevitable because Nairobi is the capital city of a strongly centralised State.

Even if we confine our attention to the provinces other than Nairobi, further strong tendencies become apparent. We see, for almost all indicators, marked differences in the degree of access for the various provinces. This is true of the quantity and quality of education (column D of table 8); of the health services (F to I); expenditure on roads (J); and of total recurrent spending per head on the part of the Government (K).

There are, moreover, systematic consistencies in these regional differences. The Coast and (especially) Central Provinces generally do well; Nyanza and, to a lesser extent, Western Province fare poorly. In order to test for the existence of statistically significant consistencies in the degree of provincial access, we measured the rank correlation between provincial income per head, on the one hand, and, on the other, the mean values of the rankings in columns B to J of table 8, obtaining a Spearmans coefficient value of 0.76. We then tested for the rank correlation between income per head and the rankings of column K, obtaining a coefficient of 0.96. Albeit rather crude, these results are certainly consistent with a conclusion that government services tend to reinforce rural regional inequalities, as well as the pro-urban bias.

This evidence of a close association between government spending patterns and regional inequalities is strongly consistent with the findings of other studies. An important example of such a study is that of Nyangira,[88] who found a strong statistical relationship, at district level, between the degree of modernisation and the allocation of public resources—a relationship, he suggests, which runs causally from the former to the latter. A less powerful but still significant finding was that, public resource allocation was statistically (and in his view, causally) associated with certain political variables, the most important being the number of ministers and assistant ministers originating in each district. A study of the distribution of educational resources,[89] similarly concludes that "educational benefits are being distributed in favour of the economically and politically powerful districts and provinces".

A moderate conclusion from this discussion is that government policies have done little to alleviate the "natural" inter-province disparities. A stronger, but more arguable, conclusion would be that the Government has actually made things worse. It should be noted, in this context, that columns (E), (I) and (J) of table 8 relate to government *capital formation* in education, health and roads respectively. Now, if the Government were seeking to use its spending powers to redress regional inequalities, we should look for a pattern of capital formation—the building of future income-generating capacity—markedly different from existing income disparities. In fact, however, it can easily be observed that (with the questionable exception of road building in Nairobi), government capital formation rather faithfully reflects existing income differentials. These facts and Nyangira's work incline us to the stronger conclusion that government efforts have in the past tended to worsen both the urban-rural imbalance and inequalities among the provinces other than Nairobi. If this is so, they have also tended to perpetuate low incomes in the western part of the country—the area where there is most poverty.

Quite apart from the quantitative and qualitative indicators discussed so far, there may be more subtle influences at work which also have an adverse influence on the relative welfare of the rural population. We particularly have in mind evidence that the type of formal schooling provided by the State, strongly geared as it is to formal-sector and white-collar job aspirations, may have a harmful effect on rural productivity. This is most likely to happen by conditioning the values of school-leavers so as to repel them from traditional occupations and induce them to migrate to the towns in search of work. Since those most likely to do so are young men, this is a serious loss of valuable labour to the rural economy.

But even among those who remain in (or return to) the rural economy, the evidence suggests that schooling could make them less productive. This, at least, is the conclusion of Hopcraft.[90] From a study of a sample of 1,500 smallholders, he concluded that the educational system tended to alienate people from the farm. He found little evidence of improved farm productivity attributable directly to schooling, and some evidence to the contrary. He found no indication that the educated had better knowledge of farming practices, although he did find that schooling broadened horizons, so that the educated person who finally commits himself to farming is more likely to be an innovator. Leonard also tested the effects of education and found that "secondary education has a uniformly detrimental effect on the work performance of agricultural extension agents...".[91] This he attributed to frustrated ambitions for higher-status work, consequential alienation from extension work and low job satisfaction. Ex-primary school pupils, with lower aspirations, were more productive.

Differential access to the Government's family planning service is another of the ways in which urban-rural inequality and rural poverty are likely to be perpetuated. Official estimates of Kenya's population growth show that, at 3.9 per cent per annum, the country has almost the fastest population growth rate in the world and a growth that has accelerated during the past two decades.[92] Rising fertility is part of the reason for this, with considerably higher fertility rates in the rural areas and with the highest rates being recorded in the western part of the country—the region which also has the highest incidence of poverty.

Rapid population growth and the associated high densities in the more fertile zones of the country serve to aggravate the land hunger and landlessness already reported. The expansion in the number of people pre-empts a large part of the country's capacity for investment, as well as adding to the difficulties of absorbing additions to the labour force in productive employment. On the basis of evidence from numerous other countries,[93] it is also probable that those already living in poverty in Kenya are the ones with the largest families, thus aggravating their condition.

Against this background, it is clear that measures to reduce fertility are likely to be crucial to a successful assault on poverty in Kenya. Kenya has, in fact, had a nominal population policy since 1967 and is the first country of tropical Africa to adopt one. However, from the simple fact that 12 years later fertility is still rising it can be deduced that the policy has been ineffective. In fact, official estimates show that only 3.5 per cent of women aged 15-49 were employing some efficient method of family planning in 1977-78—a number which includes some who have adopted birth control measures outside the official family planning programme. Moreover, a disproportionate share of the small number using family planning facilities are in the towns. For instance, Nairobi alone accounted for 38 per cent of all visits to family planning clinics in 1977, even though it contains only about 5 per cent of the national population. More generally, the programme has been found to be strongly oriented to the better-educated young women living in the towns.[94]

As one final indicator of urban bias, we should refer to the low and apparently falling share of government development spending on rural roads. During the Third Plan period, and quite contrary to the Government's stated intentions, the share of development spending on rural roads (as proxied by the category "minor and secondary" roads) fell from 33 per cent in 1973-74 to 20 per cent in 1977-78. The mileage of such roads and their share in the national roads network fell over the same period.[95]

To sum up, then, we see that the impact of the state provision of social and infrastructural facilities has been generally negative, with respect to both urban bias and inequalities between rural regions. Access to these facilities has conformed to, and quite possibly worsened, regional income disparities. There is little evidence of any *systematic* attempt by the Government to use its spending powers as instruments of regional equalisation (although there have been particular instances of this, for example in the provision of boarding schools).

Influence of industrial policies

There are at least three respects in which state policies towards industrial development impinge upon the concerns of this paper: first, the extent to which the Government *protects* local industry from foreign competition, as well as the structure of that protection, will affect the urban-rural terms of trade; second, policies towards the *location* of industry will affect the dispersion of industry into rural localities; third, policies towards the *informal sector* will affect the well-being of traditional, rural manufacturing activities.

Like many developing countries, Kenya has pursued a development strategy which favoured rapid industrialisation. By 1978 manufacturing alone contributed 16 per cent of GDP and it has long been one of the economy's fastest-growing sectors.

The effects of these industrial policies can be summed up as follows: protection policies have encouraged a pro-urban bias and have contributed to the long-term, downward trend in the terms of trade of the rural sector, although this has not been carried to extremes. Moreover, despite ambiguity in official policies, the Government has been instrumental in influencing the siting of some modern industries in rural locations, although they may have weak links with the surrounding rural economy and the great bulk of manufacturing remains in Nairobi and Mombasa. Finally, we may suggest that, on balance, it is more likely than not that the effect of government policies has been to discourage informal manufacturing in the rural economy, and that past attempts to promote small rural industries have been ineffective.

Planning rural development

Given the deep-rooted nature of the problems of rural poverty and inequality, the Government's capacity to plan effectively for the development of the rural economy is central to the concerns of this chapter. The importance of this is particularly great because the planning and execution of rural development poses special difficulties. Thus there is a need for a particularly effective machinery of planning. However, we have already referred to failings in rural planning. More recently, Killick and Kinyua have studied the extent of implementation of the Third Plan with respect to projects and policies in agriculture and roads. It may be useful at this point to summarise their results insofar as they relate to agriculture.

With regard to project implementation, they studied the extent to which actual year-by-year spending on the agricultural projects specified in the Plan had matched up to the planned amounts. The ratio of actual spending (adjusted for price changes) to planned spending was called the "implementation ratio"; it was found for the Ministry of Agriculture as a whole that this ratio was 0.53, i.e. only about half the planned spending on projects was actually undertaken. Moreover, the ratios displayed a very wide dispersion around this mean. Only 4 per cent of actual spending was within ± 10 per cent of planned amounts, and for nearly two-thirds of the project values actual spending was less than 40 per cent of planned amounts.

The authors also studied the extent to which those agricultural policy statements in the Plan had actually been put into effect. The results were mixed. In some important areas major successes were achieved: for example, in raising absolute government spending on agriculture, and in land adjudication (although this latter may not, in fact, have owed much to the existence of the Plan). Partial success was recorded in the "Kenyanisation" of large-farm ownership and, more doubtfully, in improving smallholder access to credit. Serious failures were recorded in attempts to raise the share of government spending on agriculture; on the taxation of land; on large-farm rehabilitation; and on the creation of new settlements. Taken together with the results on project implementation, the implementation record was, at best, modest.

This type of study is subject to a number of limitations, although the authors believed that, on balance, their approach was likely to have *overstated* the degree of implementation. One limitation is that the research just summarised works from the premise that there is one aspect called "policy" and another called "implementation". In reality, however, it may be impossible to draw such a sharp distinction. Policies may become distorted or transformed during execution. This, at least, was a conclusion arrived at by Holmquist in a case study of the implementation of cattle dip projects in Kisii District, Nyanza province[96]—

There is a popular conception of planning which sees the acceptance of a plan as the "big" decision which in turn determined policy outcome. But this view precludes consideration of the decision-making process of implementation and ignores the fact that fundamental policy decisions are made during, as well as prior to, implementation.

These insights into "the decision-making process of implementation" are strongly congruent with the results of American research and further underline the great difficulties any central Government will encounter in bringing about desired changes in its impact on rural development.[97] The difficulties are likely to be particularly acute if the central authority seeks to make changes that will harm existing vested interests. It is for reasons such as these that we are cautious about the speed with which effective reforms can be expected—for example in the reorientation of the extension service.

SUMMARY OF MAIN FINDINGS

It may be useful to conclude by briefly recapitulating the main findings of this study.

Extent and nature of the problems

This survey of the literature on poverty and inequality in Kenya's rural economy should be set in the context of a *national* economy generally regarded as characterised by substantial social inequities, and of a development plan that has adopted poverty alleviation as its principal objective.

Large disparities between average living standards in the urban and rural parts of the country constitute a major source of the total problem of inequality, with the great bulk of those identified as poverty groups also based in rural areas.

Even within the predominantly rural parts of the country, there are major regional disparities, both between and within the country's administrative provinces. The western parts of the country, together with North-Eastern Provinces and substantial areas of the Eastern, Coastal and Rift Valley Provinces are the most seriously affected. Central Province and those parts of Rift Valley and Eastern Provinces growing cash crops are more prosperous. At the inter-household level, there are also major disparities with poor smallholders having an average household income only one-thirtieth the size of household incomes on large farms.

A high proportion of the smallholder population and the landless live in poverty. Some 45 per cent of the smallholder population receive an annual per adult income of less than K.sh. 500 (K£25). After subsistence, this leaves only tiny cash incomes with which to buy goods other than food and with which to build up a stock of farming and household assets.

Malnutrition is generally mild, however, and only some of it is poverty-related. Rural poverty in Kenya only exceptionally kills through starvation and associated ill-health. Rather, it takes the form of a nearly complete deprivation of the benefits of economic modernisation. In other words, it provides a strong example of the failure of the "trickle-down" view of development. Pastoralists, women and the landless are particularly vulnerable to this deprivation.

Impact of government policies

Policies towards the resettlement of land formerly held by White settler farmers have helped, in a modest way, to alleviate rural poverty. But they have also widened inequalities within the African population, both between large- and small-scale farmers and within the smallholder population. While land reform policies elsewhere have, on the one hand, helped to raise rural incomes, they have also, on the other, worsened the incidence of landlessness and increased the concentration of land ownership.

The past distribution of research, extension and other agricultural services has strongly favoured the larger-scale, wealthier and more progressive farmers, leaving many poorer smallholders severely disadvantaged.

The distribution of government social services and capital formation has served to reinforce, and probably worsen, urban-rural and regional imbalances. This distortion has thus tended to perpetuate the high incidence of poverty in the western and certain other parts of the country. There is little evidence of any systematic attempts by the Government to use its spending powers as instruments of regional equalisation.

Industrial protection policies contributed to the deteriorating terms of trade of agriculture, although these have not generally been carried to extremes. The Government has been instrumental in influencing the siting of some industries in rural locations, although these may typically have few linkages with the surrounding economy and the great bulk of manufacturing remains sited in a few urban centres. It is more likely than not that government policies have discouraged informal rural manufacturing and that its attempts to promote small-scale rural industries have been ineffectual.

The past execution of Plan intentions with respect to agriculture has been modest. The constraints on implementation remain deep-seated and are likely to prove a major obstacle to the reorientation in favour of the rural poor called for in the Fourth Plan. The Government has persevered with attempts to build up decentralised planning capacities despite setbacks, but the machinery remains weak.

Notes

[1] William J. House is currently serving as an ILO technical assistance expert in Nicosia, Cyprus; and Tony Killick is currently on the research staff of the Overseas Development Institute, London. At the time of preparing this survey both were on the staff of the University of Nairobi. Neither their present nor past employers necessarily agree with the views expressed here. We are greatly indebted to the assistance of John Murugu in compiling material for this chapter.

[2] J. Heyer, J. K. Maitha and W. M. Senga (eds.): *Agricultural development in Kenya: An economic assessment* (Nairobi, Oxford University Press, 1976) is highly recommended to the reader who wishes to explore this topic in greater depth.

[3] See D. J. Pratt, P. J. Greenway and M. D. Gwyne: "A classification of East African rangelands", in *Journal of Applied Ecology,* 1966, No. 3, for a scientific classification of Kenya's land.

[4] From W. M. Senga: "Kenya's agricultural sector", table 15, in Heyer et al., op. cit.; and Kenya: *Economic Survey 1979* (Nairobi). Figures for the earlier years have been adjusted upward to make them comparable with later data.

[5] On the structure of rural employment, see Kenya: *Development Plan 1979-83* (two parts) (Nairobi, 1979), table 2.1. Data from phase I of the Integrated Rural Survey (IRS), referred to later, show non-farm operating surpluses and income from paid employment as comprising 38 per cent of rural incomes, excluding remittances and other gifts.

⁶ See L. Cliffe: "Underdevelopment or socialism? A comparative analysis of Kenya and Tanzania", in R. Harris (ed.): *The political economy of Africa* (Cambridge, Mass., 1975); and C. Leys: *Underdevelopment in Kenya* (London, Heinemann Educational Books, 1975).

⁷ H. Chenery, M. S. Ahluwalia, C. L. G. Bell, J. H. Duloy and R. Jolly (eds.): *Redistribution with growth* (London, Oxford University Press, 1974).

⁸ E. Crawford and E. Thorbecke: *Employment, income distribution, poverty alleviation and basic needs in Kenya*, Report of an ILO consulting mission, Apr. 1978 (Ithaca, NY, Cornell University, 1978; mimeographed).

⁹ ibid., p. V-8.

¹⁰ H. Rempel and W. J. House: *The Kenyan employment problem* (Nairobi, Oxford University Press, 1978), p. 21.

¹¹ A. Bigsten: *Regional inequality and development: A case study of Kenya* (University of Gothenburg, 1978).

¹² H. Rempel: *An estimate of Kenya's labour force*, Working Paper No. 159 (Institute for Development Studies, University of Nairobi, May 1974).

¹³ Crawford and Thorbecke, op. cit.

¹⁴ Integrated Rural Survey for 1974-75, tables 6.10, 8.2 and 8.6.

¹⁵ L. D. Smith: *Low income smallholder marketing and consumption patterns: Analysis and improvement, policies and programmes* (Rome, FAO, Marketing Development Project, Sep. 1978), p. 30.

¹⁶ The 6.7 per cent of holdings which reported negative incomes had attributes which suggest that they are normally relatively well off, but had suffered a substantial drop in cattle valuations for 1974-75. They were excluded from the following analysis by L. D. Smith: "An overview of agricultural development policy", in Heyer et al., op. cit., Ch. 4.

¹⁷ Smith: *Low income smallholder marketing...*, op. cit., pp. 26 and 36.

¹⁸ ibid., pp. 38-42.

¹⁹ J. Heyer: "The origins of regional inequalities in smallholder agriculture in Kenya, 1920-73", in *East African Journal of Rural Development*, 1975, Vol. 8, Nos. 1 and 2, p. 160.

²⁰ Integrated Rural Survey for 1974-75, table 6.15.

²¹ T. Kmietowicz and P. Webley: "Statistical analysis of income distribution in the Central Province of Kenya", in *Eastern Africa Economic Review*, Dec. 1975.

²² To generate this income level they assume the ratio of food to total consumption is 0.78, based on expenditure patterns of rural households with incomes of less than K.sh. 3,000.

²³ M. Shah: *Food demand projections incorporating urbanisation and income distribution, Kenya (1975-2000)* (Laxenburg (Austria), Food and Agriculture Program, ILASA).

²⁴ Integrated Rural Survey for 1974-75, table 6.16. See also D. M. Hunt: *Agricultural innovation in Mbere*, Working Paper No. 166 (Institute for Development Studies, University of Nairobi, June 1974); *Growth versus equity: An examination of the distribution of economic status and opportunity in Mbere, Eastern Kenya*, Occasional Paper No. 11 (Institute for Development Studies, University of Nairobi, 1975); *Chayanov's model of peasant household resource allocation and its relevance to Mbere Division, Eastern Kenya*, Working Paper No. 276 (Institute for Development Studies, University of Nairobi, Aug. 1976). Also F. M. Rukandema: *Some economic arithmetic of poverty: Preliminary farm data from Bukura and Shitoli sublocations of Kakamega District, Western Kenya*, Working Paper No. 253 (Institute for Development Studies, University of Nairobi, Dec. 1975).

²⁵ Crawford and Thorbecke, op. cit., table 5, p. II-11.

²⁶ Kenya: *Development Plan 1979-83* (two parts), op. cit., p. 21.

²⁷ P. Collier: *Notes on the problem of poverty in Kenya* (Sep. 1978; mimeographed), p. 1; also idem: *The rural economy of Central Province* (Apr. 1978; mimeographed).

²⁸ Audrey Chapman Smock: "Women's economic roles", in Tony Killick (ed.): *Papers on the Kenyan economy: Performance, problems and policies* (London and Nairobi, Heinemann Educational Books, 1982), pp. 219-227.

²⁹ See Kenya: *Economic Survey 1979*, op. cit., pp. 22 and 35. Also P. Mbithi and C. Barnes: *Spontaneous settlement problems in Kenya* (Kampala, East African Literature Bureau, 1975); and Kenya: *Statistical Abstract 1978*, op. cit.

[30] D. J. Campbell: *Development or decline? Resources, land use and population growth in Kajiado District,* Working Paper No. 352 (Institute for Development Studies, University of Nairobi, Mar. 1979), p. 21.

[31] Kenya: *Economic Survey 1979,* op. cit., table 5.6, p. 151.

[32] See "Rural Kenyan nutrition survey", in *Social Perspectives,* 1977, No. 4.

[33] Wanjohi et al.: *Report on nutrition status of the Mwea-Tabere irrigation scheme community* (1977; mimeographed).

[34] See Kenya, Ministry of Finance and Planning: *Food and nutritional strategy for Kenya* (n.d.; mimeographed). Also D. M. Blankhart: "Human nutrition", in Vogel et al.: *Health and diseases in Kenya* (Nairobi, East African Literature Bureau, 1970).

[35] World Bank: *World Development Report 1978* (Washington, DC, 1978), p. 104; and Kenya: *Economic Survey 1979,* op. cit.

[36] See Smith: "An overview of agricultural development policy...", op. cit., Ch. 4 on the history of agricultural policy in Kenya; also Leys, op. cit., especially Ch. 3.

[37] Smith: "An overview of agricultural development policy...", op. cit., p. 111.

[38] R. von Kaufmann: "The development of the range land areas", in Heyer et al., op. cit., Ch. 8, p. 275.

[39] R. J. M. Swynnerton: *A plan to intensify the development of African agriculture in Kenya* (Nairobi, Government Printer, 1954), p. 10.

[40] J. Heyer and J. K. Waweru: "The development of the small farm areas", in Heyer et al., op. cit., Ch. 6, p. 208-211.

[41] ILO: *Employment, incomes and equality: A strategy for increasing productive employment in Kenya* (Geneva, 1972).

[42] Chenery et al., op. cit.

[43] Tony Killick: "Strengthening Kenya's development strategy: Opportunities and constraints", in *Eastern Africa Economic Review,* Dec. 1976, pp. 1-33; also idem: "By their fruits ye shall know them: Kenya's Fourth Development Plan", in Killick: *Papers on the Kenyan economy,* op. cit., pp. 97-108.

[44] Kenya: *Economic Survey 1979,* op. cit., p. 2.

[45] Based on data provided in M. J. Dorling: *Income distribution in the small-farm sector of Kenya: Background to critical choices,* Paper presented to the Kenya Chapter of the Society for International Development (Nairobi, July 1979), p. 20.

[46] Collier: *Notes on the problem of poverty in Kenya,* op. cit.

[47] Mbithi and Barnes, op. cit.

[48] World Bank: *Kenya, into the second decade* (Baltimore, Johns Hopkins University Press, 1975), p. 468.

[49] K. Leitner: "Kenyan agricultural workers", in *Review of African Political Economy,* May-Aug. 1976, p. 34.

[50] W. J. House: *Nairobi's informal sector,* Working Paper No. 347 (Institute for Development Studies, University of Nairobi, 1978), and idem: "Nairobi's informal sector: An exploratory study", in Killick: *Papers on the Kenyan economy,* op. cit., pp. 357-368.

[51] These estimates are based on figures given in the text and on information provided by J. K. M. Kinyua: *Plan implementation in Kenya, 1974-78,* M.A. thesis (University of Nairobi, June 1978), table 4.2; see also T. Killick and J. K. M. Kinyua: "Development Plan implementation in Kenya", in Killick: *Papers on the Kenyan economy,* op. cit., pp. 109-116; Leys, op. cit., p. 89; and Kenya: *Development Plan 1979-83,* op. cit.

[52] See S. N. Hinga and J. Heyer: "The development of large farms", in Heyer et al., op. cit., Ch. 7, p. 240, for an illustrative breakdown of large-farm ownership categories for land transferred in Trans-Nzoia District.

[53] See Kenya: *Economic Survey 1979,* op. cit., p. 35 and Killick: "Strengthening Kenya's development strategy...", op. cit., pp. 19-21.

[54] See M. Fg. Scott, J. D. MacArthur and D. M. G. Newbery: *Project appraisal in practice* (London, Heinemann Educational Books, 1976), Part II; and D. Berg-Schlosser: *The distribution of income and education in Kenya: Causes and potential political consequences* (Munich, Weltforum-Verlag, 1970).

[55] G. Holtham and A. Hazlewood: *Aid and inequality in Kenya* (London, Croom Helm, 1976), p. 224; also Smith: *Low income smallholder marketing...*, op. cit., table 5; and Heyer and Waweru, op. cit., p. 215.

[56] Kenya, Central Bureau of Statistics: *Statistical Abstract 1978* (Nairobi), table 82a.

[57] See Kenya: *Development Plan 1966-70* (Nairobi, Government Printer, 1966), p. 126; idem: *Development Plan 1970-74* (Nairobi, Government Printer, 1969), p. 210; idem: *Development Plan 1974-78*, Part I (Nairobi, 1974), p. 217; idem: *Development Plan 1979-83*, two parts (Nairobi, 1979), p. 289.

[58] See Killick and Kinyua, op. cit.; and J. T. Fleming: "Tenurial reform as a prerequisite to the green revolution", in *World Development*, Jan. 1975, p. 55.

[59] W. J. Barber: "Land reform and economic change among African farmers in Kenya", in *Economic Development and Cultural Change*, Oct. 1970, table 2.

[60] H. W. O. Okoth-Ogendo: "African land tenure reform", in Heyer et al., op. cit., Ch. 5.

[61] See N. Ng'ethe et al.: *Reaching the rural poor: Lessons from the Kenyan special rural development programme*, Working Paper No. 296 (Institute for Development Studies, University of Nairobi, Jan. 1977), p. 14; also J. Ascroft et al.: "Does extension create poverty in Kenya?", in *East Africa Journal*, Mar. 1972, p. 32.

[62] Kenya: *Economic Survey 1979*, op. cit., pp. 53 and 282.

[63] From D. K. Leonard: *Reaching the peasant farmer* (Chicago, University of Chicago Press, 1977), p. 247, and Part II of the Third and Fourth Plans it can be estimated that the share of research in the Ministry's total development budget fell from 21 per cent in 1958-59, to 17 per cent in 1963-64, to about 4 per cent in the Third and Fourth Plan periods. However, it is difficult to identify all research projects with certainty, especially in the Fourth Plan, so there may be some underestimation in the latter two percentages. Heyer and Waweru, op. cit., Ch. 6, p. 202, mention the Ministry of Agriculture spending 10 per cent of its total (recurrent plus development) budget on research in 1973-74.

[64] Heyer and Waweru, op. cit., p. 202.

[65] Senga, op. cit., p. 104.

[66] Kenya: *Economic Survey 1974*, op. cit., p. 198.

[67] For similar complaints along these lines see the essays by Senga, Smith, and Heyer and Waweru, all in Heyer et al., op. cit.

[68] P. Wyeth: "Economic development in Kenyan agriculture", in Killick: *Papers on the Kenyan economy*, op. cit.

[69] J. Gerhart: *The diffusion of hybrid maize in Kenya*, Ph.D. thesis (Princeton, New Jersey, Princeton University, 1975).

[70] Kenya: *Economic Survey 1979*, op. cit., p. 209.

[71] idem: *Economic Survey 1974*, op. cit., p. 208.

[72] Barbara Jamieson: *Resource allocation to agricultural research in Kenya from 1963 to 1978*, Working Paper No. 345 (Institute for Development Studies, University of Nairobi, Oct. 1978), p. 7.

[73] From private correspondence with Ms. Jamieson.

[74] See Kenya: *Economic Survey 1979*, op. cit., p. 210; and Killick and Kinyua, op. cit., table 2.

[75] See Wyeth, op. cit.

[76] Everett Rogers: *Diffusion of innovations* (New York, Free Press, 1962).

[77] Ascroft et al., op. cit.

[78] Leonard, op. cit., p. 177.

[79] See K. A. Staudt: *Inequalities in the delivery of services to a female farm clientele: Some implications for policy*, Discussion Paper No. 247 (Institute for Development Studies, University of Nairobi, Jan. 1977); also ILO: *Employment, incomes and equality...*, op. cit., p. 153.

[80] There is an extensive literature on the SRDP. Those wishing to study it further should see two evaluations of the programme by the Institute for Development Studies, University of Nairobi: *An over-all evaluation of the SRDP*, Occasional Paper No. 8 (1972), and idem: *Second over-all evaluation of the SRDP*, Occasional Paper No. 12 (1975). I. Livingston: "Experimentation in rural development: Kenya's special rural development programme", in *Agricultural*

Administration, 1976, No. 3; M. David: *SRDP in Kenya: Can experiments in development succeed?*, Working Paper No. 254 (Institute for Development Studies, University of Nairobi, 1976); and J. W. Leach: "The Kenya special rural development programme", in *Journal of Administration Overseas*, Apr. 1974, should also be consulted.

[81] These extension experiments are discussed in Ascroft et al., op. cit.: Ng'ethe et al., op. cit.; and R. Chambers: *Managing rural development* (Uppsala, Scandinavian Institute of African Studies, 1974).

[82] See S. Schönherr and E. S. Mbuga: *New extension methods to speed up diffusion of agricultural innovations*, Discussion Paper No. 200 (Institute for Development Studies, University of Nairobi, May 1974); also Ng'ethe et al., op. cit.; and Leonard, op. cit.

[83] Gerhart, op. cit.

[84] P. Mbithi: "Going it alone", in *Development Forum*, Mar. 1979; Ascroft et al., op. cit.

[85] H. Rempel: "The labour market", table 7, in Killick: *Papers on the Kenyan economy*, op. cit.

[86] This discussion is based on estimates in J. Sharpley: "Resource transfers between agricultural and non-agricultural sectors, 1964-77", table 1, in Killick: *Papers on the Kenyan economy*, op. cit.; and Kenya: *Economic Surveys 1978* and *1979*, op. cit.

[87] Rempel: *An estimate of Kenya's labour force*, op. cit., table 4.

[88] See ILO: *Employment, incomes and equality...*, op. cit., pp. 78-81; Bigsten, op. cit., Ch. XIII; N. Nyangira: *Relative modernisation and public resource allocation* (Nairobi, East African Literature Bureau, 1975).

[89] K. Kinyanjui: *The distribution of educational resources and opportunities in Kenya*, Discussion Paper No. 208 (Institute for Development Studies, University of Nairobi, 1974).

[90] P. N. Hopcraft: *Human resources and technical skills in agricultural development: An economic evaluation of educative investments in Kenya's small farm sector*, Ph.D. thesis (Stanford University, 1974); also idem: *Does education increase farm productivity?*, Working Paper No. 279 (Institute for Development Studies, University of Nairobi, 1976).

[91] Leonard, op. cit., p. 126.

[92] See Kenya: *Economic survey 1979*, op. cit., for fuller details of the demographic situation.

[93] See S. Anad: "Aspects of poverty in Malaysia", in *Review of Income and Wealth*, Mar. 1977; also A. Fishlow: "Brazilian size distribution of income", in *American Economic Review* (Menasha, Wisconsin), May 1972.

[94] See L. Bondestam: *Population growth in Kenya* (Uppsala, Scandinavian Institute of African Studies, 1972); Kenya: *Economic Survey 1979*, op. cit., table 15.15; idem: *Statistical Abstract 1978*, op. cit., table 232.

[95] Sources: Killick and Kinyua, op. cit.; Kenya, Central Bureau of Statistics: *Statistical Abstract*, various issues; Kenya: *Sessional paper No. 4 of 1975: On economic prospects and policies* (Nairobi, 1975), p. 20.

[96] F. Holmquist: "Implementing rural development projects", in Goran Hyden, Robert Jackson and John Okumu (eds.): *Development administration: The Kenyan experience* (Nairobi, Oxford University Press, 1970), Ch. 10.

[97] See R. A. Bauer and K. J. Gerger (eds.): *The study of policy formation* (New York, Free Press, 1968); and J. L. Pressman and A. Wildowsky: *Implementation* (Berkeley, University of California Press, 1973).

GROWTH AND INEQUALITY: RURAL DEVELOPMENT IN MALAWI, 1964-78[1]   3

Dharam Ghai and Samir Radwan

The striking feature of Malawi's experience since independence was achieved in 1964 is a rapid rate of economic growth, the main source of which has been the expansion of cash crop production for exports. The growth has also been accompanied by some important structural changes. This performance appears particularly impressive when one considers the low level of development which had been attained by the country at the time of independence: very low income per head and heavy dependence on foreign assistance both for current public expenditure and for investment. This progress has been brought about mainly through the conscious growth-oriented strategy which the Government has followed. The purpose of this chapter is to examine the implications of this pattern of growth for rural development and, in particular, to assess the impact of rapid growth on income distribution and poverty in the rural areas of Malawi.

We thus begin with an analysis of growth and structural change, laying particular emphasis on the rate, pattern and sources of growth in the agrarian economy. In the next section, we attempt to estimate the distributional effects of growth and the extent to which the various socio-economic groups have benefited from it. This is followed by a brief discussion of the impact of government policies on growth and equity. The chapter concludes with an over-all assessment of the experience of rural development since independence.

GROWTH AND STRUCTURAL CHANGES, 1964-78

Background

Nyasaland (as pre-independence Malawi was known) was established as a British protectorate in 1891. Prior to its independence in 1964, it formed part of the Federation of Rhodesia and Nyasaland over the period 1953 to 1963. In pre-colonial Malawi the bulk of economic activity consisted of cultivation, hunting and livestock rearing. The main crop cultivated was maize. Agriculture was family-based and was practised on a shifting cultivation basis. The tools of production consisted of hoes and axes. Land was owned by the tribe and was allocated to individual families for use by the chief. The trade in goods was confined to the limited export of ivory, beeswax and hand-smelted iron products in exchange for cloth and guns. Subsequently, trade in slaves became increasingly important.[2]

The major changes brought about by colonialism were the abolition of slavery and the establishment of estate agriculture geared to the production of cash crops for export. The value of exports increased from negligible amounts in the late nineteenth century to a £250,000 in 1914, reaching a total of £1 million in 1940 and £9 million in 1960.[2] Over this period there were significant changes in the composition of exports: ivory and rubber gave way to coffee; when this crop failed, its place was taken by tobacco and cotton, and tea at a later stage. The bulk of the exports came from European plantations but there was a steady increase in cash crops grown by African farmers, especially in the 1920s. The early years of the colonial period also saw the recruitment of Nyasa workers for employment in the mines and farms of neighbouring countries; this was to assume increasing importance in subsequent years and became a notable feature of the Malawi economy. There was little mining or industrial development. In general, Nyasaland ranked low in the colonial priorities for development. Hence the often-quoted description of the country as a "colonial slum".

The economic structure and pattern of production remained essentially unchanged during the Federation interlude. Although it was the poorest of the three countries, Nyasaland gained little from her participation in the Federation. Thus, as will be shown later, at independence Malawi inherited a poor, underdeveloped economy, dependent largely for income generation on agricultural exports and remittances from workers in other countries, and with a backward social and economic infrastructure.

Before we discuss developments since independence, it may be useful to provide some salient features of the topography and resources of the country. Malawi is a landlocked country, bordered on the north and north-east by Tanzania, on the east and south by Mozambique and in the west by Zambia. It has an area of 118,500 square kilometers of which 94,300 square kilometers is land area and 24,200 square kilometers is lake surface. The land area consists of three topographically different regions. The Central Region consists largely of plateaux of about 1,500 metres in height, with a strip of lower-lying land along Lake Malawi. The Northern Region is mountainous, with altitudes of up to 2,500 metres. The Southern Region, which has some scattered mountains, is for a large part low-lying flat terrain about 500 metres above sea level.[3] The climate is monsoonal with a single rainy season from November to March, the rainfall averaging 30 inches (750 millimetres) with higher altitudes receiving up to 64 inches (1,000 millimetres). Other months of the year are comparatively cool and dry. There is considerable variability of rainfall from year to year.[4]

In 1977 the population was 5.6 million, estimated to be increasing by about 3 per cent per annum. Table 14 gives regional estimates of total, arable and cultivated land and population density.

About 37 per cent of the land area is classified as arable, of which 86 per cent is estimated to be cultivated in 1977. The bulk of the population is concentrated in the Southern (50 per cent) and Central (38 per cent) Regions. The cultivated area per head of the population is 2.6 acres in the Northern Region compared with 1.0 acre in the Southern. It may be pointed out that these figures do not take into account the marked differences in the quality of land among regions.

Growth, exports and employment

The period since independence has been characterised by remarkable growth and significant changes in the structure of the economy; in order to appreciate the dimensions of these changes, it is necessary to sketch the main features of the economy on the eve of independence in 1964.

The income per head in 1964 was less than K 40. Agriculture accounted for nearly 58 per cent of GDP, services for 29 per cent and industry for 13 per cent. Total subsistence output was estimated at 48 per cent of GDP at factor cost (f.c.). Within the agricultural sector, the share of monetary output was one-fourth, of which 27 per cent came from the estate sector. Exports amounted to 15 per cent of GDP (f.c.) and 29 per cent of monetary GDP (f.c.). The overwhelming proportion of exports consisted of agricultural products, tobacco and tea together accounting for two-thirds of exports. The share of estates amounted to 43 per cent of total exports. Imports accounted for 38 per cent of monetary GDP.[5]

The economy was heavily dependent upon external capital inflows. Domestic savings were negligible and national savings negative. Thus domestic investment, which amounted to 8 per cent of GDP, was financed completely through inflows of foreign resources. Only half of the current government expenditure was financed from local sources.

A similar situation prevailed with respect to the social and economic infrastructure. The country was largely dependent upon expatriates for skilled jobs. By 1965 the primary school enrolment was 44 per cent and secondary school 2 per cent. Transport and communications were grossly inadequate.

Economic growth since independence

The period since independence has witnessed remarkable economic growth. GDP, which is estimated to have risen by 4.6 per cent per annum between 1954 and 1963, grew by over 6.8 per cent annually at constant prices from 1964 to 1972, and by 6.4 per cent between 1973 and 1978 (table 15).

Table 14. Land and population, 1977

Region	Total land ('000 acres)	Potential arable land	Cultivated land	Population ('000)	Land/man ratio[1]
Northern	6 641	2 457	1 679	643	2.6
Central	8 777	3 247	2 956	2 122	1.4
Southern	7 830	2 897	2 793	2 796	1.0
Total	23 248	8 601	7 428	5 561	1.3

[1] Refers to cultivated land.
Source. FAO: *Country review paper of Malawi* (Rome, 1978).

GDP per head at constant prices grew at a rate of about 4 per cent per annum over the entire period, to reach a level of K 89.5 in 1978 (1973 prices).[6] Total investment increased rapidly from less than 8 per cent of GDP in 1964 to nearly 24 per cent in 1977-78, though with considerable annual fluctuations. At the same time, domestic savings rose from nil to over 11 per cent of GDP in 1978. Likewise the deficit on current government expenditure was steadily reduced until in 1972-73 foreign budgetary support ceased, and in most of the subsequent years a net surplus has been achieved on current account to finance a modest proportion of public development expenditure.

The over-all rapid growth has been accompanied by some significant changes in the structure of the economy. First, there has been a considerable decline in the share within the economy of agriculture, which fell from about 58 per cent in 1964 to about 52 in 1972, and further still to about 39 per cent in 1978 (all at current prices). Thus the dependence of the economy on agriculture was reduced between the early 1960s and late 1970s. The share of non-monetary output in GDP fell from 48 per cent in 1964 to around 31 per cent in 1978. Within agriculture, at 1964 constant prices, the share of estate production rose from 7 per cent (27 per cent of monetary output) in 1964 to 9.5 per cent (30 per cent of monetary output) in 1972. In 1978, at 1973 constant prices, this share amounted to 13.2 per cent, reflecting increasing dependence on estate production as a source of growth. This sector doubled its share of total agricultural output and increased its share of monetary output from just over one-fourth to 37 per cent between 1964 and 1978. It should be noted, however, that the estate share of total agricultural production is still relatively low and the main source of livelihood for the population remains the smallholder sector.

The decline in the share of agriculture is not a reflection of agricultural stagnation but of even higher rates of growth in the other sectors of the economy. Agricultural production (at constant prices) grew by 4.5 per cent between 1964 and 1972 and by a similar rate between 1973 and 1978. GDP in the industrial sector expanded by 13 per cent between 1964 and 1972 and by 7.2 per cent between 1973 and 1978, while the services sector grew by 11.7 and 8.1 per cent respectively. Growth in manufacturing was largely the result

Table 15. Selected economic indicators: annual average growth rates

Indicator	Constant 1964 prices			Constant 1973 prices			
	1964-68	1968-72	1964-72	1973-76	1976-78	1973-78[4]	1973-78[5]
Total GDP (mp)[1]	8.1	7.7	6.8	6.2	6.4	6.4	6.4
GDP (pc)[2] (mp) (K)[3]	3.4	4.9	4.2	3.6	1.1	2.6	3.6
GNP (pc) (mp) (K)	3.4	5.9	4.8	2.3	4.9	2.4	3.4
Monetary GDP (pc) (mp) (K)	6.2	7.9	7.3	5.5	2.9	4.5	5.6
Total private consumption (pc) (K)	2.4	3.8	3.0	.	.	.	.
Population	2.5	2.5	2.5	2.6	5.2	3.7	2.7

. = not available.
[1] Market prices. [2] Per head. [3] Kwachas. [4] Based on unadjusted census figures for population. [5] Based on adjusted population figures (by linear interpolation between 1966 and 1976 census years). It will be noted that this adjustment results in a higher rate of growth of GDP per head since it evens out the sudden rise in population reported by the 1976 census.
Sources. *Economic Report 1972*, tables 3.1, 2.2; *National Accounts Report 1964-1969*, table E; *Economic Report 1973*; *Malawi Statistical Yearbook 1978*.

of the emergence of processing industries and import substitution of consumer goods, notably textiles, footwear, sugar and beverages.

Expansion of exports

The rapid growth of the economy has been due principally to a dynamic expansion of exports, which increased by more than six-and-a-half times between 1964-65 and 1977-78. The dependence of the economy on exports was somewhat reduced in the early years, but by the end of the period the export orientation was greater than in 1964-65. Thus, for example, in 1977-78 exports accounted for nearly 20 per cent of GDP (current market prices) as compared with 15 per cent in 1964-65 and 14 per cent in 1971-72. This was principally due to the increasing monetarisation of total agricultural output, for as a proportion of monetary output, exports amounted to 28 per cent both at the beginning and at the end of the period, having declined to 22 per cent in 1971-72.

Between 1964-65 and 1975-76 estates accounted for over 70 per cent of the increase in agricultural exports. Thus in this sense they acted as the main "engine of growth" of the economy. The two major export crops continued to be tobacco and tea, their relative importance increasing from 65 per cent in 1964-65 to 76 per cent in 1977-78. The share of tobacco rose gradually from 37 per cent in 1964-65 to 54 per cent in 1977-78, while that of tea dropped from 28 to 22 per cent. Sugar emerged as an export crop in 1973 and accounted for 8 per cent of exports by the end of the period. The increased importance of tobacco was due in part to a sharper rise in its price as compared with tea. However, the international terms of trade, after registering an improvement in the 1960s, declined by between 10 and 20 per cent for most years in the 1970s.

Employment and migration

A notable feature of the economy during the period under review was an extraordinary expansion of wage employment in Malawi and the changing balance between employment at home and in the neighbouring countries. The 1966 population census estimated a total of 266,000 Malawians outside the country. It is not known how many of them were wage earners. On the basis of subsequent estimates of workers in South Africa and what was then Southern Rhodesia, it may be conservatively assumed that at least 200,000 were male workers. This compares with a recorded domestic wage employment in 1966 of around 130,000. The male population in the age-group 20-59 living in Malawi in 1966 was 727,000. This implies that approximately one out of five Malawian male workers were employed abroad. If we add to this the wage employees in Malawi, we get a figure of around 35 per cent of the male labour force in wage employment.[7] Even allowing for some overestimate, it appears that a relatively high proportion of adult males is in wage employment. By the same token the burden of subsistence production is disproportionately carried by women on family holdings.

In the subsequent period, especially after 1968, there was a rapid expansion of domestic wage employment. Between 1964 and 1978 it grew annually by 7.2 per cent, and for the ten years between 1968 and 1978 it expanded by nearly 9 per cent. This represents one of the fastest rates of growth in employment recorded anywhere and is quite out of line with experience in Africa and in most developing countries. Although most sectors shared in this expansion, the major contribution was made by agriculture, which accounted for 58 per cent of the increase. The share of agriculture in total employment thus rose from 38 per cent in 1964 to 50 per cent in 1978. Manufacturing also increased its share from 6 per cent in 1964 to over 10 per cent in 1978. At the same time there was a decline in the share of employment in public and some personal services. The Malawi experience is somewhat unusual in Africa also in the respect that the share of employment in the government sector fell from 33 per cent in 1968 to 25 per cent in 1976.

The expansion in wage employment was above that of GDP in real terms over the period 1964-78 but in line with the growth of monetary GDP. Although most non-agricultural sectors experienced an expansion in employment, the employment elasticity was considerably less than unity. Most of the increase in employment took place in the estate sector, which accounts for the bulk of agricultural employment.

Employment growth in the estate sector, however, shows divergent trends over the period: between 1964 and 1972 the annual increases in estate production and agricultural employment were 11.5 and 5.2 per cent respectively. The period between 1973 and 1978 shows a notable change, production and employment increasing by 9.6 and 9.2 per cent respectively. It has not been possible to determine the extent to which factors such as changing output composition, increased efficiency or changes in techniques of production are responsible for this divergent behaviour over the period. It seems highly likely, however, that relatively low wages for unskilled workers provided a strong incentive for labour-intensive patterns of production and for rapid expansion of estate production.

The question of the distribution of benefits of rapid growth is discussed in a subsequent section. Here it is useful to recall that real wages on the estates fell steadily over the period 1968 to 1976, showing a small rise in 1977. Thus the expansion of employment has taken place in the context of declining real wages for most of the period. It should be remembered that the number of workers abroad continued to increase until 1973. That real wages should have declined in the face of rapid increase in demand for labour is due principally, as explained below, to the operation of a rigorous incomes policy by the Government.

The situation took a dramatic turn in late 1974 when all recruitment of labour to South Africa was suspended following the death in an aircrash of migrant workers being taken to that country. In 1974, prior to suspension of recruitment, it was estimated that there were 250,000 Malawi workers in South Africa and Southern Rhodesia as compared with 265,000 in Malawi itself.[8] The great majority of these workers returned to Malawi in the subsequent period. The 1976 population census estimated that a total of 240,000 Malawian workers had returned from South Africa. The recruitment of workers for work in South Africa was resumed in 1977 but a ceiling was set by the South African Chamber of Mines at 20,000 workers annually. Thus, despite the extremely rapid growth of domestic employment over this period, the proportion of the total labour force in wage employment was considerably less in 1978 than at the time of independence.

The implications of such large-scale migration on income distribution and rural poverty are mentioned later, but it may be useful to indicate some macro-economic effects here. First, it is evident that despite relatively low wages in absolute terms earned by migrants, particularly up to the early 1970s, the total earnings in relation to opportunities at home were sufficiently attractive to encourage a large and continuing flow of emigrant workers. Given the sexual division of labour in agriculture, where planting and harvesting could be left to women, migration provided a convenient means of earning cash and of maximising family income.[9] The fact that migration was of a seasonal nature enabled the male workers to continue to play their role in land clearance and preparation for cultivation.

Second, while accurate data on savings and transfers by migrant workers are not available, remittances through official channels give some idea of the magnitudes involved: these rose more than tenfold between 1964 and 1975, to reach a peak of K 32 million. In the latter year the remittances were nearly equal to the total monetary output in the smallholder sector and constituted over 30 per cent of export earnings; they were thus a crucial element in financing imports needed for rapid growth. It is not surprising, therefore, that the sudden cessation of migration in 1974 has had strong adverse effects, at least in the short run, on rural incomes and foreign exchange earnings.

Third, the rapid expansion of domestic wage employment, despite low wages, particularly in agriculture, should probably be interpreted in the same manner—namely that it provided an opportunity for rural families to acquire cash earnings and to maximise family income, especially in view of the fact that subsistence output could be maintained either on family plots or on the land provided by the estates. While the existence of the family farm ensured a cheap source of labour for plantations in Malawi and for mines and farms abroad, it also enabled rural households to enhance family incomes beyond what might have been otherwise possible. One implication of such a pattern of allocation of family labour is that the wage sector and subsistence production cannot be treated in isolation from each other and, in particular, that an adequate analysis of production, consumption, rate and distribution of income at the level of the rural household must be based on an integrated view of multiple sources of income.

Conclusion

Before we proceed to a discussion of the modes of production in the rural economy, it may be useful to recapitulate the highlights of economic growth and change since independence. The rapid expansion of the economy led to an increase of somewhat less than 4 per cent in annual income per head. Although the share of agriculture fell markedly over the period, it is the rapid growth of this sector which gave impetus to the entire economy. The driving force behind agricultural growth in turn was the impressive expansion of exports, principally tobacco and tea and, at a later stage, sugar. A major contributory factor was the extraordinary growth in estate production, which also played a significant part in the very rapid growth in total wage employment. Large-scale migration provided an important source of employment, incomes and foreign exchange until 1975.

Rural economy

The two dominant modes of agricultural production in Malawi—estates and smallholders—exhibit sharp differences in size of operation, technology, productivity, yields, composition of output and labour use. At the same time, there have been significant changes in some important characteristics of these sectors during the period since independence.

Table 16 sums up the pattern of land use in the early 1970s. Of the total land area of 9.4 million hectares, half was classified as "customary land", of which only 1.9 million hectares, or 20 per cent of the total land, were under cultivation. The estate sector accounted for 1 per cent of the total land, and

Table 16. Land use

Type of land	Million hectares	% of total
Customary land	4.7	50.0
Customary land under cultivation	1.8	19.1
Customary land, fallow	2.9	30.9
Estates	0.1	1.1
Non-arable land	3.3	35.1
Other land	1.3	13.8
Total	9.4	100.0

Source. World Bank: *Agricultural Sector Review: Malawi,* Dec. 1973.

the rest represented non-arable grazing land (35 per cent) and other categories such as buildings and swamps (13.8 per cent).

Estate sector

The estate sector provided employment for approximately 130,000 workers in 1977. This compared with a probable number of smallholdings of 1.1 million in 1978.[10]

In terms of production, as indicated earlier, the large farm sector increased its share in total agriculture from 7 per cent in 1964 (27 per cent monetary agriculture) to 13.2 per cent in 1978 (36.5 per cent). Output per worker in the large farm sector in 1978 may have been of the order of K 220, as compared with approximately K 175 per holding. Assuming a total of two male adult equivalents per holding, this indicates a differential in output per man between the estate and smallholder sector of the order of 3.5.

This difference is largely a reflection of differences in techniques of production in the two sectors. The estate production is organised along the lines of classic plantations: large size of productive units, use of modern machinery and inputs such as fertilisers and improved seeds, extensive irrigation especially on sugar estates, and a full-time paid labour force. The bulk of the output is exported. The main products grown are tobacco, tea and sugar-cane, the latter especially since 1973. Whereas in 1964-65 tea accounted for 68 per cent of exports and tobacco for 25 per cent, by 1975-76 the relative share of tea had fallen to 33 per cent, that of tobacco had risen to 46 per cent and of sugar to 20 per cent.

Little information is available on the number and ownership pattern of enterprises in the estate sector. In 1971 the number of enterprises was estimated at 104, of which 61 were engaged in tobacco growing, 16 in tea and 21 in sugar-cane cultivation and fishing.[11] In 1969 it was reported that 85 per cent of estate land was in the hands of British corporate and expatriate interests.[12] Transnational corporate interests weigh heavily in this ownership pattern: thus, for example, Lonrho (United Kingdom) have extensive interests in tea, sugar and sisal estates and are in addition vertically integrated into tractors, fuel and construction enterprises; Brookers Ltd. has tea estates; while British-American Tobacco, the Imperial Tobacco Company and David Whitehead Ltd. (textiles) have processing plants using outputs from the estate and peasant sectors.

Of late, the share of foreign-owned estate lands has diminished as a result of land abandonment and incipient encroachment by landless Africans. Following these events, the Government regulated, under the Land Acquisition Act, 1970, against such independent initiatives and also purchased some abandoned estate farms for subsequent resettlement by African farmers. In consequence, large-scale freehold and long-term leasehold farming has been given a domestic rooting by new policies. These policies have included African resettlement on relatively large holdings, a small direct state participation in ownership and production, and leasehold of large- and medium-sized agricultural units by the political, bureaucratic and business élite. There has also been an increasing participation by the Malawi Congress Party's holding company (Press Holdings) which, in conjunction with the state-owned Agricultural Development and Marketing Corporation (ADMARC), inter alia owned 28 tobacco estates in 1972, accounting then for 15 per cent of the country's flue-cured tobacco crop.

Smallholder sector

The smallholder sector is the main source of income and employment for the great bulk of the population. In 1977 about 92 per cent of the population lived in rural areas, the overwhelming proportion of which were dependent directly on smallholdings for livelihood. Even for employees on estates and for emigrants, family plots represented a basic source of subsistence.

The average size of holdings is 1.5 hectares, with 1.9 hectares in the Central, 1.4 in the Northern and 1.3 in the Southern Region. The land is relatively equally distributed, especially when account is taken of the family size. In 1969 there were a total of 885,000 holdings with an average of 4.6 persons per holding, occupying a total of 1.49 million hectares of which the area devoted to crops was 1.36 million.[13] Nearly 63 per cent of the holdings were less than 1.6 hectares and about 19 per cent over 2.5 hectares. The main crops grown were maize, which occupied 78 per cent of all cultivated land (much of it in mixed stands), pulses, groundnuts, millet and cassava. Paddy, cotton and tobacco, though important sources of cash income, accounted for an insignificant proportion of cultivated land. The 1968-69 survey estimated the number of cattle at 772,000. However, only 11 per cent of holdings possessed any cattle, with the Central Region accounting for 60 per cent of the cattle owned.

Total agricultural output (at constant prices) in the smallholder sector has grown by about 3.9 per cent per annum over the period 1964 to 1978. Thus output per head has increased by 1 to 1.5 per cent per annum—a relatively good performance in comparison with countries at a comparable stage of development. Monetary output at constant prices increased rapidly at 9.1 per cent in the period 1964-72 but slowed down to 5.8 per cent between 1973 and 1978. The share of monetary output in smallholder production has risen gradually, though with considerable annual fluctuations, from around 20 per cent in 1964 to 26 per cent in 1978. Exports from the smallholder sector give some idea of the trends in production of cash crops. At the beginning of the period, three products (tobacco, groundnuts and cotton) accounted for over 80 per cent of smallholder exports, with tobacco amounting to slightly less than 50 per cent of the total. Towards the end of the period, cotton exports had practically disappeared, as it was increasingly being used domestically. The share of groundnuts also declined sharply, with tobacco emerging as the dominant crop accounting for nearly four-fifths of exports.

The commercialisation of smallholder production has been associated with the emergence of a small but growing number of "progressive" farmers. The growth of this class of farmer has been due principally to the initiation of four major development projects since 1968. The number of such farmers would, however, appear to be relatively small: one index of this is the 76,000 growers of tobacco in 1977, the number having increased from 63,000 in 1975. The total number of such "progressive" farmers is likely to be in the region of 100,000, representing perhaps around 10 per cent of smallholders. There is some evidence that they are increasingly using improved techniques of production. The 1968-69 Agriculture Survey estimated that 13 per cent of farmers used fertilisers. The consumption of fertilisers on smallholdings has increased rapidly since independence from 2,200 tons in 1964 to 26,000 tons in 1973-74 and 49,000 tons in 1977-78. The 1968-69 Survey also gave the following figures on the use of improved techniques: 41,000 oxen, 11,600 ox carts, 15,000 ploughs and 10,600 sprays. A surprising feature brought out by the Survey was that nearly 29 per cent of smallholders used hired labour on their farms.

It is interesting to note that these changes in the smallholder sector have taken place for the most part within the context of the customary land tenure system. While this reveals considerable variation, in essence the land is owned by the community or the tribe and is allocated to individual families for use by the village headman or the chief. Under the matrilineal system which prevails in the Central and Southern Regions, a husband acquires use of land through marriage and residence in his wife's village. Before being given any land for himself, the husband would have to serve a probationary period of work in the fields of his wife's parents.[14] As long as the marriage survived the husband would maintain his rights to the use of land. If, however, the marriage broke up, the husband would have to return to his own village and seek land there. As for land inheritance, the responsibility for allocating it among the children often rested with the maternal uncle.

The system is undergoing some changes: in the first place, because of growing land scarcity, the proportion of land acquired through inheritance from parents is increasing in relation to allocation of new land by the village headman or chief; second, there appears to be a decline in the proportion of husbands living in their wives' villages. Perhaps more significant for the future is the tentative attempt now being made by the Government to move towards a system of land tenure based on individual ownership with the right to sell or acquire new land. The new concept of land tenure is being applied in the Lilongwe Land Development Programme. Even there, however, it is proceeding relatively slowly: by 1977 a total of 287,500 hectares of land had been demarcated, of which 112,000 hectares were registered to different operators, most of whom were the recognised heads of extended family groups. This represents about 64 per cent of the programme area but only 2.3 per cent of the customary land under cultivation. There are no plans to give legislative title to land in other areas until the effects of the initial registration have been evaluated.[15]

The operation of the customary land tenure system has ensured land to practically all rural families: thus landlessness is virtually non-existent. At the same time, it resulted in a relatively even distribution of land, the size of the plot being effectively determined by the family labour and the dominant technology of hoe cultivation. The limited differentiation among the peasantry in the early years of independence was due principally to differences in soil fertility and incomes from wage employment, at home or abroad, received by some members of the family. This differentiation has been accentuated over the past decade and a half as a result mainly of the expansion of cash crops and use of improved tools and modern inputs, such as fertilisers and better seeds, under rural development projects. It is not clear whether the size of the operational holdings has been a significant factor in this process. However, with the rapid expansion of population, the limited opportunities for employment abroad, the introduction of a land tenure system based on individual ownership and of new technologies and the gradual disappearance of better-quality land, it is likely that, in the absence of a reorientation of rural development policy, the process of social and economic differentiation would gather increasing momentum.

GROWTH, DISTRIBUTION AND POVERTY

In the previous section it was shown that Malawi has experienced a rapid rate of economic growth over the period 1964-78. This growth has probably resulted in an increase in income per head of the order of 60 to 70 per cent. In this section we investigate the ways in which the benefits of growth have been distributed among the population. Our primary focus of interest is on the extent to which low-income groups, particularly in the rural areas, have been able to share in the increased prosperity.

It should be stressed at the outset that limitations of data preclude a comprehensive examination of the structure of income distribution at a given point in time, let alone its evolution over a period. Hence conventional analysis of changes in income distribution and incidence of poverty, based on household income and expenditure surveys over two or more periods, is not possible in the case of Malawi. Limited information on some aspects of income distribution is provided in the 1968 household income and expenditure survey for urban areas and agricultural estates and the 1968-69 national sample survey of agriculture. The former covered 2,454 households in urban areas and 1,037 on estates, while the latter covered 5,044 smallholder households. There have been no comparable surveys in the subsequent period. We shall draw on these two surveys as well as on other evidence to provide a broad assessment of the distribution of gains from growth, particularly with respect to low-income groups in the rural areas.

We begin with a brief discussion of income distribution in 1968. This is followed by a more detailed analysis of income changes among the two major groups in rural areas: estate workers and smallholders. To complete the picture, we consider likely changes in the incomes of other groups: urban unskilled workers, skilled employees and owners of enterprises; and we conclude the section with an interpretation of the main trends in income distribution since 1968.

Some features of income distribution in 1968

On the basis of the two surveys referred to earlier, Humphrey made an attempt to estimate the degree of income concentration (Gini coefficients) among the two rural groups as well as in urban areas and for the economy as a whole.[16] His results are presented in table 17. In interpreting this table, a number of factors should be borne in mind. First, the high degree of equality on the agricultural estates is the result of uniformly low wages being paid to a relatively homogeneous group of workers, and of the exclusion from the survey of expatriates and local staff earning in excess of K 20,000 per annum. Second, incomes in smallholder agriculture refer only to total cash receipts. Since subsistence production is distributed more equally and accounted for over three-quarters of total smallholder production in 1968, the concentration ratio of 0.58 exaggerates the inequality in this sector. Nevertheless, the figure serves to underline the disequalising impact of the pattern of introduction of cash crop cultivation in Malawi.

Third, the table shows that incomes in the urban sector are much less evenly distributed than in rural areas. This is a reflection of the low wages paid to unskilled workers, and the extremely high incomes of expatriates and local skilled employees and businessmen. Finally, the high figure of inequality in over-all income distribution results directly from large inequalities in intra-urban and rural-urban income distribution. However, in view of what has been said earlier regarding the exclusion of subsistence production in smallholder agriculture, the figure for over-all inequality has only limited significance. Unfortunately, it is not possible to obtain a quantitative esti-

Table 17. Concentration ratios, 1968

Area covered	Ratio
Total urban	0.67
Agricultural estates	0.28
Smallholder agriculture	0.58
Nationwide	0.76

Source. David H. Humphrey: "Malawi's economic progress and prospects", in *Eastern Africa Economic Review*, Dec. 1973.

mate of changes in income distribution since 1968. Nevertheless, on the basis of some partial indicators, an attempt will be made to give some idea of trends in the incomes of the main economic groups.

Agricultural estate workers

The numbers of employees in agriculture rose rapidly from 48,300 in 1969 to nearly 134,000 in 1977, at an annual rate of 13.6 per cent.[17] On the basis of the ratio of employees to households in the 1968-69 survey, this implies about 70,000 households in 1977, or somewhat less than 7 per cent of all rural households. In 1968 the average household income was nearly K 140 per annum, 68 per cent of which was derived from wages, 16 per cent from free housing and other income in kind, 10 per cent from self-employment, and the rest from other sources (table 18). There was a general tendency for the household size to increase with income, though less than proportionately.

As indicated earlier, no comparable estate household income survey has been carried out since 1968-69. However, it is possible to get an approximate idea of the trends in the incomes of estate employees from movements in agricultural wages; these are shown in table 19. It will be noted that, while money wages increased by 50 per cent between 1968 and 1977, the real wages showed a practically continuous decline over the period, having fallen by about 20 per cent by 1976-77. The decline is likely to be even sharper for the lowest income groups with proportionately greater expenditure on food, the price of maize having risen faster than the over-all price index.

Wages, together with free housing and other income in kind, constituted nearly 85 per cent of average estate household income. There is no reason to suppose that the other sources of income rose rapidly over this period. Thus it is virtually certain that the real incomes of the estate households fell continuously between 1968 and 1977.

The decline in real wages on the estates was accompanied by a massive expansion of employment. A number of factors explain this apparent paradox: on the demand side, the declining real wages provided an incentive for

Table 18. Distribution of households by income class and source of income (agricultural estates only)

Income group (shillings per year)	Households		Average income per household (shillings)	Source of income (percentages)				
	No.	%		Wages and salaries	Self-employment	Rent, interest, pensions	Remittances	Other[1]
0- 1 000	647	62.4	1 070.8	64.3	12.3	3.0	0.9	19.5
1 001- 1 400	192	18.5	1 420.7	66.1	9.7	5.0	1.3	17.9
1 401- 2 000	110	10.6	1 797.2	68.9	11.2	5.5	0.8	13.6
2 001- 2 800	44	4.2	2 441.7	74.8	4.8	6.6	1.5	12.3
2 801- 4 000	24	2.3	3 339.6	80.0	3.1	5.8	2.5	8.6
4 001- 6 000	13	1.3	4 052.8	72.4	5.4	14.7	3.2	4.3
6 001-10 000	6	0.6	6 929.4	82.3	2.6	12.7	0.2	2.2
10 001-20 000	1	0.1	9 768.0	98.8	–	–	–	1.2
20 001+	–	–	–	–	–	–	–	–
Total	1 037	100.0	1 402.9	68.0	10.0	4.8	1.1	16.1

– = magnitude nil.
[1] Mainly free housing and other income in kind.
Source. Household income and expenditure survey, 1968.

Table 19. Money and real wages in agriculture, 1968 to 1977

Year	Money		Real wages[1]		Price of maize	
	Kwacha per month	Index	Kwacha per month	Index	Tambala per lb.	Index
1968	8.09	100.0	8.09	100.0	1.64	100.0
1969	8.15	100.7	8.03	99.3	1.62	98.7
1970	8.82	109.0	7.98	98.6	2.25	137.2
1971	9.22	114.0	7.71	95.3	2.37	144.5
1972	9.51	117.6	7.68	94.9	2.60	158.5
1973	9.52	117.7	7.37	91.1	2.55	155.4
1974	10.90	134.7	7.26	89.7	2.81	171.3
1975	10.65	131.6	6.14	75.9	4.29	261.6
1976	10.98	135.7	6.07	75.0	4.63	282.3
1977	12.69	156.9	6.73	83.2	4.13	251.8

[1] Real wages calculated by deflating money wages by the "low-income consumer price index".
Sources. ILO: Year Book of Labour Statistics, various issues; Malawi Statistical Yearbook 1978, table 17.

expansion of plantations; on the supply side, as shown later, the operation of the incomes policy by the Government prevented a rise in real wages. The lack of difficulty experienced by the plantations in recruiting labour is due to the fact that even the relatively low wages for agricultural workers represented an opportunity for increased cash incomes for the poorer strata of the smallholder sector. The supply of labour was further enhanced suddenly by the cessation of migration to South Africa. Thus, while the old employees suffered an absolute decline in their living standards, the new entrants to the estate labour force experienced a clear improvement in their incomes. Nevertheless, the majority of estate households—certainly the 63 per cent with annual incomes below K100—continue to live in poverty.

Smallholders

Smallholders constitute over 90 per cent of the rural population. It is not possible, on the basis of the existing data, to draw any precise conclusions regarding the trends in the level and distribution of income among smallholders. As indicated earlier, the 1968-69 survey provides information only on the cash receipts of smallholders. This information is summarised in table 20, which brings out some interesting features of smallholder cash incomes. It is surprising to find that current farm cash income accounts for less than 30 per cent of total cash receipts. The major source of household cash receipts (57 per cent) is income from employment in agriculture and in other sectors, as well as from the operation of non-agricultural enterprises. Transfers, credit and savings withdrawals constitute less than 14 per cent of total cash receipts. There are, however, significant differences in this respect among the various income groups: the poorest 43 per cent of households derive nearly 68 per cent of their cash income from remittances, etc., while the richest 14 per cent obtain nearly 70 per cent of their cash receipts from employment and operation of non-agricultural enterprises. Net farm cash incomes are relatively important only for the middle income groups: here the differentiation is largely associated with the introduction of cash crops, especially cotton and tobacco, which experienced a trend growth over the period 1964-78 of 10.7 and 7.2 per cent per annum respectively (table 21).

These features of the smallholders' cash incomes highlight the importance of internal and external migration as a means of supplementing the incomes of the poorer households. The recorded remittances from abroad rose dramatically from K4.3 million in 1968 to K32.1 million in 1975; they must therefore have played a major role in reducing inequalities among smallholders (the Gini coefficient of distribution of remittances, etc., in 1968 was 0.176). By the same token, the sharp decline in remittances to K2.4 million in 1977 must have greatly aggravated rural poverty and inequalities. The relatively great importance of non-farming activities as a source of cash income for better-off smallholders lends support to the proposition that these activities represent more profitable investment opportunities than farming for smallholders with surplus cash, and constitute a basis for further accumulation and differentiation (the Gini coefficient of other current income was 0.734).

It is difficult to draw any conclusions regarding over-all inequality among smallholders, in view of the fact that 75 to 80 per cent of total income was represented by subsistence output and little is known about its distribution.

Malawi

Table 20. Distribution of smallholders' cash income, 1968-69 [1]

Income class (1)	Households (2) No.	%	Average household size	Current farm cash income [2] (3) Kwacha	%	Other current income [3] (4) Kwacha	%	Other receipts [4] (5) Kwacha	%	(3) + (4) (6) Kwacha	%	(5) + (6) (7) Kwacha	% by income class
nil	469	9.3	4.0	389	6.7	206	3.5	5 216	89.8	595	10.2	5 811	2.7
0.1-10	1 680	33.3	4.2	4 888	28.6	4 401	25.7	7 830	45.7	9 289	54.3	17 119	7.9
10.1-20	974	19.3	4.8	8 474	37.5	8 328	36.9	5 775	25.6	16 802	74.4	22 577	10.5
20.1-30	535	10.6	4.6	7 848	44.4	7 148	40.4	2 696	15.2	14 996	84.8	17 692	8.2
30.1-40	333	6.6	4.9	6 713	43.6	6 600	42.9	2 081	13.5	13 313	86.5	15 394	7.1
40.1-60	343	6.8	5.2	7 604	35.5	10 774	50.3	3 035	14.2	18 378	85.8	21 413	9.9
60.1-80	262	5.2	5.0	7 254	34.5	12 772	60.7	1 007	4.8	20 026	95.2	21 033	9.7
80.1-120	176	3.5	5.3	6 519	32.5	12 631	63.0	914	4.5	19 150	95.5	20 064	9.3
120.1-200	156	3.1	5.3	8 346	32.4	17 445	67.6	–	–	25 791	100.0	25 791	11.9
200 +	116	2.3	5.5	5 667	11.6	42 060	85.8	1 295	2.6	47 727	97.4	49 022	22.8
Total	5 044			63 702	29.5	122 365	56.7	29 849	13.8	186 067	86.2	215 916	100.0
Gini coefficient				0.522		0.734		0.176		0.672		0.563	

– = magnitude nil.

[1] Aggregate incomes for certain income classes do not fall within the designated class limits. This is due to the fact that classes were defined by income net of expenditures in the original data. Gross income is used here. [2] Includes all cash crops, livestock and poultry. [3] Includes all types of wages and income from forestry, fishing, quarrying and mining and distribution. [4] Includes all transfers, credit and savings withdrawals.

Source. Malawi: *National sample survey of agriculture, 1968-69* (Zomba, 1970), table R6.12.

Table 21. Trend rates of growth of area, production and yields of some smallholder crops, 1964-78 [1]

Item	Rates of growth (%)		
	Groundnuts	Tobacco	Cotton
Area	0.3	1.9	1.0
Production	−2.1	7.2	10.7
Yield	−1.8	5.2	9.7

[1] Trend rates of growth were estimated by the formula Log $Y = \alpha + \beta T$ where T = time.
Source. FAO: *Production Yearbook*, various issues.

What can be stated with certainly is that the over-all income differences must have been substantially less than those indicated by the Gini coefficient of 0.58 for smallholder cash receipts. A rough approximation to distribution of farm income (both cash and subsistence) may be obtained from the size distribution of land among smallholders; this is shown in table 22. The Gini coefficient of land distribution was 0.407, while that for the cropped area 0.385. It is quite likely that these ratios overstate the degree of inequality from farm incomes. This is so for two reasons: first, there is a tendency for the household size to increase with the size of holdings; second, table 22 makes no allowances for differences in the quality of land: in general, the average farm size is smaller in the more fertile areas.

Our discussion so far has been concerned with the structure and distribution of cash incomes among smallholders. Can anything be stated concerning the changes in smallholder income since 1968? One approach to this issue is to estimate production per smallholder (cash crops and subsistence) from the GDP tables. The results for 1969 and 1978 are shown in table 23. The figures in that table for 1969 and 1978 are not strictly comparable as the latter are expressed in constant 1973 prices. Allowance for price increases between 1969 and 1973 would somewhat reduce the growth figures shown in the table. Adequate data are not available to determine changes in output distribution among smallholders over the period, but it is highly likely that inequalities have increased. Some evidence to that effect was presented above. As shown in the next section, the process of differentiation was greatly accelerated by the establishment of four major rural development projects over the period, and the concentration of resources on the progressive farmers. Furthermore, it would appear that land distribution has worsened over the period: while there is little overt landlessness, the growing population pressure is leading to the exhaustion of the land frontier, resulting in shorter fallow periods and cultivation of poorer-quality land. In some areas where estate agriculture in tobacco has grown at fast rates, there is evidence of loss of land by smallholders and emigration of the landless to Zambia.

Table 22. Size distribution of land among smallholders, 1968-69

Size class	No. of households ('000s)	%	Cumulative %	Area of land-holdings ('000 hectares)	%	Cumulative %	Average size (hectares)
Total landholding							
1.<0.8	254	28.7	28.7	132.08	8.9	8.9	0.52
2. 0.8-1.6	302	34.1	62.8	362.40	24.3	33.2	1.20
3. 1.6-2.4	163	18.4	81.2	326.00	21.9	55.0	2.00
4. 2.4-4.9	148	16.7	98.0	540.20	36.2	91.3	3.65
5. >4.9	18	2.0	100.0	129.96	8.7	100.0	7.22
Total	885	100.0		1 490.64	100.0		

Gini coefficient = 0.407
Decile distribution:

1	2	3	4	5	6	7	8	9	10
1.8	3.3	4.5	5.7	6.8	8.3	10.4	12.5	16.3	30.4

Cropped area

Size class	No. of households ('000s)	%	Cumulative %	Area of land-holdings ('000 hectares)	%	Cumulative %	Average size (hectares)
1.<0.8	254	28.7	28.7	122.60	9.0	9.0	0.48
2. 0.8-1.6	302	34.1	62.8	352.20	25.9	34.9	1.17
3. 1.6-2.4	163	18.4	81.2	314.60	23.1	58.0	1.93
4. 2.4-4.9	148	16.7	98.0	464.20	34.1	92.1	3.14
5. >4.9	18	2.0	100.0	107.80	7.9	100.0	6.00
Total	885	100.0		1 361.40	100.0		

Gini coefficient = 0.385
Decile distribution:

1	2	3	4	5	6	7	8	9	10
1.9	3.4	4.6	6.1	7.2	8.9	11.0	13.2	15.6	28.3

Source. Malawi: *National sample survey of agriculture, 1968-69.*

Table 23. Average production per smallholder, 1969 and 1978 (Kwachas)[1]

Type of production	1969	1978	Percentage increase
Subsistence	90.6	130	43
Cash crops	27.3	46	70
Total	117.9	176	49

[1] Calculated on the basis of 885,000 holdings in 1969 from the agricultural survey; assumed annual increase of 2.5 per cent since 1969.
Source. Malawi Statistical Yearbook 1978.

It may be asked to what extent the increases in production noted above have been translated in increased real incomes for the smallholders. Some evidence on this issue is presented in table 24. It will be seen that the terms of trade faced by smallholders showed a limited improvement up to 1971 but have been declining since. A more appropriate index for smallholders based on exclusion of food prices is given in the last column, which shows an improvement of about 10 per cent between 1969 and 1971 and a gradual decline thereafter (except for 1975). Between 1969 and 1978 the index declined by about 12 per cent. In interpreting this table it should be borne in mind that cash agricultural output in 1978 amounted to about a quarter of the total smallholder production and that not all of it is marketed through the state marketing agency—ADMARC—which fixes producer prices. On the basis of the above figures, it may be tentatively concluded that, despite a deterioration in their terms of trade between 1969 and 1978, the purchasing power of the average cash income of smallholders was probably increased by more than 30 per cent over the period. It should, however, be recalled that most of the cash crop production is concentrated in the hands of a relatively small proportion of smallholders—probably no more than 10 per cent. As for the bulk of smallholders, in view of their heavy reliance on remittances from abroad, which declined sharply after 1975, it is highly likely that their total cash receipts in 1978 were considerably lower than in 1969.

The smallholder terms of trade have been depressed by the procurement policy of the state marketing agency. The profits made by ADMARC in 1971-72 are shown in table 25. It is apparent that the profit margins, especially on cash crops, were very high: 141 per cent of the purchase value for tobacco, 54 for groundnuts and 33 for cotton. The total profits earned in 1971-72 represented 22 per cent of the smallholder cash crop production. The ADMARC profits have continued to show a strong upward trend in the subsequent period. Since most of these profits are invested in projects unrelated to smallholder production, it is clear that the ADMARC operations have resulted in a major transfer of surplus from the smallholder sector to finance urban development and other government expenditure.

Some concluding remarks on growth and distribution

To complete our discussion of income distribution, the experience of three other groups will be briefly considered—urban unskilled workers, high-level employees and owners of corporate businesses. Some relevant statistics are presented in table 26. The numbers of non-agricultural employees increased by 44 per cent between 1969 and 1977, but the real wages of unskilled employees, as represented by minimum urban wages, fell by 30 per cent over the period. The skilled employees may be roughly represented by individuals assessed for income tax. Their numbers increased by 60 per cent over the period but their real wages declined. However, more significant from the point of view of income distribution—and indicative of the staggering dimensions of over-all inequalities in this area—is the fact that, in 1977, their

Table 24. Terms of trade indices for smallholders, 1968-78

Year	Index of producers' prices, major crops (1)	Low-income price index total (2)	Price index excluding food (2A)	Terms of trade index (3) = (1) ÷ (2)	Terms of trade index (4) = (1) ÷ (2A)
1968	100	100	100	100	100
1969	104	101	101	103	103
1970	115	111	106	104	108
1971	126	120	112	105	112
1972	126	124	115	102	110
1973	131	131	119	100	110
1974	149	150	136	99	110
1975	181	174	152	104	119
1976	172	181	162	95	106
1977	167	189	173	88	97
1978	169	205	191	82	88

Sources. Col. (1): based on data on producers' prices paid to smallholders by ADMARC for the six major crops (tobacco, groundnuts, cotton, rice, maize and pulses) published in *Malawi Statistical Yearbook 1978*, tables 8.15 and 8.16, pp. 91 and 92. The index is based on a weighted average of these crops; col. (2): index of Blantyre retail prices for low incomes with base shifted to 1965, ibid., table 17.3, p. 174.

Table 25. ADMARC profitability from trading, 1971-72

Crop	Purchase value	Buying costs	Percentage of cost to value	Change in stocks	Sales value	Sales costs	Percentage of cost to value	Overhead	Profits	Percentage of profits to purchase value
Cotton	2 466	896	36	+316	4 250	102	2	198	804	33
Groundnuts	3 907	904	23	+429	7 065	372	5	200	2 111	54
Tobacco	3 933	1 415	36	0	11 456	183	2	312	5 553	141
Maize	884	413	47[1]	−208	1 599	46	3	91	−42	−5
Paddy	1 223	228	19	+145	1 173	2	0	51	−185	−15
Other	1 486	261	18	+184	2 168	73	3	58	474	32
Total	13 899	4 117	29	+866	27 711	778	3	910	8 715	63

[1] Includes storage costs.

Source. ADMARC annual accounts, 1971-72. Estimated by World Bank: *Malawi agricultural sectoral review* (1973).

average remuneration was 20 times the urban and 32 times the rural minimum wages. The data on companies are even more telling, for they show that while the number of companies taxed rose by a mere 15 per cent, their average profits increased by 300 per cent. It should clearly be borne in mind that these are declared profits.

Table 26. Some indices of income changes for main socio-economic groups

	1969	1977	% increase
Average smallholder production (Kwacha p.a.)	117.9	176[1]	–
Subsistence	90.6	130[1]	–
Cash	27.3	46[1]	–
Average agricultural wages (Kwacha p.a.)	97.8	152.4	56
Minimum rural wages (Kwacha p.a.)	82.8	90.0	9
Number of agricultural employees ('000)	48.3	133.8	168
Number of non-agricultural employees ('000)	98.2	141.8	44
Minimum urban wages (Blantyre) (Kwacha p.a.)	137	144	5.1
Number of individuals assessed for tax income	9 537	15 310[2]	60
Average income (Kwacha p.a.)	2 225	2 863	29
Number of companies assessed for tax	495	570[2]	15
Average company profits (Kwacha p.a.)	25 438	100 101	297
Low-income cost-of-living index (1970 = 100)	91.3	170.7	87
Low-income cost-of-living index excluding food (1970 = 100)	95.7	163.4	71
High-income cost-of-living index (1970 = 100)	94.7	179.8[2]	90

– = not applicable.
[1] 1978 figures in 1973 prices, taken from table 23. [2] 1976.
Source. Malawi Statistical Yearbook 1978.

It is now possible to draw upon our earlier discussion in order to present an over-all picture of the gains from growth in Malawi since 1969. While the average smallholder incomes have risen in real terms, most of the gains have accrued to a small minority of progressive farmers. For the bulk of smallholders, the subsistence levels of production have been maintained; however, the modest gains in cash resulting from crop sales have been completely swamped by large losses consequent upon the cessation of emigration to neighbouring countries. Those employed on estates and as urban unskilled workers have experienced a sharp fall in their already very low incomes, with the decline being slightly greater for the latter category. However, the vast expansion of employment, especially on the estate sector, has resulted in higher income opportunities for members of rural households, the new employees having thus been enabled to increase both their own and their total family incomes. The opposite trends in the average incomes of smallholders and employees have resulted in a reduction in income differences between those two categories, and to a lesser extent between estate workers and unskilled urban employees.

Another group of beneficiaries from growth is constituted by skilled employees. Their members have increased rapidly and, in view of their relatively extremely high incomes, they are likely to have absorbed a substantial proportion of the gains from growth. It is, however, the owners of corporate enterprises who have proportionately benefited the most. Many of these enterprises are foreign-owned, but in recent years there has been some increase in participation by local interests.

We may conclude from the foregoing that there must have been a significant increase in income inequalities since 1969. The evidence for this is to be found in the concentration of gains among progressive smallholders, the large expansion in the number of highly paid employees and an increase in the share of profits in the value added of the corporate sector. Given the extremely low incomes of estate workers, urban unskilled employees and the great majority of smallholders, poverty is widespread. This is further corroborated by such other indicators as infant mortality (142 per 1,000), average life expectancy (46 years) and adult literacy rate (25 per cent).

It is, however, difficult to argue that the pattern of growth *per se* has increased poverty in the sense that a higher proportion or numbers of households were living below a certain specified income in 1977 as compared with 1969. Nevertheless, the loss of jobs for some 200,000 workers abroad, and the impact of such loss on the incomes of smallholders, clearly aggravated poverty after 1975.

STATE POLICIES

Since independence, Malawi has pursued what might be broadly described as a growth-oriented, export-led strategy of development. The key elements of this strategy have been reliance on private enterprise in agriculture and industry, encouragement of foreign private and public capital inflows, and stimulation of exports. Little systematic attempt has been made to alter the pattern of growth in favour of the poverty groups, either in urban or rural areas. There have been no major alterations in patterns of asset ownership and almost no institutional innovation. It has been assumed that the achievement of high rates of economic growth will automatically lead to reduction in poverty and rising living standards.[18]

Unlike many other Third World countries, particularly in Africa, Malawi has emphasised agricultural production in its development strategy. As indicated earlier, this has been done primarily through encouragement of investment by plantations. The distribution of public expenditure also to some extent reflects this orientation. A growing proportion of the expenditure on economic infrastructure has been geared to the agricultural sector. Annual current outlays on the Revenue Account rose for "Agriculture and national resources" from K2.8 million to K8.9 million over the period 1964-77, representing a rise from 7.5 to 9.5 per cent of total current expenditure. On the Development Account, outlays grew slightly more rapidly in absolute terms (from K2.5 million to K12.0 million over the period 1964-77), though their share fell from 29 per cent in 1969 to 20 per cent in 1977.

Within agriculture, government policy has concentrated on the estate sector and the *Achikumbe* or "progressive farmers" in the smallholder sector.

Foreign private investment in plantations and agro-industry has been encouraged by a favourable tax system and a liberal policy on imports and repatriation of profits and capital. A major incentive to investment in plantations has been the existence of cheap labour. As will be shown later, this has been the main objective of the Government's incomes policy.

The second plank of agricultural policy has been the growth of cash crops and exports through assistance to *Achikumbe* farmers. Through this policy,

... smallholders who persistently achieve certain high standards of husbandry and market orientation receive an *Achikumbe* certificate signed by the President, and other tangible benefits such as priority in marketing of crops through ADMARC and permission to take certain crops such as tobacco straight to the auction floor rather than being compelled to sell to ADMARC from whom they would receive only a quarter or one-third of the auction price. Other privileges include priority in the supply of farm inputs, and advice with farm planning and organisation.[19]

The main instrument for promoting smallholder production has consisted of four major integrated development projects—funded mainly by the World Bank and the European Development Bank, and initiated between 1968 and 1972—covering an area of 23,760 square kilometres and containing about 279,000 farm families, or about 20 per cent of the rural population. These projects are aimed at increasing agricultural production and rural incomes through the intensive and co-ordinated supply of improved extension services, credit, infrastructure, storage and marketing facilities, health, housing and rural water, as well as through the promotion of irrigation and land improvement. Only about 30 per cent of households show evidence of benefiting directly from project policies in the chosen areas. The costs for each farmer have been high, varying between K 500 in Lilongwe to over K 1,000 in Karonga.[20] The projects, which have met with a limited measure of success, have certainly accentuated rural stratification, the concentration of resources thereon having resulted in a neglect of the majority of farmers in the country. The relatively high costs of the projects and their limited coverage have led the Government since 1976 to embark on a new National Rural Development Programme (NRDP) under which it is hoped to reach greater numbers of producers more quickly. The NRDP would concentrate on the provision of inputs and services, with a lesser emphasis on infrastructure. It is anticipated that 60 per cent of farm families will be participating in the NRDP by 1985.

An important aspect of state policy relates to the internal terms of trade faced by producers. It was indicated earlier that while these terms improved slightly (or stayed constant) between 1968 and 1975, they deteriorated sharply between 1975 and 1978. To the extent that these trends reflect export and import prices, they are largely beyond the control of the Government. However, as indicated earlier, the operations of ADMARC have depressed these terms to levels lower than they might have been otherwise. Since only smallholders are required to market through ADMARC, a significant discriminatory measure has thus been introduced against them in relation to estates, which must in part be held responsible for the relatively low growth of smallholder production.

The role of ADMARC has widened beyond that of price policy formation and domestic and export marketing for smallholders. It now embraces supply of input and investment in agro-industrial production in the form of estates and processing plants (e.g. fruit canning and tung processing). Indeed, its assets have grown remarkably in value from K 13.6 million in 1970-71 to K 115.5 million in 1977-78.

ADMARC has been well placed to regulate both the size of the inputs going to smallholders and the redeployment of surpluses acquired. In the 1970s ADMARC crop marketing profits were substantial: the net profits of the organisation rose from K 8.9 million in 1971-72 to K 17.4 million in 1977-78. Although not all profits had been obtained from crop marketing, they amounted to a sizeable share of the value of marketed smallholder output (76 per cent in 1975). Moreover, an increasing element of investments has had no connection with the smallholder sector, whilst through its loan and equity policies it would seem apparent that a net transfer of funds has been effected to either non-smallholder agriculture or non-agricultural ventures. The extraction and deployment pattern of the relatively high profits made on crop marketing by ADMARC is likely to have had one important effect: available surpluses for re-investment directly by peasant households have been reduced by depleting the latter sector of that growth in capital stock which it might otherwise have had.

The Government has also sought to influence income distribution through its incomes policy. The urban and rural minimum wage rates have been kept at relatively low levels, and fell in real terms between 1966 and 1977. As shown earlier the real wages in agriculture have also fallen. The incomes policy, which does not appear to control higher wage incomes, profits or dividends, was formulated to maintain economic stability and growth.[21] It sought to control rural-urban migration through reducing income differentials between peasants and wage earners and to promote wage employment. Thus, it became "government policy to maintain the present level of minimum wages and to make no general increases in the foreseeable future".[21] Given the weak trade union structure and the non-existence of a trade union for agricultural or estate workers, it has been relatively easy to implement this policy. The existence of a surplus supply of labour, even at the existing wage rates, especially after the cessation of large-scale migration, and the small number and effective organisation of employers, both in estate agriculture and manufacturing, further facilitated the implementation of this policy.

It would seem that this policy has been an important factor in the expansion of estate production and exports, as well as in the impressive growth of wage employment in agriculture and manufacturing. As shown earlier, it also reduced differentials between average peasant incomes and the wages of estate workers and of urban unskilled employees. At the same time, however, it has contributed to a widening of the already vast income differences between unskilled wage earners and the high income groups deriving their income from salaries, professional practice, rent, dividends and profits.

CONCLUSION

The rapid growth sustained by Malawi has facilited the transition into a more commercialised and, to some extent, more diversified economy. A number of factors combined to make this possible. First of all, Malawi was able to achieve a rate of agricultural expansion well in excess of the population growth. The smallholder sector was able, through higher yields and by bringing new land under cultivation, to produce enough food for the growing population, in addition to cash crops for exports. Second, the small but expanding estate sector, enjoying full government support and drawing on cheap labour, has been instrumental in achieving a fast growth of cash-crop production and export. Third, the focus of government policies and foreign aid has also been on rapid growth of cash crops either from the estate or smallholder sector. Finally, as a labour reserve economy, Malawi was able to mitigate the effect of population pressure through the emigration, at least until the mid 1970s, of a sizeable proportion of its labour force to South Africa. The export orientation of the economy increased over the period and its growth became more dependent on the world markets for exports.

The growth strategy pursued by the country has been associated with increasing inequalities. The major beneficiaries of growth have been the corporate enterprises and the increasing number of skilled employees. Within the smallholder sector, the bulk of the benefits have accrued to a small minority of "progressive farmers" who have enjoyed the greater part of government support to agriculture. This has resulted in what some Malawians call the "new class" of private farmers.[22] On the other hand, the real wages of agricultural and unskilled urban workers have declined somewhat over the period under consideration. There is awareness of these problems, and the Government is attempting a broad-based policy package for the smallholder sector; it is too early, however, to judge the impact that this is likely to have.

Notes

[1] This chapter draws on material prepared by D. G. Clarke under the title *Rural poverty in Malawi: 1964-77* (ILO, 1980; mimeographed).

[2] John G. Pike: *Malawi: A political and economic history* (London, Pall Mall Press, 1968).

[3] World Bank: *Memorandum on the economy of Malawi* (Sep. 1977).

⁴ R. W. Kettlewell: *Agricultural change in Nyasaland, 1945-1960,* Food Research Institute Studies (Palo Alto, California, Stanford University Press, 1965).

⁵ For a useful account of the economic situation at independence, see David H. Humphrey: "Malawi's economic progress and prospects", in *Eastern Africa Economic Review,* Dec. 1973.

⁶ The periods 1964-72 and 1973-78 are analysed separately because of a break in the constant price GDP series in 1973.

⁷ On the assumption that approximately 10 per cent of wage employees in smaller enterprises are not covered by recorded wage employment and that women form 10 per cent of the wage employees.

⁸ World Bank: *Memorandum on the economy of Malawi* (30 Sep. 1977), pp. 12-13.

⁹ William Barber: *The economy of British Central Africa* (Palo Alto, California, Stanford University Press, 1961).

¹⁰ Assuming an increase of 2.5 per cent between 1968-69 and 1978.

¹¹ *Malawian Agricultural Statistics 1971.*

¹² H. Dequin: *Agricultural development in Malawi* (Munich, Institut für Wirtschaftsforschung, 1969).

¹³ *National sample survey of agriculture, 1968-69.*

¹⁴ B. H. Kinsey: *Rural development in Malawi: A review of the Lilongwe land development program* (World Bank, 1974).

¹⁵ FAO: *Country review paper of Malawi* (Rome, 1978).

¹⁶ See Humphrey, op. cit., for a detailed discussion.

¹⁷ These numbers refer to employees in enterprises with more than 20 workers.

¹⁸ See Malawi: *Statement of development policies, 1971-1980* (Zomba, Government Press, 1971).

¹⁹ Humphrey, op. cit.

²⁰ FAO: *Country review paper of Malawi,* op. cit.

²¹ Ministry of Labour: *A national wages and salaries policy for Malawi* (Limbe, 1969).

²² Alifeyo Chilivumbo: "On rural development: A note on Malawi's programmes of development for exploitation", in *Africa Development,* No. 2, 1978, p. 43.

EXPORT-LED RURAL DEVELOPMENT: THE IVORY COAST

4

Eddy Lee

The Ivory Coast is a relatively small country in western Africa which, in recent decades, has achieved rapid economic growth. In 1975 it was estimated to have a population of 6.7 million and a GNP per head of US$ 500.[1] In spite of a high rate of population growth (3.8 per cent per annum between 1960 and 1975) it still remains a land-abundant country.

The growth performance of the Ivory Coast—frequently described in hyperbolic terms as "the economic miracle"—has been well above the average for African developing countries. The rate of growth in GNP between 1960 and 1975 was 7.4 per cent per annum in constant prices and the corresponding rate per head was 3.6 per cent per annum.

Export of agricultural products was the main engine of this growth; total exports grew almost *pari passu* with GNP at 7.2 per cent per annum between 1960 and 1975. Exports were 40 per cent of total GNP in 1960 and maintained this same proportion in 1975. With huge untapped reserves of arable land, economic growth was fuelled by the rapid extension of the land frontier to produce agricultural products for the world market. The rate of expansion was also accelerated above what was feasible on the basis of natural population increase by resort to large inflows of unskilled labour from neighbouring countries. In essence, therefore, the pattern of growth in the Ivory Coast was similar to that experienced in the earlier phases of the economic history of "plantation economies" in south-east Asia and the Caribbean,[2] and of neighbouring western African economies such as Ghana and Nigeria.

In the rest of this chapter we shall attempt to document the nature of this "export-led rural development" and its implications for rural incomes and poverty in the Ivory Coast. The sources of growth in the rural economy will be considered, as a prelude to the discussion of changes in rural incomes and their distribution. We then go on to discuss policy issues deriving from the analysis, and conclude with an over-all evaluation of the Ivory Coast's experience with rural development.

SOURCES AND PATTERN OF GROWTH

To understand the sources and pattern of growth in the rural economy of the Ivory Coast it is necessary to consider the basis features of the agrarian system prior to the introduction of cash crops as well as its subsequent evolution. While no accurate historical records or data exist, it has been possible to draw some inferences about the pre-cash economy agrarian system from the more recent observations that are available.

The dominant characteristic of the original agrarian system in the Ivory Coast was that it was a subsistence agriculture carried out in the context of land abundance. The technology was extremely simple, consisting of no more than rudimentary hand tools, and hence labour supply was the effective constraint on the expansion of agricultural output. The system was essentially based on the cultivation of roots and tubers; except for some regions in the north, where cereals such as rice, sorghum and millet were grown, manioc, yams and cocoyams were the predominant crops. In addition, plantains were also an important food crop in some regions in the south. Since tubers and root crops (and also plantains) yield high calorie equivalents of food per unit of land, the agrarian system generally provided adequate food supplies for the population.[3]

Within this basic system of subsistence agriculture, however, some significant regional differences could be identified. Cropping patterns varied across regions, each with a different dominant food crop and constituting distinct regional food systems.[4] There was also a sharp ecological difference between the savannah north and the rain forest regions of the south, and systems of land tenure varied between the ethnic groups which occupied different parts of the country. These land tenure systems ranged from the centralised and hierarchical system of the Akan group of tribes[5] to the simple system of shifting cultivation where notions of proprietary or usufructuary rights over land were little developed. An example of the former group is the Agni, where the traditional socio-economic structure, to quote from the study by Stavenhagen "was a class structure in which nobles, free men and slaves occupied three distinct positions in a network of social relationships".[6] In spite of this socio-economic structure, private ownership of land did not exist. "Instead land tenure rights (were) held collectively. In principle, the entire Agni society, represented by a king whose authority is sanctioned by religion, collectively holds the territory under its control."[7] Access to land was free for members of the tribe but, with their strong sense of territorial imperative, this was not true for outsiders. The latter fact is of some significance in considering the impact of the introduction of cash crops, and the migration pattern it set in motion, on the type of agrarian relationships which evolved.

The introduction of cash crops brought about profound changes in the simple agrarian system that has been described in the preceding paragraphs. Cash crops were first introduced in the forest zone of the south-east and this region remains to this day the one where commercial farming is the most highly developed. Since this is the region where the Akan peoples predominate, a fair idea can be gained of the types of changes that have occurred by considering the study on the Agni carried out by Stavenhagen, to which reference was made above.

In the Agni region traditional subsistence agriculture was of the slash-and-burn type. In contrast, the cash crops that were introduced (coffee and cocoa) are standing perennials which require higher labour inputs, especially during the seasonal peak from October to February. Although land was abundant enough for cash cropping and traditional agriculture to coexist, the supply of family labour was an effective constraint on total output. The immigration of labour thus became a necessary condition for the expansion of cash cropping. This immigration took various forms (seasonal workers, permanent labourers and even sharecroppers) and agrarian institutions adapted by giving rise to different forms of labour contracts. One was the *abusan*, a type of sharecropping arrangement whereby the immigrant paid a one-third share, while another involved wages being paid at the end of the year. Yet another form was for payment to be made for specified tasks, performed under the supervision of the owners.

This influx of immigrant workers and the consequential changes in agrarian institutions led to growing differentiation in agricultural production. Land acquired a value beyond the mere provision of subsistence and came increasingly to be viewed as private property. Chiefs and heads of "lineages" or extended families were well placed to mobilise immigrant labour to expand the area under cash crops, and it has been claimed that a "plantocracy" has evolved in some regions with the most intensive development of cash cropping.[8] In general, differential access to money incomes arising from different timing of adoption and extent of cash cropping provided the basis for emerging differentiation in the agrarian structure.

Two further aspects of the changing agrarian system should be noted. The first is the existence of a large potential labour supply in the savannah belt in the north of the Ivory Coast, as well as in the Sahelian countries of Mali and the Upper Volta. Due to the difficult ecological conditions to be found there, agricultural productivity and incomes are significantly lower than that in the cash cropping areas of the forest zone. There was thus a pool of labour, with a low supply price, to be tapped by the emerging cash crop economy of the south. The second aspect to note is that the adoption of cash cropping typically has not meant complete specialisation in such crops. Food crops have continued to be grown for subsistence, and a mixed cropping pattern, even mixed crops in a single field, is the general rule.

It is within the above framework of agrarian change that the following statistics on growth in the agricultural sector of the Ivory Coast have to be evaluated. By 1950 the harvested areas[9] of coffee and cocoa were 179,000 and 158,000 hectares respectively. The corresponding figures for total output were 63,500 and 60,000 tons. In 1977 the figures for harvested area under coffee had increased more than fivefold to 921,000 hectares, while that under cocoa had increased by more than three times to 526,000 hectares. Over the same period (1950-77) the production of coffee has increased to 291,000 tons and that of cocoa to 228,000 tons. Direct figures on the numbers employed in the cultivation of these crops are not available, but it can be safely inferred from population figures that there was a large inflow of foreign labour into these activities. "Population growth appears to have accelerated from a rate of 1.4 per cent a year for the period 1920-45 to 2.8 per cent for 1955-65 and to 4.0 per cent for 1965-75. The increase in the indigenous population is estimated at 2.5 per cent, the balance coming from heavy and apparently accelerating immigration. In 1973 it was estimated that about 25 per cent of the total population was foreign born."[10] Most of this immigration was of males of working age entering the export crop sector in the south of the country. Writing of the period up to 1965, Samir Amin estimated that:

> the foreign population of the Ivory Coast, negligible in 1920, amounted to about 100,000 persons in 1950. But the period of great immigration from the north extended from 1950 to 1965, with the accelerated growth of the south and the development of plantations for coffee, cocoa, and bananas. This immigration ... was to bring about 850,000 persons to the Ivory Coast ... composed of young working men.[11]

Over and above this immigration of foreign labour, one must also add the internal migration, especially from the northern savannah belt of the country, and the re-allocation of labour from subsistence to cash crops in the south itself, to obtain an estimate of the total increase in labour input into the export crop sector. On the former, a World Bank report in 1970 had this to say: "Originally the savannah had a greater and denser population than the forest; but once the forest area was opened up ... a continuous flow of migrants began. In 1965 there was a net balance of 100,000 internal migrants to the forest areas."[12] On the latter, Samir Amin has estimated that "very roughly, the population of the southern regions who were well integrated in the plantation economy increased from 50 to 80 per cent of the total rural population of the south between 1950 and 1965".[13] In summary, these admittedly fragmentary items of information on labour supply do point to a massive inflow of labour into the plantation sector.

The preceding paragraph, while indicating the broad orders of magnitude of the increase in land and labour devoted to the plantation sector and in production, yields little insight into the nature of this expansion. To ascertain this we have estimated the trend rates of growth of output, area and yields for the major export crops and these are presented in table 27. It will be seen that both coffee and cocoa showed high trend rates of growth in physical output of 6.27 and 6.37 per cent per annum over the period between 1947-48 and 1976-77. The corresponding figures for total harvested area were also high, being 4.38 and 5.52 per cent per annum. The value of R^2 for all these trend equations was also extremely high—an indication of steady growth in these magnitudes over the relevant period. Implicit in these figures is the fact that yields increased only modestly, as is borne out by the relevant estimates of

Table 27. Trend growth rates for production, area and yields of principal export crops

Crop	α		β	R^2	N	Period covered
Cocoa						
Production	10.535 (0.0785)	+	0.0627 (0.0044)	0.880	30	1947-48 to 1976-77
Area	5.0215 (0.0157)	+	0.0438 (0.0009)	0.987	28	1949-50 to 1976-77
Coffee						
Production	10.900 (0.0929)	+	0.0637 (0.005)	0.841	30	1947-48 to 1976-77
Area	5.377 (0.0587)	+	0.0552 (0.0035)	0.903	28	1949-50 to 1976-77

Notes. (1) α and β are the coefficients derived from estimating trend growth rates in the form Log $Y = α + β T$, where T = time.
Figures in parentheses are standard errors.
N refers to the number of observations.
"Production" refers to total output in tons.
(2) The yield function can be derived from the production and area functions. The yield trend growth rate can be determined by the difference between β for production and β for area.
Source. Ministry of Agriculture: *Statistiques agricoles: Memento 1947-1977.*

trend growth rates. While cocoa yields per hectare increased by 1.78 per cent per annum, they were much lower, at 0.75 per cent, in the case of coffee. Moreoever, the relatively poor fit of the trend equation, especially with regard to coffee, indicates a high degree of variability in yields from year to year.

In the case of coffee, yields were very low in comparison with other producing countries in the initial period of our study, and the virtual stagnation in yields since then has resulted in a considerably widened productivity gap between the Ivory Coast and other producers. This is shown in the figures of comparative yields given in table 28. It will be seen that in the period 1948-52 yields were significantly lower than in western African countries with comparable agro-climatic conditions, only about 60 per cent of the levels in Kenya and Uganda, and less than half of those in Latin American countries, such as El Salvador and Colombia. By 1973-75 the gap in yields between the Ivory Coast and other producers in eastern Africa and Latin America had widened very sharply.

For cocoa, the productivity differentials are less stark. Comparative yield figures are as given in table 29. It will be seen that yields in the Ivory Coast in the 1950s were not unfavourable in comparison with the few other producing countries for which data are available, but that there was no increase in yields between this period and that of 1961-65. The greater number of observations available for this latter period shows, however, that yields in the Ivory Coast were lower than in Nigeria, less than half of those in Central American countries such as El Salvador and Honduras, and only a fraction of those in Haiti. By 1973-75 there had been a sharp increase in yields in the Ivory Coast,

Table 28. Coffee yields in selected countries (kg per hectare)

Country	1948-52	1973-75
Ivory Coast	264	278
Angola	332	420[1]
United Republic of Cameroon	356	314
Kenya	452	950
Uganda	413	718
Brazil	407	571
Colombia	543	627
Costa Rica	454	977
El Salvador	665	1 051

[1] 1973 figure.
Source. FAO: *Production Yearbook*.

Table 29. Cocoa yields in selected countries (kg per hectare)

Country	1952/53-1956/57	1961-65	1973-75
Ivory Coast	344	329	478
Ghana	151	250	249
Nigeria	.	395	297
Bolivia	.	336	526
Brazil	425	317	448
Colombia	.	445	400
Cuba	325	503	712
El Salvador	.	750	750
Haiti	212	2 295	2 063
Honduras	.	781	989
Malaysia (Sabah)	.	232	1 211

. = not available.
Source. FAO: *Production Yearbook*.

but these were still low in comparison with all Latin American countries except Brazil and Colombia.

It would appear from these figures that the main source of growth was the extension of cultivated area, with increases in the physical productivity of land making only a relatively minor contribution, especially in the case of coffee. There is little reason therefore to challenge earlier characterisations of the growth of the Ivory Coast export sector as the expansion of an extensive mode of agricultural production with relatively little intensification of production techniques. This would be particularly true in view of the large gap in yields that still prevails between the Ivory Coast and other producers of coffee. Moreover, despite progress since the mid-1960s, cocoa yields are still substantially lower than in the most "advanced" producing countries in the Caribbean and Central America.

It is unfortunate that adequate data do not exist to estimate the change in output per man in export crop production. Although yields per unit of land have not increased significantly from their low initial levels, it is, in theory, possible that output per man may have increased. For this to happen it would be necessary to have had labour-saving technical progress which resulted in only very moderate increases in the productivity of land. Such a possibility is, however, an extremely remote one in the context of the Ivory Coast. Given the indications of a generally stagnant level of agricultural technology and the large inflows of labour into the plantation sector, we are inclined to agree with the assessment that "the over-all increases in incomes and in output per head in agriculture in recent years have been due not so much to improvements in technology as to the migration (from regions of low productivity) of people —partly coming from outside the Ivory Coast—to the relatively more productive forest areas".[14]

While the expansion of the export crop sector has been the main source of growth in the rural economy, it is also necessary to evaluate changes in the food crop sector, since this has an important bearing on subsistence levels and rural welfare. Accurate figures on the change in the area under food crops are not available, the first agricultural census having been conducted only in 1974-75. According to estimates issued by the Ministry of Agriculture,[15] the area cultivated with the principal food crops (rice, manioc, yams, sorghum, maize, millet and groundnuts) increased by 1.92 times between 1950 and 1977—a margin of increase considerably lower than that for export crops. The proportion of total cultivated area under these food crops decreased from 61 to 43 per cent over the same period. Over approximately the same period—namely between 1950 and 1975—the rural population increased 2.25 times,[16] indicating a lag in the expansion of the cultivated area under food crops behind the growth of rural population. In terms of growth rates, cultivated area increased by 2.56 per cent per annum, while rural population increased by 3.45 per cent per annum.

Estimates have been made of the trend growth rates for total physical output and yields per hectare for six principal food crops. Table 30 shows that, with the exception of rice and maize, growth in output of food crops has been significantly lower than that for export crops and rural population. This was particularly pronounced in the case of millet and the root crops. For millet and yams, there was no statistically significant trend in production, whereas for sorghum and manioc the growth rates were 2.4 and 1.4 per cent respectively, but with high variability around the trends. With regard to yields, there was no statistically significant trend for sorghum, yams or manioc, and only millet showed a positive trend of 1 per cent per annum. These four crops, while constituting almost 50 per cent of the total area devoted to

Agrarian policies and rural poverty in Africa

Table 30. Trend growth rates for production and yields of major food crops

Crop	Estimated growth rates	Period covered
Millet		
Production	+ 0.0033	1949-50 to 1976-77
Yield	+ 0.0103	1950-51 to 1976-77
Maize		
Production	+ 0.0722	1949-50 to 1976-77
Yield	+ 0.0252	1950-51 to 1976-77
Paddy		
Production	+ 0.0659	1949-50 to 1976-77
Yield	+ 0.0395	1949-50 to 1976-77
Sorghum		
Production	+ 0.0244	1949-50 to 1976-77
Yield	+ 0.00005	1949-50 to 1976-77
Yams		
Production	− 0.0084	1949-50 to 1976-77
Yield	+ 0.0080	1949-50 to 1976-77
Manioc		
Production	+ 0.0142	1949-50 to 1976-77
Yield	− 0.0023	1949-50 to 1976-77

Notes and source. As for table 27.

food crops in 1959, made up a considerably higher proportion of total food production measured in cereal equivalents.[17] This is accounted for by the fact that the yields of the root crops are considerably higher than for cereals (more than 5 tons per hectare for yams and manioc compared with approximately 0.5 tons for cereals) and this difference remains large even when converted into cereal equivalents. In 1950 these two crops constituted 68 per cent of the total production of the six food crops discussed above, and these two together with millet and sorghum accounted for 75 per cent of the total production. Thus the general stagnation in the production conditions for yams, manioc, millet and sorghum that we have documented has a significant bearing on the state of total food production.

As mentioned earlier, rice and maize were the two exceptions to the over-all stagnation in the food crop sector. Total production increased steadily at 7.2 and 6.6 per cent per annum for maize and rice respectively, and the corresponding figures for yields were 2.5 and 3.9 per cent. The significance of these figures, however, has to be tempered by remembering that the initial levels of yields were very low by international standards—an average of 0.48 tons per hectare for rice and 0.34 for maize for the period 1949-51.[18] Even by the terminal period the yields remained low by international standards in spite of the intervening growth—an average of 1.2 tons per hectare for rice and 0.49 for maize for the period 1975-77.[19] Moreover, by 1975 rice and maize still accounted for only 32 per cent of the estimated consumption of carbohydrates per head[20] and imports of rice increased from 43,000 tons to 148,000 (approximately one-quarter of the total of imports and domestic production in that year) between 1962 and 1973.[21] Thus, although the shift in the composition of food output in favour of rice and maize was a necessary response to changing food consumption habits with increasing urbanisation and incomes, there was a long way to go in terms of intensification of production and self-sufficiency in rice.

Two interesting questions may be posed in the interpretation of these changes in the food crop sector. The first is the extent to which the stagnation in the production of traditional food crops is explained by shifts in demand arising from rising income levels. If a shift in demand were the sole explanation of stagnation in production, it could be argued that the changes we observed indicated a flexible supply response to changing conditions in demand, rather than defective supply conditions giving rise to an inelastic supply. We do not have the necessary data to resolve this issue, but it is clear that, given the rise in incomes per head that has occurred in the Ivory Coast, the shift in preferences towards cereals would have played an important part in explaining the stagnation in the traditional food crop sector. But this still begs a second important question, namely, what happened to food availability per head in the rural areas, especially in the case of the rural poor? It could be argued, on *a priori* grounds, that the changes we have described in the food crop sector probably resulted in a fall in food availability per head in the rural areas. Given that two-thirds of consumption of carbohydrates per head come from the traditional crops—millet, sorghum, yams and manioc—in the terminal year (and hence a much higher proportion in the initial period), the gap between the growth rates of production of traditional food crops and those of the rural population is likely to have implied declining food availability per head for a large part of the rural population. In contrast with a growth of 3.45 per cent per annum in the rural population, annual growth rates of output were 0.3 per cent for millet, 2.4 per cent for sorghum, 0.8 per cent for yams and 1.4 per cent for manioc. Due to the small base, in terms of total food consumption, from which this growth occurred, it is unlikely that the faster growth of rice and maize production would have compensated for this shortfall in the supply of traditional food crops. Furthermore, the increase in rice imports that we noted indicates the inadequacy of the response in domestic supply. It could further be argued that since rice and maize feature more prominently in the urban than in the rural diet—on account, inter alia, of higher urban incomes and greater effective demand—the bulk of the increase in rice availability (domestic production plus imports) would have been diverted for urban consumption. Moreover, given the relative proportions involved, it is unlikely that the consumption of traditional food crops "released" by the shift to rice and maize could have fully compensated for the shortfall between production and growth of rural population. Thus, while there exists some suggestive evidence that rural food consumption per head declined, it is impossible to reach any firmer conclusion with the available data.

It should be noted, none the less, that this possibility of a decline in rural consumption per head constitutes an argument against interpreting the changes in the food crop sector as being merely a "normal" response to changes in demand. An equally plausible interpretation would be that rising incomes in the urban population have bid away the increasing supply of "new" food crops, shifting demand away from the producers of traditional food crops and leading to a stagnation in production, thus reducing food consumption per head among the majority of the rural population. Such a reduction in food consumption would have affected particularly those who were not self-sufficient producers of traditional food crops—for example, wage labourers in the rural areas [22] and those without sufficient land—as well as others with a marketable surplus in traditional food crops who experienced a fall in demand for these products.

In summary, the growth in the agricultural sector in the Ivory Coast has come predominantly from expansion of an extensive mode of production. High rates of growth of production were achieved in the export crop sector through this type of expansion. In the food crop sector, however, production and yields were generally stagnant, except for rice and maize.

RURAL INCOME AND ITS DISTRIBUTION

The pattern of growth in the agricultural sector described in the previous section gave rise to the impressive over-all growth record for which the economy of the Ivory Coast has been lauded. It is not, however, the level of these growth indicators that is our chief concern; rather, it is the implications of this growth process on rural incomes, on differentiation in rural society and on rural poverty. To document this, however, is a daunting task since the usual caveats about inadequate data apply with greater than usual force to the Ivory Coast. No time-series data exist on producer incomes, employment or cost of living in the rural areas; no comprehensive household survey data exist for rural income distribution or consumption patterns, and the first agricultural census was carried out only in 1974. In the following discussion, therefore, we shall be relying on a number of surrogate measures to construct an—admittedly—imprecise profile of changes in rural income and its distribution. While not all the procedures that we have resorted to can be defended in terms of methodological rigour, we hope nevertheless that we have avoided inconsistencies and not stretched the bounds of credibility too far.

Our point of departure is to establish what has been happening to producer incomes during the rapid expansion of export crop production. In table 31 we present the estimates of changes in gross revenue per hectare for

Table 31. Trend growth rates of gross revenue per hectare and of producer prices for cocoa and coffee

Crop	Period	α	β	R^2	N
Cocoa					
Gross revenue per hectare at current prices	1956-57 to 1976-77	9.574	+ 0.0729	0.79	21
Gross revenue per hectare at constant 1960 prices	1960-61 to 1976-77	20.48	+ 0.03	0.33	17
Producer price	1956-57 to 1976-77	53.32	+ 0.04	0.52	21
Coffee					
Gross revenue per hectare at current prices	1961-62 to 1976-77	22.06	+ 0.05	0.88	16
Gross revenue per hectare at constant 1960 prices	1961-62 to 1976-77	23.84	+ 0.0018	0.0012	16
Producer price	1961-62 to 1976-77	64.18	+ 0.05	0.52	16

Notes and source. As for table 27.

coffee and cocoa—the predominant export crops and, in any case, the only ones for which we have reliable data. These estimates were arrived at by multiplying the total output by the producer price and then dividing this by the harvested area for each year. It is obvious that this is a very crude measure of producer incomes; since no annual data on average costs of production are available it is not possible to state anything concerning net revenue, nor can any idea be gained regarding differences in earnings between different subgroups of producers. What these estimates do indicate, however, is the direction and magnitude of change in average producer incomes if we assume that production costs and average farm size have remained constant; in other words, they convey the effect of changes in producer prices and output on producer incomes per unit of land.

The figures in table 31 show that gross revenue per hectare at current prices in cocoa and coffee production increased by 7 and 5 per cent respectively for the periods for which data are available. The main factor responsible was the increase in producer prices, at a trend rate of 4 and 5 per cent per annum for cocoa and coffee respectively over the corresponding periods. According to this indicator, therefore, it would appear that the gains from export expansion have been adequately transmitted to producers. Such an inference would, however, be premature since it is necessary to get an estimate of changes in gross revenue per hectare in constant prices before anything can be said about the real incomes of producers. Unfortunately, there is no rural cost-of-living index to be used as a deflator, and we have therefore resorted to the expedient of using the Abidjan price index for "consommation de type africain"[23] as the deflator. Although there can be a wide absolute difference in rural and urban costs of living, it is likely that the magnitude and direction of *change* in cost-of-living indices cannot be greatly at variance. Since a very high proportion of the consumption of urban Africans would consist of rural-supplied food items, this would constitute a link between movements in urban and rural price indices. If this (admittedly very crude) deflator is used, the picture is altered significantly: the rate of increase in gross revenue per hectare for cocoa drops to 3 per cent per annum, while it becomes insignificant in the case of coffee. Thus in comparison with the *real* rates of growth of over 7 per cent per annum of GDP and of the value of total exports, our proxy for real producer incomes shows that these gains have been inadequately transmitted to producers in the case of cocoa, and hardly at all in the case of coffee. This conclusion needs, of course, to be qualified by the fact that low growth in producer *incomes per hectare* does not rule out the possibility that producer incomes increased whether as a result of expanding the land frontier because new land was acquired, or because an increased proportion of land was devoted to cocoa and coffee, or through a combination of both these causes. How real these latter possibilities are would depend on the nature of the process of expansion of land under cocoa and coffee cultivation. One polar case would be where there was a fixed number of producers of these crops over the period in question and where all the increase in cultivated area was distributed, in some fashion, among this fixed group. In this case, low growth or stagnation in producer incomes per hectare would not imply that income per producer behaved in the same manner. Those among the fixed number of producers who acquired new land or shifted a higher proportion of their land to coffee and cocoa would indeed have experienced increasing incomes. The other polar case would be where all new land was acquired by new entrants into coffee and cocoa cultivation (either by transferring land from other crops or by clearing new land) with the average farm size of the *original* group of producers remaining constant. In this case, producer incomes per hectare would be an accurate index of income movements of the original group of producers. The actual position would, of course, be in between these polar cases; but it could be argued that it was closer to the latter of the two.

First, given the prevailing absence of labour-saving technological aids, the area under cocoa and coffee would be limited by the availability of family labour, unless the producer were able to afford hired labour. Thus a generalised expansion of cultivated area by existing producers would be possible only if there was a corresponding increase in the proportion of farms using hired labour. However, as will be seen later from table 34, even by 1974-75 it was only farms of above 10 hectares, comprising 11 per cent of all farms, that used hired labour to any significant extent. Thus the argument of generalised expansion of cultivated area could be sustained only if we are prepared to argue that the majority of the original producers had, in the initial period, cocoa and coffee farms *below* the potential area permitted by the availability of family labour, and that the expansion in area was confined to range between this low initial level and the upper limit set by the availability of family labour. This appears an implausible picture of change to be applied to a majority of producers and would, in any case, imply that very little cocoa and coffee were grown by the average producer in the initial year.

Second, the figures on the large inflow of population into the plantation regions, when viewed in conjunction with the low proportion of farms using significant amounts of hired labour cited above, indicate clearly that many of these "immigrants" into the plantation zone became *new entrants* to cash crop production. Land abundance made it possible for migrants to clear new land in order to enter cash crop production, and the available evidence on adaptations of traditional land tenure systems to accommodate this trend [24] provides confirmation that this was a significant phenomenon.

It could be argued, therefore, that while average incomes per hectare provide an accurate proxy for total incomes of the majority of original producers, they do not of course do so for the minority who expanded cultivated area through hired labour. The latter fact is related to the dynamics of growing differentiation which we shall be discussing shortly. It should further be noted that average income per hectare also serves as an accurate proxy for incomes of new entrants: once a fixed area is acquired by them, average income per hectare is a valid proxy unless we assume that they have been continuously expanding their cultivated area. However, the same arguments against assuming that this occurred for the majority of original producers would equally apply to the new entrants.

In the case of producers of food crops, similar calculations of incomes per hectare cannot be made since no data on producer prices are available; they would, however, be of less significance than for producers of export crops on account of the high proportion of such production that is for auto-consumption. All that can be said in respect of producer incomes for these crops is that, except for rice and maize, there was little sign of productivity growth being a source of income growth and that the latter probably showed little change.

In assessing over-all changes in rural incomes it needs to be borne in mind, especially in the light of our earlier discussion, that the major means of obtaining an increase in income in the rural economy of the Ivory Coast during this period was by *entering* export crop production. Thus, although on the average income increases may not have been very great in each activity separately, there was scope for a farmer to obtain an increase in income by adopting cash crop cultivation (if he was not previously engaged in such cultivation) or by expanding the area under such cultivation. These migrants from the north of the country who managed to obtain land and start cultivation of a cash crop would be clearly one category who benefited from the growth process. Another would be those already resident in the south who managed to expand export crop production through the use of immigrant labour.

Ideally, one way of measuring the nature of the gains from the growth process and its distribution would be to estimate the changes in the relative size of the above subgroups of farmers. Again, we have no data to enable us to do anything of the sort. The best that can be done is to examine the cropping pattern of farmholdings in the south from the 1973-74 agricultural census in order to see how widespread the cultivation of cash crops had become towards the terminal period of our study. The figures in table 32 show that only 15.6 per cent of all farms grew neither cocoa nor coffee, an indication of how far the cash crop economy had penetrated. The table also shows that mixed cropping was very common: 45 per cent of all holdings grew both coffee and cocoa. It is also of interest to note that the rate of "adoption" of coffee and cocoa is correlated with the size of landholdings: over 90 per cent of farms with less than 0.5 hectare cultivated neither coffee nor cocoa, while the percentage drops to below 8 per cent for farms with between 2 and 5 hectares. Farms less than 2 hectares in size accounted for 23.5 per cent of the total number of farms but they comprised almost two-thirds of all farms which grew neither of the principal cash crops.

These figures suggest that the benefits from growth went largely to farmers possessing between 5 and 20 hectares of land; it is likely that those who were able to expand their landholdings were drawn largely from these size categories in the initial distribution of landholdings, since the farmers concerned would have been better placed in terms of operating surpluses to finance the expansion of cultivated area and to employ hired labour.

So far we have dealt only with those who have access to land and are operating a farm. There is, however, a significant subgroup who are dependent on wage employment in agriculture: those who have supplied the labour for the expansion of farms beyond the limits that can be cultivated by family labour. In 1973-74 there were an estimated 177,318 permanent workers, comprising 10.5 per cent of the total employed labour force in agriculture,[25] and virtually all of these were concentrated in the plantation economy of the south. In table 33 we have presented data on minimum wages for such workers over the period 1960-76. It should be noted that large-scale coffee and cocoa plantations[26] of over 100 hectares are exempted from the provisions of the minimum wage legislation[27] and that in 1968 it was estimated that one-sixth of the agricultural wage employees worked on such plantations. It is also likely that these minimum wages represent actual or even maximum wages in view of the labour supply situation and are not fictitious legislative figures which are below the market wage.

It will be seen from table 33 that the real minimum wage has been falling steadily between 1960 and 1976. The minimum wage of 156 CFA francs per day remained unchanged for a 16-year period between 1956 and 1972, despite increases in the cost of living. Although since 1972 the minimum wage has been raised three times, the increases have been just sufficient to keep up with the higher rates of inflation in this period. Thus this sector of the rural population experienced a steady reduction in their living standards during this period of rapid growth. It should be noted, however, that notwithstand-

Ivory Coast

Table 32. Extent of cash crop cultivation by size of landholding, 1974-75 (Ivory Coast, south)

Size of landholding (hectares)	Total number of holdings	Number of holdings growing both coffee and cocoa	Number of holdings growing only coffee	Number of holdings growing only cocoa	Number of holdings growing neither coffee nor cocoa	(5)+(1)
	(1)	(2)	(3)	(4)	(5)	(6)
<0.5	15 497	179	1 018	231	14 069	0.908
0.5- 0.99	24 248	1 454	6 023	1 407	15 364	0.634
1.0- 1.99	64 473	13 014	24 742	4 089	22 628	0.351
2.0- 4.99	161 897	64 092	73 019	11 101	13 678	0.084
5.0- 9.99	120 159	76 658	35 428	5 835	2 238	0.019
10.0-19.99	47 586	35 276	8 236	3 171	903	0.019
20.0-49.99	9 157	8 071	911	21	154	0.017
50.0-99.50	285	157	9	17	102	0.357
Total	443 295	198 901	149 386	25 872	69 136	0.156

Source. Census of agriculture, 1974-75, table 6-2-5.

Table 33. Minimum wages in agriculture, 1956-76

Year	Wage rate per day in CFA francs (coffee, cotton, cocoa, rice) (1)	Abidjan African consumer price index (2)	= (1) at 1960 prices (3)	Index of real wages (4)
1956	156	.	.	.
1957	156	.	.	.
1958	156	.	.	.
1959	156	.	.	.
1960	156	100.0	156	100
1961	156	102.9	152	97
1962	156	112.7	138	88
1963	156	112.4	138	88
1964	156	113.9	137	88
1965	156	117.0	133	85
1966	156	121.9	128	82
1967	156	124.6	125	80
1968	156	131.4	119	76
1969	156	135.6	115	74
1970	156	148.9	105	67
1971	156	147.7	106	68
1972	156	148.2	105	67
1973	160	164.1	98	63
1974	200	193.1	104	67
1975	200	211.4	95	61
1976	250	240.6	104	67

. = not available.
Source. Ministry of Agriculture: *Statistiques agricoles*.

ing the fall in real wages, new entrants to this group of wage employees would probably have experienced an increase in income as compared with their previous position. This would apply to migrants both from the north and from neighbouring countries who have entered wage employment, since it can be assumed that even the declining real minimum wage was sufficiently above the supply price of labour in the areas of origin to induce a continuing inflow of workers. But it still remains true that all those who were in this group at the initial period as well as all new entrants experienced a decline in their initial wage levels, although the magnitude of the fall is less the more recent is the date of entry.

Thus far, we have concentrated on indicators of total and average changes in income levels in the rural areas. It remains for us to calibrate the extent of inequality that exists and to indicate the main sources of differentiation.

It has been claimed that inequality in the distribution of incomes in the Ivory Coast is relatively low in relation to average values derived from cross-section data on income distribution in developing countries.[28] This particular claim is, however, inconclusive. There are no reliable household survey data on which over-all measures of inequality can be based and the method that was used in this particular estimate yields, at the very best, an estimate of unknown validity. The estimate for 1973-74 was derived from figures on average income per head in the 24 departments into which the Ivory Coast is divided;[29] thus it ignores income differences between households within each department and, if anything, yields information only on interdepartmental differences in income. It is totally invalid to treat a measure of over-all inequality derived from this procedure as being equivalent to the size distribution of household incomes, and to use it for comparisons with the size distribution of incomes in other countries. Equally baseless is the attempt to compare these figures with an earlier estimate in 1970, derived on a totally different basis, and to draw conclusions to the effect that the distribution of incomes had improved. Furthermore, the "explanations" advanced to support this contention—namely, rising farm incomes and agricultural wages—are unconvincing. Our earlier discussion shows that farm incomes cannot be assumed to have increased all round, while real wages stagnated or even fell over this period.

Rather than attempt to construct dubious measures of over-all inequality, we shall examine the partial indicators that are available on inequality in rural areas of the Ivory Coast. One such indicator is the size distribution of landholdings obtained from the 1973-74 census of agriculture. Table 34 shows that the Gini coefficient for the size distribution of landholdings was 0.446, a high figure compared with those of most sub-Saharan African countries for which estimates are available, as well as some Asian countries (Thailand and Sri Lanka).[30] Furthermore, this figure refers only to landholdings in the traditional sector (defined as holdings of less than 100 hectares) and is thus an underestimate. According to the census, there were 550 holdings of more than 100 hectares; however, information on the total area cov-

Ivory Coast

Table 34. Size distribution of landholdings, 1974-75

Size group (hectares)	Ivory Coast, south				Ivory Coast, north				Whole country			
	Number of farms	%	Total area (hectares)	%	Number of farms	%	Total area (hectares)	%	Number of farms	%	Total area (hectares)	%
<0.5	15 497	3.5	4 101	0.2	3 922	3.7	1 042	0.3	19 419	3.5	5 143	0.2
0.5- 0.99	24 248	5.5	18 462	0.8	6 754	8.2	6 722	1.8	33 002	6.0	25 184	0.9
1.0- 1.99	64 473	14.5	96 338	4.1	25 183	23.7	38 714	10.3	89 656	16.3	135 052	4.9
2.0- 2.99	62 688	14.1	155 533	6.5	20 151	18.9	50 798	13.5	82 839	15.1	206 331	7.5
3.0- 3.99	53 865	12.2	187 805	7.9	18 329	17.2	63 962	17.0	72 194	13.1	251 767	9.1
4.0- 4.99	45 337	10.2	202 640	8.5	9 044	8.5	40 390	10.7	54 381	9.9	243 030	8.8
5.0- 7.49	81 948	18.5	499 092	21.0	11 495	10.8	70 381	18.7	93 443	17.1	569 474	20.8
7.5- 9.99	38 211	8.6	327 085	13.8	5 246	4.9	45 499	12.1	43 457	7.9	372 485	13.5
10.0-14.99	37 126	8.4	446 275	18.8	3 198	3.0	37 854	10.0	40 324	7.3	484 129	17.6
15.0-19.99	10 460	2.4	178 376	7.5	684	0.6	11 546	3.1	11 144	2.0	189 922	6.9
20.0-29.99	6 902	1.6	162 840	6.9	352	0.3	8 217	2.2	7 254	1.3	171 057	6.2
30.0-39.99	1 954	0.4	67 006	2.8	55	0.1	1 964	0.5	2 009	0.4	68 970	2.5
40.0-49.99	301	0.1	13 149	0.6	–	–	–	–	301	0.1	13 149	0.5
50.0-59.99	164	–	8 763	0.4	–	–	–	–	164	–	8 763	0.3
60.0-99.99	121	–	8 937	0.4	–	–	–	–	121	–	8 937	0.3
Total	443 295	100.0	2 368 402	100.0	106 413	100.0	377 090	100.0	549 708	100.0	2 753 591	100.0
Gini coefficient		0.450				0.411				0.446		

– = not applicable.
Source. Census of agriculture, 1974-75, tables 2-1/1, 2-1/B, 2-1-1.

ered by this modern sector has not been published. As expected, land concentration is greater in the cash-cropping south than in the north, with Gini coefficients of 0.45 and 0.41 respectively.

Of the total of 549,000 holdings, 11 per cent were of more than 10 hectares in size; most of these were concentrated in the south, where 13 per cent of holdings exceeded 10 hectares and 40 per cent were of more than 5 hectares. The north had correspondingly fewer large farms: only 4 per cent were of more than 10 hectares and only 19.7 per cent exceeded 5 hectares.

Referring to 1965, Samir Amin claimed that "there is at present in the Ivory Coast ... a rural bourgeoisie of 20,000 planters, a landless proletariat of 120,000 labourers. These figures are not negligible compared to the 200,000 small and medium plantations, since one-tenth of the farmers own about 30 per cent of the land under cultivation."[31] The basis of this claim is not specified, nor is any information given regarding the definition of the rural bourgeoisie. It can, however, be stated that even if his estimate of the degree of inequality was not true of 1965, the situation in 1973-74 now conforms to it. The top 11 per cent of landholders operated 34.3 per cent of total cultivated land in 1973-74. Furthermore, if we consider those operating more than 10 hectares as constituting a rural bourgeoisie, we find that their numbers have increased to over 60,000. The average size of the 89 per cent of total holdings which were less than 10 hectares in size was 3.8 hectares, whereas the average size of the top 11 per cent of holdings was 15.5 hectares. There were 20,000 holdings of between 15 and 40 hectares in size, almost 400 of between 40 and 100 hectares in size and, as mentioned earlier, 550 holdings of over 100 hectares which were not included in the statistics for the traditional sector.

These facts on land distribution show that, since the introduction of cash cropping, differentiation has proceeded to a very substantial degree. There is moreover a high degree of congruence between the distribution of land and other aspects of differentiation. It will be recalled that the extent of adoption of cash cropping was correlated with size of landholding, with small farms of under 1 hectare showing only very limited adoption of cash cropping. This difference in cropping pattern is reflected in the differences in sources of income shown in table 35. Of the total income of farms of less than 2 hectares in size, 70.5 per cent comes from the imputed value of own consumption, while the proportion drops sharply as farm size increases and is only 16.7 per cent of total income for farms of over 20 hectares. The variations in total cash earnings per farm are huge: the range is almost 35 times between farms of over 20 hectares and those below 2 hectares in size, the large farms earning 1.5 million CFA francs in cash per year. In contrast, the range in total income between these two groups of farms is much smaller, being only 12.3.

Other aspects of differentiation can be seen in table 36. The proportion of farms owning farm machinery such as "sprayers", "pulverisers" and "huskers" is negligible for farms up to 5 hectares. Thereafter the proportion rises, reaching significant levels for farms of over 20 hectares. Similarly, while the

Table 35. Distribution of agricultural income—Ivory Coast, south, 1978

Size of landholding (hectares)	Number of farms	Average number of residents per farm	Percentage of total income in the form of auto-consumption	Percentage of total income coming from receipts of cash cropping	Percentage of total income derived from cash sale of food crops
0.0- 1.99	130 800	4.5	70.5	15.1	14.4
2.0- 4.99	203 200	5.7	47.9	38.1	13.9
5.0- 9.99	150 200	7.1	34.0	55.3	10.6
10.0-19.99	58 400	10.6	27.9	65.8	6.2
20.0-99.99	13 900	13.6	16.7	81.7	1.6
Total	556 500				

Size of landholding (hectares)	Value of cash income per head (CFA francs)	Value of cash income per farm (CFA francs)	Total income per head (CFA francs)	Total income per farm (CFA francs)
0.0- 1.99	9 622	43 300	32 622	146 800
2.0- 4.99	25 000	142 500	48 000	273 600
5.0- 9.99	44 633	316 900	67 633	480 200
10.0-19.99	59 236	627 900	82 235	871 700
20.0-99.99	114 824	1 504 200	137 824	1 805 500

Source. Unpublished estimates, Direction du Plan.

Table 36. Aspects of differentiation, 1974-75

Size of landholding (hectares)	Percentage of farms owning sprayers	Percentage of farms owning pulverisers	Percentage of farms owning coffee huskers	Percentage using tractors for land preparation	Average number of permanent workers per farm
<0.5	–	–	–	1.8	0.1
0.5- 0.99	–	1.0	1.0	1.6	0.1
1.0- 1.99	1.0	1.0	1.0	1.2	0.1
2.0- 4.99	2.0	1.0	1.0	1.7	0.1
5.0- 9.99	11.0	2.0	3.0	0.7	0.4
10.0-19.99	22.0	6.0	6.0	1.7	1.2
20.0-49.99	37.0	18.0	17.0	1.1	3.2
50.0-99.99	40.0	21.0	33.0	33.0	7.5

– = not applicable.
Source. Census of agriculture, 1974-75.

proportion of farms using tractors for land preparation is negligible in the case of farms of less than 50 hectares, one-third of farms above 50 hectares use tractors. The same is largely true with respect to the use of hired labour: farms of less than 5 hectares have no more than an average 0.1 permanent workers per farm, a figure which implies that no more than 10 per cent of such farms employ permanent workers. Farms of over 10 hectares, on the other hand, have an average of 1.2 permanent workers, this figure rising to 7.5 for farms of over 50 hectares in size. The proportion of farms using temporary workers is also closely related to farm size.

The above cross-section picture of differentiation allows us to draw some inferences concerning the dynamics of growth and inequality in the rural economy of the Ivory Coast. It is clear that the large differences in landownership and extent of cash cropping lead to large differences in cash income. This in turn implies huge differences in the capacity to accumulate and invest, that would be translated into greater inequality in asset distribution and incomes: only those with a monetary surplus will have the capacity to invest in opening up new land and in extending their farms through purchase, and to acquire the necessary labour and machinery to expand their farming operations. Furthermore, given the structure of land distribution and the cropping pattern, the benefits from any changes in producer prices for export crops will be largely appropriated by larger farmers. It is probable that the rise in production and producer prices that we documented earlier would have conferred disproportionate benefits on the larger farmers, given their greater specialisation in—and larger holdings of—cash crops. Similarly, the declining real wages that have been documented would have benefited this group of farmers, who were the most intensive users of labour. Declining wages together with rising producer prices would have meant declining costs and rising profit margins, since agricultural production is still based on simple, labour-intensive methods. At the same time, income polarisation between the landless agricultural workers and farmers with significant participation in cash crop production would have intensified. Some idea of this polarisation can be obtained from estimates of the earnings of wage labourers in agriculture and different groups of farmers. Labourers in agriculture were estimated to earn an average of 6,100 CFA francs per month in 1974,[32] whereas the figures on total earnings per farm for 1978 show that, even after allowing for the intervening inflation, average earnings in the smallest farms (<2 hectares) would be almost twice as high. Earnings per farm in farms of over 20 hectares would be more than 20 times greater than the average wage of labourers.

It might be argued that the measures of inequality that we have used are overstated in that household size is positively correlated with size of landholding. While this is strictly true, it will be seen from table 37 that adjustments for this factor still leave a considerable degree of inequality. Although the average size of household ranges from 3.7 to 30 between the smallest and largest group of farms, adjustment for this factor still leaves a range in land availability per head of 30 : 1 between these two groups. Rather than being a liability, it could be argued that large family size in a situation of land abundance is an enabling factor which allows entry into cash crop cultivation. Given the mode of agricultural production in the Ivory Coast, it is only those

Table 37. Demographic characteristics of farm households by size of landholding (Ivory Coast, south)

Size of landholding (hectares)	Average size of land-holding (hectares)	Average size of household	(1) + (2) Average size per head	Average number of agricultural active household members	Average number of permanent labourers	(4) + (5)	(4) + (6) Hectares per agricultural worker
	(1)	(2)	(3)	(4)	(5)	(6)	(7)
<0.5	0.26	3.7	0.07	1.9	0.1	2.0	0.13
0.5- 0.99	0.76	4.8	0.16	2.3	0.1	2.4	0.32
1.0- 1.99	1.49	5.6	0.27	2.6	0.1	2.7	0.55
2.0- 4.99	3.37	6.5	0.52	3.0	0.1	3.1	1.09
5.0- 9.99	6.88	8.1	0.85	3.6	0.4	4.0	1.72
10.0-19.99	13.10	11.9	1.10	5.5	1.2	6.7	1.70
20.0-49.99	26.50	17.2	1.54	9.1	3.2	12.3	2.15
50.0-99.99	62.10	30.0	2.07	15.5	7.5	23.0	2.70

Source. Census of agriculture, 1974-75.

with a supply of family labour above that required to provide for family subsistence who can break into cash crop production and hence acquire the cash income for further expansion through the use of hired labour. Thus the positive correlation between family size and farm size does not merely reflect an egalitarian expansion of farm size to meet the subsistence needs of larger families. Instead, since average farm size increases by an overwhelmingly larger proportion than average family size, the distribution of family size could itself be seen as an important factor in creating increasing differentiation.

Two further aspects of income inequality are significant in the Ivory Coast: first, the wide regional inequality between south and north; and, second, the income differences between nationals and the large foreign population. As regards the former, the gap in average incomes between south and north has recently been estimated to be 7 : 1.[33] As we have noted earlier, adverse ecological conditions in the north have ruled out export crop cultivation and have limited productivity increases in traditional crops. The north has thus been a labour exporter to the south, which by now accounts for 80 per cent of the total number of farms and 89 per cent of the total population of the Ivory Coast.[34] Despite recent governmental concern over this trend, reflected in rising levels of public investment per head being allocated to the north,[35] there has been as yet no significant diminution of the income gap. A growth strategy has naturally tended to concentrate investments in the south where more profitable opportunities existed, and it will be interesting to observe whether a shift in allocation of investments towards the north can be sustained. The limits to this would appear to be set by the availability of productive investment opportunities in the north in relation to the competition from the south. This would be particularly true of private investment.

The large foreign population in the Ivory Coast complicates the measurement of income distribution and the interpretation of the usual indices of inequality. If data existed for constructing a complete size distribution of incomes in the country, such data would probably yield a measure of inequality which would exaggerate the degree of its incidence among the native population. This is due to the fact that the extremes of the income scale are largely occupied by low-paid unskilled immigrant workers from other African countries, on the one hand, and the high-income Europeans, Syrians and Lebanese, on the other. The former group accounts for almost 30 per cent of the population (and probably a higher percentage of the labour force) while the latter numbered about 150,000 (or about 2.5 per cent of the population) in 1975.[36]

Through a massive reliance on foreign factors of production, the Ivory Coast has succeeded in shifting the burden of proletarianisation on to neighbouring countries; however, they have also drifted into a high degree of dependence on foreign capital and skills. Present policies are now directed at "Ivorianisation" at the top in the interests of an emerging native bourgeoisie and the skilled or educated. At the bottom, however, no such pressures are at work. Rather, the aim is to preserve a continuing supply of cheap labour in the face of a growing awareness among labour-exporting countries that they have a right to be compensated for their subsidisation, through labour exports, of the costs of the Ivory Coast "economic miracle".

Within the rural economy the availability of foreign labour probably contributed to the ending of forced labour and spared the local Ivorian peasantry from the policies of forced proletarianisation that was applied elsewhere in Africa to generate the necessary labour supplies. In this sense, therefore, foreign labour benefited the majority of the local Ivorian peasantry while, at the same time, as we have argued, being an important contributing factor towards growing differentiation. Such differentiation is now very advanced and probably constitutes a serious obstacle to the future developmental prospects of the mass of the Ivorian peasantry within the present agrarian system. Furthermore, it could be held that the availability of cheap foreign labour served to keep the incomes of local Ivorian peasants low by removing the pressures to increase productivity through intensification of agricultural production.

IMPACT OF GOVERNMENT POLICIES

In the preceding sections an analysis has been made of the basic economic forces which underlay the growth process and the distribution of the benefits of growth. The role of the State was, admittedly, implicit in the discussion, in so far as it opted for the pattern of growth that actually prevailed. Its laissez-faire policies allowed for the massive immigration of foreign labour and the expansion of export crop production.[37] Moreover, there were no significant attempts to intensify agricultural production or to moderate the extent of differentiation among the peasantry. None the less, its agricultural pricing, fiscal and investment policies during this period had important repercussions on rural incomes. These policies will be discussed in the present section.

The most striking feature of government policies during this period was the transfer of a high proportion of the agricultural surplus to finance the expansion of the urban economy. This was effected through setting producer prices of export crops well below world market prices, and also through export taxes. It should be noted that, although the Caisse de stabilisation et de soutien des prix des produits agricoles (CSSPPA), which fixes producer prices, could, in principle, pay subsidies to producers during years of low prices, this has not occurred at all during the period covered by our study. The CSSPPA has had a net operating surplus in every year[38] and has in fact acted as a mechanism for transferring the agricultural surplus.

Table 38 shows the level of producer prices in relation to f.o.b. prices for coffee and cocoa beans. It will be seen that for the period 1960-75 producers of coffee and cocoa were paid an average of only 54 to 55 per cent of the f.o.b. price. The world market price of these products increased steadily over this period, yet producer prices were generally adjusted upwards only with a lag. There has also been a fall since the early 1960s in the ratio of producer prices to f.o.b. price, from an average of 0.61 for coffee and 0.67 for cocoa between 1960 and 1965, to 0.52 for the period 1971-75.

The producer prices for coffee, cocoa and five other export crops are fixed by the CSSPPA. The method of operation is as follows:

For each season covering the period from 1 October to 30 September, the CSSPPA publishes a list of prices for coffee and cocoa called the "differential". This list does not only fix the price guaranteed to the producer, but also provides a detailed breakdown of all the costs included in the f.o.b. value guaranteed to exporters as well as the c.i.f. values for the principal destinations. All the intermediaries along the chain, from production to exportation, are private entrepreneurs to whom the CSSPPA guarantees a remunerative unit price.[39]

In addition to this marketing system an export tax is also levied.

The net result of these operations yields the distribution of the export earnings from cocoa and coffee that is shown in table 39. About one-third of the total export proceeds of these two crops goes to the public sector. Since the rate of surplus extraction is basically the same for the other export crops,[40] a very high proportion of the benefits from export growth described earlier has been appropriated by the State.

It should be noted that this extraction of the agricultural surplus has not been balanced to any significant degree by a flow of state investments into agriculture. Although the share of agriculture in public investments had increased to 26 per cent in the 1971-75 Plan period, there was still a large net transfer of resources out of agriculture. The total value of public investments

Table 38. Producer and export f.o.b. prices

Year	Coffee beans				Cocoa beans			
	Producer price (1)	Price f.o.b. (2)	(1)+(2) (3)		Producer price (4)	Price f.o.b. (5)	(4)+(5) (6)	
1960	98	126.4	0.775		85	138.3	0.615	
1961	.	.	.		89	110.7	0.804	
1962	73	132.5	0.551	0.605	64	104.8	0.611	0.668
1963	73	134.5	0.543		64	113.3	0.565	
1964	80	155.2	0.515		70	116.7	0.599	
1965	90	139.5	0.645		70	86.2	0.812	
1966	75	166.4	0.451		55	105.4	0.522	
1967	90	170.5	0.528		70	132.1	0.530	
1968	90	167.4	0.538	0.496	70	159.7	0.438	0.449
1969	95	169.4	0.531		70	222.0	0.315	
1970	95	221.0	0.430		80	182.3	0.439	
1971	105	228.2	0.460		85	147.7	0.575	
1972	105	195.2	0.538		85	142.1	0.598	
1973	105	206.0	0.510	0.525	85	194.4	0.437	0.519
1974	120	242.0	0.496		110	303.3	0.363	
1975	150	242.0	0.619		175	280.5	0.624	
Average 1960-75			0.542				0.553	

. = not available.
Sources. Ministry of Agriculture: *Statistiques agricoles*, op. cit., p. 14; World Bank: *Ivory Coast: The challenge of success* (Baltimore and London, Johns Hopkins University Press, 1978), table SA 18; Ministry of Agriculture: *Statistiques agricoles: Memento 1947-1977*, op. cit., p. 17.

Table 39. Distribution of the export earnings from cocoa and coffee (percentages)

Distribution	Average 1965-66 to 1974-75	
	Cocoa	Coffee
Farmers	50	54
Traders, transporters	12	15
Export tax	22	16
Stabilisation fund	16	15
f.o.b. price	100	100
Total public sector share	38	31

Source. World Bank, op. cit., p. 81.

in agriculture during this period was 110,000 million CFA francs; however, the total value of export taxes was 110,500 million CFA francs and the total surplus of the CSSPPA, a government institution, 79,000 million CFA francs.[41] Furthermore, the public investment in agriculture did not make a very substantial contribution to increasing rural incomes. There has been a proliferation of government agencies, each assigned the responsibility for promoting the expansion of production in a particular crop. There has, however, been insufficient control over the expenditures of these organisations and there have been cost over-runs and delays in the implementation of projects.[42] In the case of coffee and cocoa, only 13 and 23 per cent respectively of the total increase in planted acreage between 1971 and 1975 was accounted for by public investments. "The coffee replanting program has been only about 50 per cent achieved, at more than twice the cost in real terms foreseen in the Plan."[43] Although no information is available on the costs of private planting and replanting, it is probable that these problems with the statutory bodies meant that the use of investment funds in the public sector was relatively inefficient. There was moreover a tendency for investments to be concentrated on high-cost projects with low employment creation. This was clearly the case with investments in the sugar industry, which absorbed 37 per cent of the total investment in the plantation crop sector, and in which the investment costs per hectare were almost 40 times higher than that for cocoa. The high costs were due to "the choice of what seems to be an oversophisticated production technology".[44] The agricultural development plan for 1976-80 also "tentatively advocates a large-scale programme to develop motorised medium-scale modern farms in the savannah region ... (with individual family farms of about 20 hectares each grouped together in a block of 100 farms (2,000 hectares)".[45] Such a scheme, when viewed in the context of the average farm size of 3.5 hectares prevailing in 1974-75, would increase inequality in landownership by conferring disproportionate benefits to the small minority of families that could be absorbed by it.

One important consequence of this pattern of investment was the low level of replanting with high-yielding varieties which was reflected in the stagnation in yields per hectare that was observed in an earlier section. Very little was done to shift from the extensive mode of expansion of export production to a more intensive, higher-productivity agriculture relying less on the import of unskilled labour. The existing pattern of agricultural growth would appear, however, to serve the interests of the present ruling group very well. The large planters of the south dominate political power[46] and the extensive pattern of growth, as was argued in the previous section, would have conferred disproportionate benefits to this group. Furthermore, a high proportion of the native peasantry (as opposed to landless immigrant workers) would have experienced income gains despite the high rate of surplus extraction from the export sector and the probable increase in the degree of differentiation.

The use of the agricultural surplus to support a very high-cost urbanisation (*vide* the high concentration of ultra-modern skyscrapers in Abidjan), a relatively inefficient industrialisation through import substitution, and a growing service sector also accorded with the interests of the dominant political groups. The World Bank has pointed out that profits made in agriculture were used to create jobs for skilled Ivory Coast nationals, and that these jobs were created by buying modern plants and technology abroad at relatively high cost.[47] It would be a reasonable conjecture that "skilled Ivory Coast nationals" are more likely to emerge from the economically and politically dominant groups than from the rural masses. Since it is the large planters who dominate political power, it is likely that they have acquiesced to, if not actively promoted, these changes. Thus the extraction of the agricultural surplus need not indicate a conflict of interest between the large planters and the urban rich. In spite of the transfer of agricultural surplus, the large planters benefited disproportionately from the pattern of growth and, given their political dominance, they could also have been the main beneficiaries from urbanisation and industrialisation. This could have occurred either through a shift of a part of the investible surplus of large planters into urban activities or in gaining preferential access to high-paying urban jobs for themselves or their relatives.

It should be noted, however, that since industrialisation is still at an early stage and relies heavily on foreign capital, no vigorous indigenous capitalist class can be said to have emerged. This must account to some extent for the lack of a sharp contradiction between plantation and industrial interests.

The particular pattern of structural change in the Ivory Coast economy has led to enormous gaps between the value added per worker in agriculture and that in the industrial and service sectors. In 1975 the value added per worker in agriculture (which accounted for 82.4 per cent of total employment) was the equivalent of US$ 409, whereas that in industry (9.3 per cent of total employment) and services was US$ 3,074 and US$ 7,500 respectively.[48] It has been argued that these figures exaggerate intersectoral differences in productivity since "agricultural production is valued only at producer prices (and) taxes, trade margins and transport are all attributed to other sectors".[49] While this is true in so far as intersectoral productivity differences are concerned, this particular method of measuring value added per worker makes it reflect more closely intersectoral *income* differences. Valuation of agricultural production at producer prices reflects the incomes of farmers and also captures the effect of the large surplus transfer out of agriculture.

CONCLUSION

The "economic miracle" of the Ivory Coast was essentially based on export-led growth of its rural sector by expansion through an extensive mode of agricultural production. This has severely limited qualitative improvements in the rural economy and has led to growing differentiation in rural society. State policies have also resulted in a high level of extraction of the agricultural surplus which reduced the extent of growth in rural incomes and led to vast intersectoral differences in productivity and incomes. The "open door" policies have also led to growing dependence on foreign unskilled labour in agriculture, and on foreign skills and capital in the modern sector.

In spite of the dazzling over-all growth statistics, the standard of living in the rural areas remains poor. While basic food requirements appear, on the average, to be met,[50] there is the possibility that, as discussed earlier, for a large part of the population food consumption per head may have fallen. Life expectancy at birth, however, was only 43.5 years in 1975. This is a low level for countries in the Ivory Coast's range of incomes per head and is probably a good reflection of the low level of satisfaction of basic needs other than brute subsistence. The level of educational development likewise remains low: in 1975 only 50 per cent of children between 7 and 12 years of age were enrolled in primary schools. Despite an increase to 71 per cent in 1978, this figure is undoubtedly lower for the rural areas. Moreover, the literacy rate was only 20 per cent in the mid-1970s. The "economic miracle" does not, therefore, appear to have been transmitted into a general increase in the level of basic-needs satisfaction in the rural areas, where more than 80 per cent of the labour force continue to be employed.

It is also relevant to note that the pattern of growth was only possible with the abundant land frontier that was available and also that it was greatly facilitated by favourable world market prices for coffee and tea and favoured access to the European Community market as the result of the French connection. The Ivory Coast also enjoyed a certain regional economic hegemony in relation to other francophone West African States such as Mali and the Upper Volta, which supplied unskilled labour to it and provided a market for its industrial sector.

Notes

[1] World Bank: *Ivory Coast: The challenge of success* (Baltimore and London, Johns Hopkins University Press, 1978).

[2] Samir Amin: *Le développement du capitalisme en Côte-d'Ivoire* (Paris, Les Editions de Minuit, 1967).

[3] ibid.

[4] ibid.

[5] Rodolfo Stavenhagen: "Commercial farming and class relations in the Ivory Coast", in Rodolfo Stavenhagen (ed.): *Social classes in agrarian societies* (New York, Doubleday, 1975).

[6] ibid., p. 122.

[7] ibid., p. 125.

[8] Amin, op. cit.

[9] Since both coffee and cocoa are perennials, the harvested area is less than the total cultivated area by the extent of the area under immature plants.

[10] World Bank: *Ivory Coast ...*, op. cit., p. 124.

[11] Amin, op. cit., p. 32 (extract translated by editors).

[12] World Bank: *Economic growth and prospects of the Ivory Coast* (Washington, DC, 1970), p. 19.

[13] Amin, op. cit., p. 84 (extract translated by editors).
[14] World Bank: *Economic growth...*, op. cit., p. 19.
[15] Ministry of Agriculture: *Statistiques agricoles: Memento 1947-77* (Abidjan, 1979).
[16] World Bank: *Ivory Coast...*, op. cit.; Amin, op. cit.
[17] Using as a rough approximation, converting roots, etc., to grain using a conversion factor of 1/3. See *mimeographed version* of World Bank: *Ivory Coast...*, op. cit., Appendix table 7.
[18] Ministry of Agriculture, op. cit.
[19] ibid. Comparative international figures for rice and maize yields were as follows:

Country	1948-52 (in kg per hectare)		1973-75 (in kg per hectare)	
	Rice	Maize	Rice	Maize
Ivory Coast	510	350	1 259	535
Ghana	1 150	1 180	1 044	1 118
Nigeria	1 460	890	1 232	635
Senegal	900	950	1 440	809
Indonesia	1 610	760	2 636	1 117
Republic of Korea	3 620	–	5 135	1 666
Malaysia	1 930	–	3 043	–
Thailand	1 310	910	1 840	2 328
Brazil	1 580	1 261	1 480	1 521

Source. FAO: *Production Yearbook.*

[20] World Bank: *Ivory Coast...*, op. cit.
[21] ibid.
[22] It should be noted that real wages fell during this period. See table 33.
[23] As opposed to the "French" pattern of consumption of expatriates and Africans in the higher income brackets.
[24] Stavenhagen, op. cit.
[25] Census of agriculture, 1974-75.
[26] ibid. There were estimated to be 550 of these large-scale plantations, defined as holdings of over 100 hectares in size.
[27] World Bank: *Economic growth...*, op. cit.
[28] idem: *Ivory Coast...*, op. cit.; Yolande Bresson and Bruno Ponson: *Repartition personnelle des revenus et croissance en Côte-d'Ivoire* (Geneva, ILO, 1978; mimeographed).
[29] World Bank: *Ivory Coast...*, op. cit., pp. 380-382.
[30] D. Ghai, E. Lee and S. Radwan: *Rural poverty in the Third World* (Geneva, ILO, 1979; mimeographed World Employment Programme research working paper; restricted), table IVa.
[31] Amin, op. cit., p. 107 (extract translated by editors).
[32] Ministère du Plan: *Le secteur privé en Côte-d'Ivoire 1974,* p. 98.
[33] R. Dumont et al.: *Pauvreté et inégalités rurales en Afrique de l'ouest francophone* (Geneva, ILO, 1981), pp. 50 ff.
[34] Census of agriculture, 1974-75; World Bank: *Ivory Coast...,*, op. cit.
[35] World Bank: *Ivory Coast...*, op. cit., p. 151.
[36] ibid., p. 124.
[37] ibid., p. 155.
[38] ibid., pp. 327-373.
[39] Michel Benoit-Cattin: *Le café et le cacao dans l'économie de la Côte-d'Ivoire* (University of Montpellier I, Oct. 1973), pp. 7-8 (extract translated by editors).
[40] World Bank: *Economic growth...*, op. cit., p. 23.

[41] All in current prices, calculated from World Bank: *Ivory Coast ...*, op. cit., *mimeographed version,* Annex 1.

[42] idem: *Ivory Coast ...*, op. cit.

[43] ibid.

[44] ibid., p. 211.

[45] ibid., p. 244.

[46] See Bonnie Campbell: "Social change and class formation in a French West African State", in *Canadian Journal of African Studies,* 1974, No. 2, pp. 285-306, for a discussion of the dominance of the planter class in the political structure of the Ivory Coast.

[47] World Bank: *Economic growth ...*, op. cit., p. 5.

[48] idem: *Ivory Coast ...*, op. cit., *mimeographed version,* p. 1.

[49] ibid., p. 31.

[50] See Ghai, Lee and Radwan, op. cit., p. 15.

RURAL POVERTY IN BOTSWANA: DIMENSIONS, CAUSES AND CONSTRAINTS

5

Christopher Colclough and Peter Fallon

Botswana is a large, though sparsely populated, country occupying a central position in southern Africa. The Kgalagadi (Kalahari) Desert covers about four-fifths of the area of this flat, semi-arid tableland. Surface water is scant, except in an isolated region in the north-west, where the Okavango river flows into Botswana from Angola to form a large inland delta. The only other perennial surface waters are those of the Chobe river, forming part of the northern border with Namibia, and of the Limpopo, which, in the south-east, divides Botswana from South Africa. Most of the population, which currently comprises about 750,000 people, live along this eastern border, where soil fertility and rainfall levels are more favourable for cattle and arable production than elsewhere in the country.

Until 1966 Botswana was a British Protectorate, under the name of Bechuanaland. During the colonial period the population had been almost entirely dependent upon cattle farming, subsistence crop production and wage remittances from migrant workers employed in South Africa. Since independence, however, important discoveries of diamonds and copper-nickel have been made, whose exploitation has brought remarkably high levels of economic growth, rapid urbanisation and substantial increases in formal employment. Despite these changes, most people continue to live outside the towns, and the incomes of many of them have remained relatively untouched by this process of modernisation. This chapter examines the dimensions of rural poverty in Botswana and the economic constraints faced by rural people. It analyses the reasons why modern sector development, and the range of policies which has supported it, have not yet had a fundamental effect upon the lives and livelihoods of the poorest half of the population.

ECONOMIC TRENDS SINCE INDEPENDENCE

At independence the new Government in Botswana, led by Sir Seretse Khama, faced a range of particularly critical problems. Income per head was estimated to be about P60 per year[1] (then equivalent to about US$80), which placed Botswana amongst the poorest 20 countries in the world. A series of drought years had reduced the cattle herd by about one-third between 1961 and 1966, and in the latter year one-fifth of the population was receiving famine relief provided by international agencies. In addition, almost all capital expenditure and about 40 per cent of current expenditure of the Government were financed directly by aid from the United Kingdom. Thus, at that time the Government was unable to raise sufficient revenue from domestic sources to pay even the wages and salaries of all its own employees.

The subsequent economic history of Botswana, however, has been heavily influenced by two major changes. The first of these was an improvement in the weather cycle. Independence coincided with the beginning of a period of good rains, which enabled both cattle and crop incomes to increase substantially. The national cattle herd increased steadily from around 1 million head in 1966, to about 3 million ten years later. Although crop production has been more variable, the annual tonnage of sorghum and maize, which are the two main staples, increased sevenfold over the same period. The second major change has been the development of a minerals sector. In 1967 De Beers, which had been prospecting in the territory since the early 1950s, announced the discovery of a major diamondiferous kimberlite pipe in the north-central part of the Kgalagadi. And in the same year, Bamangwato Concessions Ltd.—a locally incorporated subsidiary of the Roan Selection Trust (RST) mining company—proved 33 million tonnes of copper-nickel ore 200 kilometres to the south-east of the diamond deposit. The investment undertaken for the development of these two prospects between 1969 and 1973 transformed the output of the modern sector in Botswana, in addition to having a major impact on the Government's domestic revenue.

The aggregate picture is summarised by the national accounts data shown in table 40, from which it will be seen that during the first three years after independence there was a healthy increase in total output. This, however, was entirely a result of gains made in two sectors: agriculture and government services. In the former case, growth was heavily dominated by increases in the number of cattle on the hoof. Indeed, the tradition of allowing the herd to build up during years of good rainfall (also common elsewhere in Africa) caused the annual slaughter to fall back from 140,000 head to about 100,000 head between 1965 and 1968. This fall in rates of slaughter led to an increase in the value added by agriculture at this time, due to the fact that a growth of the cattle herd is treated as capital accumulation in the national accounts. At the same time, the lower throughput of cattle at the Botswana Meat Commission (which dominates the manufacturing sector) accounts for the decline in manufacturing activity over the period. The enhanced level of government spending was made possible mainly through increased financial support for the recurrent budget from the United Kingdom. In 1966 the administrative headquarters had been transferred from Mafeking, in South Africa, to Gabor-

Table 40. Industrial origin of the GDP, current market prices (P million), and aggregate real growth estimates, 1965 to 1976-77[1]

Industrial sector	1965	1968-69	1973-74	1976-77[2]
Agriculture, forestry, hunting, fishing	11.1	23.2	69.2	70.8
Mining, quarrying, prospecting	0.2	0.2	15.9	43.3
Manufacturing	3.8	2.8	10.1	21.0
Water, electricity	0.2	0.3	3.3	10.9
Building, construction	2.1	1.9	21.1	17.5
Wholesale and retail trade, restaurants, hotels	6.2	5.1	35.1	57.9
Transport, storage, communications	2.7	3.4	5.3	1.7
Banking, insurance, real estate, ownership of dwellings, business services	2.1	3.5	13.2	23.4
General government	4.5	9.5	18.2	47.3
Other services		1.3	6.8	12.1
Errors and omissions	–	–	–0.7	–6.8
GDP at current market prices	32.9	51.2	197.5	299.2
GDP at constant 1974-75 prices	59.0	84.0	227.0	235.0
Average annual rate of growth (%)	.	10.6	22.0	1.2

. = not available. – = not applicable.
[1] Data for 1965 refer to the calendar year. All other data are for the period July to June of the years shown. Owing to the lack of adequate price series, it is not possible to produce estimates for sectoral contributions to value added at constant prices. [2] Preliminary figures.
Sources. Central Statistical Office: *National Accounts of Botswana, 1974-75* (1976), tables 0.2.c and 0.2.5; idem: *National Accounts of Botswana, 1976-77* (1978), table III.

one, a new capital town established in 1963. The institution of new departments resulted in an increase in government employment of 70 per cent between 1964 and 1969. In the absence of the agricultural recovery and the enhanced level of aid from the United Kingdom, total output would have remained stagnant at this time.

During the following five years, however, there was a dramatic change to very high rates of economic growth embracing all sectors of the economy. The weather remained kind, which allowed the growth of the cattle herd to be maintained; in addition, the annual offtake doubled to about 200,000 head in 1973. This, together with the strong international beef price—which similarly doubled over this period—accounts for most of the growth in the manufacturing sector during these years. The major impetus, however, came from the two mining projects already mentioned. Both deposits were some distance from existing population centres. Thus, their development required not only the establishment of mining plant and infrastructure, but also the construction of the two new towns of Orapa (diamonds) and Selebi-Phikwe (copper-nickel). The funds for these projects, most of which came from abroad, involved a total investment which, in real terms, was equivalent to more than three times the value of the GDP in 1968-69. As a result, value added from construction increased tenfold over the five years, and these investments also stimulated the trade, transport and services sectors. The Orapa mine came on stream in 1971-72, and the net output from minerals increased from negligible amounts to about P16 millions by 1973-74. Overall, the rate of growth in total output was, in real terms, in excess of 20 per cent per year over this five-year period.

These rates of increase in total output, however, were not sustained. Although initial losses at the copper-nickel mine were rectified in 1976, the output of several other sectors declined after 1973-74. Once the infrastructure work for the mines had been completed, investment spending in both the public and the private sectors fell heavily. The main direct impact was felt in the construction sector which, measured in real terms, was operating in 1976-77 at scarcely more than half its capacity of three years earlier. Moreover, in 1977 the railway experienced a large operating loss as a result of the situation in Zimbabwe, which caused value added from transport to decline substantially in that year. As a result of these difficulties, growth rates of total output in the mid-1970s were much lower than those of the boom period at the beginning of the decade.

In spite of recent setbacks, however, the transformation of domestic production in Botswana over the whole period is remarkable. Real income per head tripled over the decade ending in 1976, which, outside the oil-producing nations, represented unusual progress and was mainly the result of large investment in the minerals sector, diversification of the economy, and rapid economic growth. The share of minerals output rose from less than 1 per cent to more than 15 per cent of GDP, and the infrastructure requirements stimulated—if only temporarily—activity in other sectors. The imports needed for the mining investments also had an expansionary effect upon government revenues through the working of the Customs Union Agreement with South Africa. Under this Agreement, which had been renegotiated in 1969, Botswana received, in common with Lesotho and Swaziland, a revenue transfer from South Africa equal to about one-fifth of the value of annual imports from all sources at the rates of duty then in force. As a result, Botswana's customs receipts reached P30 million by 1974-75, which was more than 20 times greater than their value had been six years earlier. These enhanced revenues enabled the Government to balance its recurrent budget in 1972-73—much earlier than had seemed possible at independence—and, when later supplemented by taxes and royalties from diamonds, they generated considerable surpluses that could be made available for expenditure on development.

This aggregate picture therefore suggests that major progress has been made; however, it provides only one side of the story. On the other side, it must be stated that the main beneficiaries of the enhanced levels of non-agricultural output have so far been a fairly small proportion of the population; that although crop production has increased, the major part of the agricultural recovery has come from increases in the cattle herd, which could easily be lost with the onset of another drought; that within the agricultural sector, those who have mainly profited from its recovery include only about half of the rural population; and that, although there are signs that the Government is genuinely committed to rural development, only a small proportion of the enhanced revenues have as yet directly benefited the rural population. However, in order to substantiate these statements about the trend of rural incomes, it is first necessary to examine the characteristics of rural income distribution in Botswana, and to identify the main variables which affect the standard of living of rural households.

RURAL INCOME DISTRIBUTION: STATISTICAL ANALYSIS

Our data are taken from the Rural Income Distribution Survey, 1974-75 (RIDS), published by the Central Statistics Office in 1976. This sample survey was carried out in Botswana by the Central Statistics Office between February 1974 and March 1975 with the aim of measuring the statistical distribution of annual household incomes in rural areas. A stratified sample of 3,093 households was investigated, giving an over-all sampling fraction for the rural areas of 3.5 per cent. Some 1,115 households were interviewed in the field, the data on the remainder being taken from income tax returns and from those for the 1974 Employment Survey. The data have been cleaned by the Population and Human Resources Division at the World Bank and consist of 1,484 households including those taken from the 1974 Employment Survey. The data base generated by RIDS is very detailed, providing a wealth of information on a range of household characteristics, including the education, health, labour time input and assets of family members, as well as the size and sources of annual household incomes.

In this chapter we analyse data from the survey made available to us on magnetic tape by the Botswana authorities. Our investigation is restricted to an examination of the data relating to the size distribution and sources of annual income for rural households. All the relevant data collected in RIDS are used in the following analysis, with the exception of some confidential information relating to a small number of wealthy households in the top percentile of the distribution.

OVER-ALL DISTRIBUTION AND MEASURES OF INEQUALITY

Table 41 shows the shares of total income that accrue to various percentile groups. Total income in this table, and throughout the subsequent analysis, is defined as value added plus net interest, dividends and transfers received less income tax. It is clear from the table that the distribution of total income in rural Botswana is very uneven. Whereas the poorest decile of rural house-

Table 41. Distribution of total rural incomes

Percentile income group	Percentage of total income received	Mean annual income (Pula)
Bottom 10%	1.5	129
10-20%	2.5	219
20-30%	3.4	298
30-40%	4.4	385
40-50%	5.5	483
50-60%	7.1	616
60-70%	8.9	777
70-80%	11.1	970
80-90%	16.8	1 468
Top 10%	38.8	3 390
Top 5%	26.2	4 485
Top 1%	10.2	8 940

Source. Computer analysis of data from RIDS, 1974-75.

holds receive only 1.5 per cent of total rural income, the richest decile receive nearly 39 per cent. Furthermore, the richest 1 per cent receive nearly as high a percentage of total income as do the poorest 40 per cent. Admittedly, part of this inequality in total income received may be due to inequality in household size: in other words, those households with high total income may also tend to be larger than the average. However, as we show below, standardising for household size appears to have little impact upon levels of inequality.

This aggregate picture also hides significant differences between incomes earned in various parts of rural Botswana. It is useful to distinguish between residents in the eight larger villages with populations of between 10,000 and 30,000, and the rest of the rural population. People who live in these centres tend to be better off (often maintaining an additional dwelling outside the village at the place where they grow their crops or graze their cattle) and have easier access to schools, services and wage employment. It can be seen from table 42 that the incomes of residents of large villages are considerably higher, on average, than those of small-village dwellers.

The table also makes special mention of three other rural groups which, although small, have sufficiently different circumstances from the rest of the rural population to merit separate treatment. First, the farmers in the fertile lands of the Barolong region in southern Botswana have practised an individual tenure system since the end of the nineteenth century; they tend to be more prosperous, and have much higher crop incomes than farmers in other parts of the country. Second, the refugee settlements in the north-west tended, at least at the time of the survey, to be poorer than other rural dwellers. This group, coming mainly from Angola, were not integrated into the local population, nor were they either cattle farmers or wage earners, and their subsist-

Table 42. Mean and median annual household incomes in rural areas

Category	Mean annual income per household (Pula)	Median annual income per household (Pula)	No. of households
Small villages	826	550	72 883
Large villages	1 243	818	13 356
Barolong[1]	2 266	1 300	1 920
Refugees	226	205	640
Freehold farm employees	528	410	3 022
All households	903	575	91 821

[1] Refers to the farmers in the Barolong region of southern Botswana who have their own land tenure system and tend to be more prosperous than Botswana farmers elsewhere.
Source. Computer analysis of data from RIDS, 1974-75.

ence needs were met mainly from handicraft manufacture and from growing a limited range of crops. Finally, freehold farm employees also constitute a special group: they are made up of families living in one of the five freehold farming blocks which were originally granted to Europeans as concessions at the end of the nineteenth century. They are the only group for whom wages—from wage employment on one of these privately owned farms—constitute the main source of household income. For all these groups it can be seen that median incomes are less than mean incomes: in other words, the majority of households in each group earn less than the average, thus illustrating the positive skewness in the income distribution.

Table 43 illustrates some aspects of the incidence of poverty in the rural areas. The Central Statistics Office in Botswana has calculated a Poverty Datum Line (PDL) on the basis of minimum food and other requirements according to the age and sex composition of each household.[2] Thus, large households with a high proportion of adults will necessarily have a higher PDL value than small households with a low proportion of adults. In this sense, the PDL can be thought of as a minimum consumption needs index, and a household with an income of less than its PDL can thus be regarded as suffering from absolute poverty.[3] The first column in table 43 shows that, on average, households in rural Botswana have incomes almost 50 per cent higher than the PDL. In general, freehold farm employees almost reach the poverty line, defined in these terms, and only refugees are a long way below it. However, column 2, showing the proportion of households with incomes less than PDL, considerably modifies this apparently favourable picture, for it can be seen that over one-half of rural households are suffering from absolute poverty by our definition. Only in the relatively prosperous Barolong region are a minor proportion of households in these circumstances. The problem, as has been indicated above, is particularly severe in the refugee settlements and among freehold farm employees.

Table 43. Poverty in rural Botswana

Category	Mean ratio income to poverty datum line	Percentage with income below poverty datum line	Income-gap ratio
Small villages	1.39	54.5	0.44
Large villages	1.85	46.5	0.43
Barolong	2.94	22.5	0.25
Refugees	0.53	92.5	0.53
Freehold farm employees	0.97	72.4	0.33
All households	1.47	53.6	0.44

Source. Computer analysis of data from RIDS, 1974-75.

Clearly, a simple head-count of the poor does not tell us how near to the poverty line they actually are. If all 54 per cent "poor" rural households are very close to their PDL, this is obviously a much better situation than if they are all well below it. Column 3, therefore, presents the income-gap ratio, which expresses the mean shortfall between income and PDL as a percentage of PDL for all those suffering from absolute poverty. Thus, for example, an income-gap ratio of 0.30 would mean that the incomes of those in absolute poverty were on average 30 per cent less than their PDL. It can be seen from table 43 that poor households in both small and large villages are in general about the same relative distance from the PDL. Poor Barolong households are nearest to the PDL of all the groups, while poor refugees are furthest away. It is interesting to note, however, that although freehold farm employees contain the second highest proportion of poor households, they are nevertheless closer to the PDL than most other groups. The degree of poverty among poor freehold farm employees is therefore less severe than that amongst the other rural groups.

Table 44 presents three different measures of inequality for both total household income and total household income divided by PDL. The latter can for present purposes be roughly regarded as an index of the distribution of household income per adult male equivalent family member.[4] The three measures of inequality chosen are perhaps the commonest used in empirical work: the coefficient of variation is simply the standard deviation divided by the mean; the variance of logs refers in this case to variance of the natural logarithm, while the Gini coefficient is the familiar measure computed from the area between a Lorenz curve plotting the distribution of household incomes and the corresponding diagonal. All three measures are independent of the units in which income is measured, and will increase in value as income is transferred from poorer to richer households. They differ, however, in terms of their sensitivity to transfers at different income levels. The coefficient of variation, for example, attaches the same weight to a small redis-

Table 44. Income inequality in rural Botswana

Category	Coefficient of variation	Variance of logs	Gini coefficient	Mean (Pula)
Inequality in total income				
Small villages	1.27	0.72	0.49	826
Large villages	1.35	0.88	0.51	1 243
Barolong	1.38	0.60	0.49	2 266
Refugees	0.45	0.16	0.23	226
Freehold farm employees	0.86	0.31	0.30	528
All households	1.39	0.77	0.49	903
Inequality in income as ratio to PDL				
Small villages	1.38	0.63	0.48	1.39
Large villages	1.23	0.83	0.51	1.85
Barolong	1.39	0.69	0.51	2.94
Refugees	0.51	0.19	0.25	0.53
Freehold farm employees	0.86	0.34	0.31	0.97
All households	1.39	0.67	0.49	1.47

Source. Computer analysis of data from RIDS, 1974-75.

tribution between two individuals with a given income difference independently of their income levels. If one wished to give more weight to transfers at the lower end of the distribution, the variance of logs is a more appropriate measure. The Gini coefficient, on the other hand, attaches more weight to transfers at middle income levels.

The high values obtained for the Gini coefficient both for the total and for small villages, large villages and Barolong households taken separately show that income inequality is high in rural Botswana even by the standards of least developed countries. Although the Gini coefficients are roughly similar between small villages, large villages and Barolong households, the variance of logs is highest for large villages, indicating that income is particularly uneven among poorer households in the latter group. In the sample as a whole, total income is positively correlated with PDL at low income levels, although the opposite is true at higher income levels. Thus, with regard to the differences shown in the two halves of table 44, standardising for PDL reduces income inequality among all households as measured by the variance of logs, but has no clear effect when the other criteria are applied. Among households within each group, inequality in total income is, however, in some cases higher and sometimes lower than inequality in income as a ratio to PDL, depending on the measure used. This lack of a consistent pattern reflects differences both in the relationship between total household incomes and PDL within different groups, as well as in the shape and, in particular, the skewness of the various total income distributions. Notwithstanding these differences, however, income inequality by all measures is least among freehold farm employees—a not surprising result, given the relative uniformity of wage incomes as compared with incomes from other types of work.

SOURCES OF INCOME AND CHARACTERISTICS ASSOCIATED WITH POVERTY

So far, rural income distribution in Botswana has been examined in terms of total incomes accruing to households in different (non-homogeneous) categories. Although the "location" variable serves as a proxy for a range of different socio-economic circumstances which act as constraints on household incomes, within each category a wide variety of economic activities are practised. It is therefore necessary to examine the sources of household income in greater detail, in order that we may identify the economic conditions most usually associated with rural poverty in Botswana. What it has been sought to establish is whether a household's relative position on the income ladder is related to the main sources from which it derives its income. For example, are poorer households more likely to depend on non-agricultural activities? Are wage earners typically to be found among the richer or the poorer households?

Table 45 shows the average proportions of income derived from each source for households in various income groups. Since income relative to PDL is a more appropriate measure of household welfare than total income *per se*, we have defined our income groups, as in table 41, on the basis of the former rather than upon the latter measure. Table 45 indicates that, as we move from poorer to richer households, while the proportion of income received from both animals and wages rises, there is a decline in that received from manufacturing (predominantly traditional beer brewing), transfers (mostly income received from family members working in South Africa) and gathering (a simple subsistence activity). Proportions of income received from crops, trading, hunting and other sources are not strongly related to income level. Table 45 also shows the simple (or zero order) correlation coefficients between both measures of household income and the proportions of income derived from each source. The associations described above are in general confirmed at an extremely high level of statistical significance, namely 0.01 per cent. Moreover, there appears to be a weaker positive relationship between the share of income derived from trading and income levels; this is significant at the 5 per cent level.

The information conveyed in table 46, showing the breakdown of income by source for each of the locational household groupings, demonstrates the logic for treating these groups separately. While both large and small villages reveal similar proportions of income derived from crops, large village house-

Botswana

Table 45. Income profiles for the rural population[1]

Percentile income groupings[2]	Crops	Animals	Wages	Manufacturing	Trading	Services	Hunting	Gathering	Housing	Transfers	Other	Total
Poorest 10%	9	14	9	10	–	2	9	22	11	19	−5	100
10%–30%	10	13	17	6	1	3	1	16	10	19	4	100
30%–50%	10	19	21	5	2	2	1	12	9	14	5	100
50%–70%	11	31	17	6	1	2	1	7	6	11	7	100
70%–90%	8	39	20	4	1	1	3	5	5	8	6	100
Richest 10%	5	40	38	1	1	2	4	3	2	5	−1	100
Total	9	27	20	5	1	2	3	10	7	13	4	
Correlation coefficient[3] between income/PDL and share in income	−0.13*	0.19*	0.16*	−0.14*	0.05	−0.02	0.01	−0.28*	−0.32*	−0.16*	−0.03	
Correlation coefficient[3] between income and share in income	−0.02	0.32*	0.04	−0.13*	0.07	0.01	−0.01	−0.29*	−0.32*	−0.18*	0.04	

[1] The proportions shown are simple unweighted averages of proportions of income derived from each source by the households in each income range, not the proportions of income from each source. [2] Income groupings are defined in terms of total income expressed as a ratio to PDL. [3] * denotes significance at the 0.01 per cent level.
Source. Computer analysis of data from RIDS, 1974-75.

Table 46. Sources of rural incomes by location of household groups (percentages)

Category	Crops	Animals	Wages	Manufacturing	Trading	Services	Hunting	Gathering	Housing	Transfers	Other
Small villages	9	29	16	5	1	2	3	11	7	12	5
Large villages	8	22	32	5	1	3	1	5	7	14	2
Barolong	37	33	4	1	3	2	0	4	3	6	7
Refugees	39	4	3	20	0	3	1	19	5	5	2
Freehold employees	1	4	85	0	0	0	0	7	5	0	0
Total	9	27	20	5	1	2	3	10	7	13	4

Source. Computer analysis of data from RIDS, 1974-75.

holds are less dependent upon income from animals. Barolong farmers have the highest total proportion of income derived from agricultural activity of any group, while refugees are heavily dependent upon crops. The table demonstrates that freehold farm employees receive the great majority of their income from wages, although wage income is also important in large villages and to a lesser extent in small villages. Income from transfers is most important in large villages, which in part reflects the greater proximity of these centres to recruiting points for mine employment in South Africa. The most important pattern revealed by table 46, however, is that in the case of all groups, the bulk of household income is accounted for by animals, crops, wages and transfers. Thus, if the variation in income from these sources can be adequately explained, much will have been done towards explaining rural income inequality as a whole.

The following methodology has been applied: the number of animals (largely cattle) owned by each household is taken as given; if a constant return is assumed per Pula-worth of animals owned, this can be proxied by total income derived from animals. Returns to cattle rearing are generally believed to be high in Botswana; hence cattle ownership is the dominant form of wealth holding and may therefore be used as a proxy for total wealth. It is further assumed that both the location of the household and number of adult equivalent household members (as before, PDL may be taken as a proxy) are given.

If a household's endowment of animals is increased, this will raise income in two ways: first, it will raise income directly through higher total returns from animals; and, second, it will raise total income by increasing income from other sources, due to the fact that a large cattle herd makes ploughing and planting easier, thus raising income from crops. Equally, a higher cattle income also allows individual household members greater time and opportunity to invest in human capital by acquiring more education or training. Household wage incomes will thus clearly be raised, owing to the links between the amount of education received and eligibility for regular paid employment. On the other hand, higher returns from those remaining on the farm should make mine work in South Africa less attractive, thus reducing income from transfers. Our basic hypothesis is therefore that a household's endowment of animals, together with its total available labour input and its location, determine the other major components of income and hence total income.

We start from a model of the form

$$Y_i = \alpha_1 L_i + \alpha_2 K_i + \alpha_3 L_i^2 + \sum_{j=1}^{T} B_j D_j L_i$$

where Y_i is income (Pula)
K_i is cattle (as proxied by income from animals) (Pula)
L_i is household size as measured by PDL per male adult equivalent (Pula) [5]
D_j is the dummy on the jth household group—small village is taken as the base

and the i subscript refers to the ith observation.

We thus include the L^2 term to pick up any scale effects arising from differences in household size, and the household group dummies to standardise for absolute differences in marginal returns to household size.

Dividing through by L_i we obtain

$$\frac{Y}{L_i} = \alpha_1 + \alpha_2 \frac{K_i}{L_i} + \alpha_3 L_i + \sum_{j=1}^{T} B_j D_j$$

which is the form of equation presented in table 47. The coefficients can therefore be given the following straightforward interpretation: α gives the impact on income/PDL of a unit increase in animals/PDL holding household size constant; α_3 indicates whether there are scale effects arising from household size; the combination of α_1 (constant term) and α_3 provide an estimate of average income per adult equivalent in small village households of varying sizes—for example, $\alpha_1 + 3\alpha_3$ would give average income per adult among small village households comprising three adults or their equivalent. Finally, in each case B_j measures differences in income/PDL between the other household groups and small villages after holding animals/PDL constant. A similar interpretation follows if one substitutes crop income and other typical sources of income for total income (Y).

The results are presented in the first column of table 47. The over-all importance of cattle ownership is brought clearly into focus by the size of the coefficient for animals/PDL: a one-unit increase in animal income, holding family size and composition constant, creates a 1.22-unit increase in total income. As expected, this coefficient is very significantly greater than unity. Thus, income from cattle ownership has an "amplifier" effect on total

Table 47. Determinants of rural income and income by source (weighted[1] ordinary least squares)[2]

Independent variables	Dependent variables			
	Income/PDL	Crop/PDL	Wages/PDL	Transfers/PDL
Animals/PDL	1.22 (317.8)	0.02 (23.0)	0.08 (20.5)	0.004 (3.76)
PDL	−21.9 (80.7)	0.36 (5.5)	−17.9 (68.2)	−2.3 (32.9)
Large village	88.7 (45.2)	2.9 (6.2)	97.5 (51.3)	3.9 (7.9)
Barolong	213.7 (44.7)	173.6 (152.0)	−12.0 (2.6)	−5.8 (4.7)
Refugee	−64.3 (7.9)	21.3 (10.9)	−60.7 (7.6)	−17.5 (8.4)
Freehold farm	10.2 (2.7)	−9.3 (10.1)	71.1 (19.1)	−21.8 (22.2)
Constant	211.3	9.3	123.9	30.2
F	18 692	4 078	1 252	276
R^2	0.55	0.21	0.08	0.02

[1] Each observation in the sample is weighted by the inverse of the probability of that household being selected from the population. [2] t statistics are presented in parentheses.

income: any increase in the former leads to an even larger increase in the latter. The implications of this important result are discussed in the next section of the chapter.

Household size (as proxied by PDL) would appear to have a negative effect on income per adult equivalent. This might reflect a tendency for larger households to have a relatively greater number of non-participants as, for example, children; on the other hand, there may be a genuine negative scale effect to household labour input if outside employment opportunities are limited. The locational differences are as one would have been led to expect, on the basis of earlier tables, although it is interesting to note that while freehold farm employees have lower incomes on average than households in small villages, this is not true once animal ownership is taken into account.

The three other main sources of income—wages, crops and transfers—are each significantly related to animal ownership. The positive relation between animals/PDL and transfers/PDL is surprising, as one would have expected households that were well endowed with animals to have been less inclined to send family members to work in the South African mines or to the urban centres in Botswana. There are, however, a number of possible explanations for this result. First, those households with migrants may tend to be smaller than the average as defined by RIDS (migrant workers were excluded in the survey from the definition of household size); if so, this would tend to raise the apparent value of animals/PDL for such households, and thereby contribute to the above results. On the other hand, however, it is also likely that the wages earned by migrants from richer households—namely those with more cattle—tend to be greater than those from poorer families, owing to easier access to schooling and better urban contacts for the children of these groups. Whatever the explanation, the magnitude of the coefficient in question is very small.

It can be seen that larger households have a relatively greater income from crops and a relatively lower one from wages and transfers. Other data show that family size is negatively related to the average years of schooling of family members, which probably explains why households with a larger PDL have less wage income relative to size. On the other hand, the observed negative relation between PDL and transfer income may reflect a problem arising from the model we have used. Households who send family members to South Africa will tend, all other things being equal, to have a lower observed PDL. Likewise, if family members stay at home, households will be observed to have a higher PDL. However, in the absence of any information about the number of people from each household currently remitting funds home—such data not having been collected in RIDS—it is impossible to disentangle the effect of household size upon transfers from that of receiving transfers upon household size.

Our results therefore tend to confirm the view that animal ownership is the key determinant of household income: not only does a larger cattle herd directly increase income from animal ownership, but it also raises income from the other major sources. However, other exogenous variables such as household size and composition and household location also play a significant, if smaller, role in the determination of total household income.

RURAL INCOME DISTRIBUTION: DYNAMIC ANALYSIS

The analysis in the previous section has demonstrated three important facts: first, that income distribution in rural Botswana is extremely unequal; second, that the single most important determinant of household incomes per head is income from animals; and, third, that income from animals is itself positively correlated with the other major constituents of rural household incomes, namely income from crops, wages and transfers.

The main objective of this section is to examine the changing influences upon the shape of rural income distribution over time. In the first section it was shown that total incomes, including those from the agricultural sector, have grown very considerably since the early 1960s. The important question that needs to be asked therefore is: has the standard of living of every sector of the community been raised by the same amount, or have the rich captured proportionately more of the benefits of this growth? Given the link between the ownership of cattle and rural household income, the crucial issue in this context is how the distribution of cattle ownership has been changing over time. Subsidiary issues concern the extent to which improved opportunities for crop, wage or transfer incomes have been made available for poor, as opposed to richer households.

Cattle

The size of the national herd of cattle in Botswana has increased roughly tenfold since the beginning of the century, to about 3 million head in 1978. This growth, however, has not been uninterrupted. Periodic droughts have caused stock losses, often of a dramatic nature—most notably around 1914, 1933, the late 1940s and the mid-1960s. These droughts, together with occasional outbreaks of foot-and-mouth disease, particularly affected the farmers on marginal lands where the pasture was thin, and those who owned only a few head of cattle. Furthermore, the introduction of cash levies which were paid throughout the century to the Government in the form of poll taxes and in some cases also to chiefs to finance public undertakings in the tribal areas, often necessitated the sale of cattle in order to raise the necessary funds. Evidence from the available sources suggests that a combination of these factors resulted in an increasingly concentrated ownership of the total national stock of cattle during the colonial period.

There were, and still are, a number of traditional mechanisms which somewhat weakened the forces making for greater concentration of cattle ownership. The most notable of these has been the *mafisa* system, whereby large-scale cattle owners allocate part of their herd to relatives or friends, thereby providing the owners with a responsible source of cattle care, while the recipient is able to use the cattle for ploughing and for milk, and is also sometimes allowed to retain calves from the cattle he tends.

The mitigating effects of this institution, however, are fairly small. Indeed, since the early 1960s the distribution of cattle ownership has become far more skewed, mainly as a result of the prolonged drought which lasted from 1961 to 1966. During this period of drought, in order to relieve pressure upon badly depleted grazing areas, the colonial administration drilled new boreholes in areas which had previously lacked a water source. These new lands were then allocated to the larger cattle owners who could afford to repay the costs over the long term. As a result, however, at the end of the drought many of these owners of large herds had acquired *de facto* exclusive rights to extensive areas of new grazing land while, on the other hand, many of the smaller cattle-owners had lost all their herd. During the five years of severe drought, about 400,000 cattle died, representing about one-third of the total national herd in 1960. A large number of households lost all their cattle, and, at independence in 1966, one-fifth of the population of the country were in receipt of famine relief following the succession of crop failures.

The drought itself was therefore an important factor in increasing the unequal distribution of cattle ownership. Since no taxation or other measures were introduced by the Government which could secure even a partial redistribution of the increments to the herd when the good rains returned, the benefits of the subsequent recovery accrued to those who had not lost all their cattle. Thus, there was a *permanent* worsening of the distribution of cattle ownership at this time. On the basis of our earlier analysis of the links between cattle and household incomes, this obviously implies a movement in the distribution of total rural income in the same direction.

In addition, however, when the total national stock of cattle began to build up again in the early 1970s, there arose acute competition for good grazing lands. The convention of allowing those who could afford to install boreholes the exclusive rights to the use of their water, and thus to the surrounding grazing land, meant that the richer farmers could effectively pre-empt the better grazing lands for their own use. Until this "land grab" was stopped by the Government in 1975, when borehole applications were frozen, poorer farmers were being pushed on to the more marginal lands, thereby increasing their vulnerability to water shortage and further promoting the tendencies towards polarisation in herd size and in patterns of ownership. These trends may have been exacerbated by the additional economic demands made upon rural families since independence: with greater opportunities for urban work, many more people left the villages in an endeavour to find a job. Typically, rural-urban transfers of food were needed in order to sustain migrants during their period of job search.[6] In addition, head taxes and other charges to rural (and urban) families tended to rise. These costs may well have increased the economic burden on poorer rural households (at least in the short term) and led them to sell their cattle more readily than in earlier days. Finally, social and economic changes within Batswana families appear to have led to some divergence away from the traditional extended-family system: households are becoming smaller, and this appears to be happening more quickly than any corresponding fragmentation of cattle herds.

Its is difficult to assess the relative importance of these various economic and social factors. However, it is clear that their combined effects have seriously worsened the unequal distribution of cattle ownership which existed at the end of the drought. The results are indicated in table 48: it can be seen that of those households engaged in agricultural production, the proportion neither holding nor owning cattle has increased significantly in recent years. In addition to these "agricultural" households, there is a small but growing rural group which does not engage in agricultural production. These are often amongst the very poorest households in the community, many headed by women and for the most part neither owning nor having access to cattle. Taking the rural population as a whole, then, it is estimated that the proportion of households without cattle increased from roughly one-third to one-half of the total between 1960 and 1975[7]—a period during which the number of people living in the rural areas increased by about 20 per cent. Thus, the apparent progress in the agricultural sector indicated by the national accounts data in table 40 cannot be interpreted to mean that the rural population *as a whole* is now much better off than it was at independence. On

Table 48. Proportion of agricultural households without cattle

Period	Number of agricultural households[1] ('000s)	Percentage owning no cattle	Percentage holding[2] no cattle
1968-69	48	30[3]	29
1970-71	52	25	23
1971-72	64	32	30
1974-75	80	37	32
1975-76	80	39[3]	36

[1] Estimates of the number of households engaged in agricultural production in each year shown. Based upon agricultural surveys. [2] Difference between proportion of households "holding" and "owning" cattle represents the *mafisa* system. [3] Rough estimates, based upon trend shown in final column.

Source. Based upon Christopher Colclough and Stephen McCarthy: *The political economy of Botswana: A study of growth and distribution* (Oxford, Oxford University Press, 1980), table 5.1.

---

the contrary, the growth in the cattle herd, which these output data mainly represent, has directly benefited only half of the rural population and, even within this category, the gains have been disproportionately captured by the richer groups.

Crops

As our analysis of the data from RIDS has shown, this trend in the distribution of cattle ownership will also have worsened the distribution of income from crops, mainly on account of the fact that households who own no cattle experience considerable difficulties in having their lands ploughed. While it is true that cattle can be borrowed, yet all too often the borrower has to wait until the owner himself no longer needs them for his own ploughing. Moreover, those who own no cattle, being poorer, are often short of agricultural implements; and if there are males of working age attached to the household, these are more likely to be absent in the mines of South Africa, or in one of Botswana's urban centres, causing a shortage of labour to work the lands. This range of disadvantages implies that households without cattle plough later (if they do so at all) than others, working a smaller area, and, since the rains are usually already advanced, producing lower yields. In 1971–72, for example, of those undertaking arable production, households which did not own cattle planted only two-thirds of the area, and received less than one-quarter the yields of households owning cattle.[8]

Employment

We know from tables 45 and 47 that in 1974 households with larger herds of cattle also received both a higher absolute *and* proportional amount of their total incomes from wages than those with few cattle. In other words, wealth in the form of cattle ownership and access to more or higher income-earning opportunities from wage employment are strongly positively correlated in the rural areas of Botswana. The question which now arises is whether and how this static picture has been changing since independence. This is less easy to analyse than the trends in cattle and crop incomes already discussed. Formal employment in Botswana has increased much more rapidly than the labour force, at an average rate of about 10 per cent year over the first decade of independence, and the proportion of the rural population with regular jobs doubled from about 2.5 per cent in 1964 to about 5 per cent in 1975. However, the distributional effects of this development depend upon whether these jobs have been taken by persons from poorer or from richer households. There is some evidence, however, to suggest that here too the main benefits have been captured by the richer groups.

In the first place, the very poorest rural households often comprise single women with children and are without males of working age. Such families would be unable to benefit from employment opportunities, even if they were available. Second, the incidence of job creation in the rural areas has been mainly concentrated in the larger villages, which are the main centres for local government and district trading activities. It is the richer households who can afford to maintain additional dwellings on their lands or cattle posts who live in these centres, and they are therefore likely to have had more favoured access to new jobs than those not living so close at hand.

Some evidence for this is given by our analysis of the RIDS data in tables 44 and 46. The mean income of households in large villages in 1974 (15 per cent of all households) was 50 per cent greater than that for households in small villages (80 per cent of all households). On the other hand, the mean income from wages was three times greater for households in large villages than for small-village residents. This is not only a function of a greater incidence of wage employment opportunities in large villages, but also of a difference in wage rates. Many of the wage incomes in small villages are derived from informal or seasonal work in agriculture, including cattle herding, where the wages paid are very low. The growth of the national cattle herd, at a rate of some 10 per cent since the late 1960s, will have increased the number of such jobs. On the other hand, the increased concentration of ownership, together with increases in average herd size, will also have brought negative employment effects through scale economies. It is probable that the over-all impact of these factors on job creation has been much less than that of the growth of formal employment, to which the richer groups have had more favoured access. Thus, the distributional implications of increases in wage employment opportunities have probably been neutral, or even negative, in the rural areas, and they have probably done little to alleviate directly the poverty of the poorest groups.

Transfers

Transfers to rural households in Botswana come from two main sources:

first, transfer payments are made by—for the most part—unskilled and semi-skilled migrant workers in South Africa, approximately half of whom work in the South African mines; second, such payments are also made by members of rural families working in Botswana's urban centres. Payments from each of these sources will now be briefly discussed.

Migration for work in South Africa has increased substantially since the Second World War. By the mid-1960s the number of such absentees had increased to 50,000 persons (compared with about 10,000 during the 1930s), and had reached about 70,000 ten years later. At present, about one-third of Botswana's male labour force typically work in South Africa. The most important explanation for this recent trend is the increase in the levels of wages paid in South Africa. For a range of domestic and international reasons, minimum wages in the mines began to move forward in 1970, but the increases accelerated rapidly after 1972. Over the seven years from 1969 to 1976 minimum wages in the gold-mines of South Africa multiplied more than sevenfold. As a result, transfer payments from South Africa to Botswana households have increased considerably. Estimates suggest that the inflow from mineworkers alone increased from about P2 million per year in the late 1960s to about P10 million in 1976. Total transfers from migrant workers in South Africa in all sectors, are today probably twice this figure.

Wages in Botswana's urban centres have similarly increased since independence, more particularly between 1973 and 1976, when unskilled wages almost doubled in real terms. Higher up the earnings ladder, wages since independence have not kept abreast of inflation, although substantially larger earnings were given to the higher paid than to unskilled groups in 1978. The effect of these changes on remittances is not entirely clear. Quantitatively, the unskilled wage rates are the most relevant for absentee workers from rural families. The number of skilled workers is, over-all, smaller, and many of these tend to become permanent urban residents. On the other hand, rural-urban transfers are significant, particularly for migrants who have recently arrived in the towns, and one recent survey suggested that at least from the peri-urban areas, net transfers of money to villages was insignificant.[9]

Nevertheless, taking account of wage movements in both South Africa and urban Botswana, it is likely that many of the poorer rural Botswana households have benefited from increased real transfers since 1973. As shown in table 45, in 1974 transfers represented about one-fifth of household incomes for the poorest third of the rural population. If the assumption is made that remittances were maintained as a constant share of the incomes of those workers sending them over the following three years, this would have increased the real incomes of the poorest third of rural households by about 20 per cent, assuming that the real value of the other items of their incomes were maintained.

It should be noted that this conclusion does not conflict with the results of our earlier regression analysis, which showed that transfer income was positively correlated with cattle ownership: whilst it is true that those with more cattle tend to receive higher transfers, per head, than those with less (possibly, as mentioned earlier, because migrants from rich households are more likely to secure clerical or professional jobs, and can *afford* to send more money back to their respective villages), the *proportional* importance of transfer income is four times greater for the poorer than for the richer groups (table 45). Thus, increases in the real value of transfers to the rural areas will recently have had a much greater proportional impact upon the standard of living of the poorer than of the richer groups.

The role of government

So far, nothing has been said concerning the extent to which government policy has encouraged or weakened the trends in rural incomes discussed above. A full analysis of this issue would need to include a review of the taxation, subsidy, incomes and expenditure policies of the Government, together with an analysis of more detailed sectoral programmes in agriculture, health, education and other services. Here we shall focus only upon the Government's investment pattern since independence. The Government's capital budget is extremely important, particularly in small economies where domestic private investment constitutes a small proportion of total capital formation. In Botswana it has played a crucial role in determining the major shifts in economic structure and in productive capacity which were outlined in the introductory sections of this chapter.

The over-all distributional impact of public capital expenditure is summarised in table 49. These data, calculated from government accounts for each of the years shown, classify capital expenditure according to whether it involved the construction of rural or urban facilities. The "urban" category is strictly limited to expenditure in the five main towns of Botswana, the definition "rural" covering all other expenditure. In addition, however, a separate category—"communications"— is included, which is heavily dominated by road construction between the main towns. Although expenditure under the "communications" heading does bring benefits to rural dwellers, the main beneficiaries tend to be urban residents (particularly those who own cars or urban businesses) and the justification for such projects in Botswana has normally been based upon a high benefit/cost ratio with regard to output from the formal sector of the economy. A number of trends indicated by this table are important.

First, it can be seen that there was a dramatic build-up in urban investment until 1973. This was heavily dominated by the construction of Selebi-Phikwe, the new copper-mining town in the east-central part of the country. Although foreign investors provided the mining facilities, the rest of the urban infrastructure was the responsibility of the Government. By 1972-73, fully 85 per cent of public capital expenditure was devoted to urban construction—almost all of it at Selebi-Phikwe in that year.

Second, although urban investment fell back during the following years,

Table 49. Government capital budget expenditure by location and/or main beneficiary group, 1966-67 to 1976-77
(P millions (percentages))

Year	Rural	Urban	Communications	Total
1966-67	1.44(26)	2.79(49)	1.43(25)	5.66
1967-68	1.48(39)	1.69(45)	0.62(16)	3.79
1968-69	0.85(25)	2.12(64)	0.37(11)	3.34
1969-70	1.00(20)	2.19(45)	1.71(35)	4.90
1970-71	1.39(16)	5.34(63)	1.80(21)	8.53
1971-72	1.00(8)	9.03(74)	2.22(18)	12.25
1972-73	1.63(5)	25.39(85)	2.86(10)	29.88
1973-74	3.32(10)	23.86(70)	6.93(20)	34.11
1974-75	12.31(33)	13.67(37)	11.00(30)	37.00
1975-76	13.40(32)	17.69(42)	11.00(26)	42.09
1976-77	11.70(27)	14.15(33)	17.22(40)	43.07

Source. Colclough and McCarthy, op. cit., table 4.4.

after 1973-74 this was to a large extent compensated for by an increased emphasis upon communications facilities. This was almost entirely a reflection of increased investment in roads, which, by 1976-77, accounted for almost 40 per cent of the capital budget. Following the attempted—and, in the event, largely unsuccessful—Arab oil embargo of South Africa in 1973, the Botswana Government realised how acutely vulnerable the economy was to the variability of supplies from that source, not only of oil but of all other commodities. Accordingly a decision was taken to mount a "lifeline" programme involving the tarring of the existing road running along the eastern part of the country between Gaborone and Francistown, and the construction of a new road from Francistown to link with Zambia in the far north. This project, which dominated the capital budget in the late 1970s, involved an investment of over P50 million, a very high cost for a scheme which was based entirely upon strategic considerations, and which has brought few real development benefits, at least in the short term, particularly in view of the recent political changes in Zimbabwe.

The counterpart to the above trends is the changing emphasis that has been given to rural investment. It can be seen that this remained stagnant between 1966 and 1973 which, allowing for inflation, represents a significant drop in expenditure measured in real terms over those years. By 1972-73 investment in rural development had dropped to only 5 per cent of the Government's capital budget, almost all the Government's development capacity being currently devoted to the construction of the new copper-mining township.

Nevertheless, it would be unjust to say that the Government has completely ignored the needs of the rural areas. The President and Cabinet had constantly reaffirmed in political speeches and government documents the pledges for broad-based development that had been given at independence. The explicit strategy adopted by the Government at that time had been to concentrate initially on infrastructure and mining projects which would secure budgetary independence from the United Kingdom and which would generate financial surpluses which could subsequently be used for rural development.[10] By 1973 these initial aims had been achieved. Accordingly, in that year an Accelerated Rural Development Programme (ARDP) was initiated. This focused upon the provision of infrastructure in the rural areas (schools, clinics, water supplies, rural roads) and was rather designed to satisfy basic needs than to stimulate rural production directly. The results are shown in table 49. The ARDP increased capital provision in the rural areas more than fourfold between 1973-74 and 1975-76. During these three years more than P20 million was spent on ARDP projects,[11] which itself was more than two-and-a-half times the total amount of capital expenditure on rural facilities undertaken by the Government since independence.

CONCLUSION

We have argued that changes in rural income distribution were probably greatest during the immediate post-independence years. Between 1965 and 1973-74 the average real growth of agricultural output was close to 16 per cent per annum. These increases, however, were mainly captured by cattle owners, who constituted the richest half of the rural population. Over these years real wages for unskilled workers increased only slightly, both in Botswana and South Africa, probably implying that the real value of transfers to the rural areas remained approximately unchanged. Moreover, public expenditure on rural projects declined significantly in real terms. During this period, therefore, the distribution of rural incomes almost certainly deteriorated sharply. Since that date, however, the situation has changed to some extent, real output in agriculture having made slow progress whilst transfers have increased at a much faster rate. As a result, the growth in household incomes of the poorer families between 1973-74 and 1977 was probably faster than those of the richer groups.

It is as yet too early to tell whether there will be a change of emphasis, in public expenditure and other aspects of development policy towards promoting rural development, on the part of the Government. Whilst ARDP is perhaps a good omen, it was introduced at a time when the Government felt itself to be under political pressure, and although it did succeed in raising rural expenditure to a higher level, table 49 has shown that the proportion of capital investment devoted to such provision at independence (26 per cent) had scarcely been exceeded ten years later. Moreover, because the initial benefits from the high rates of economic growth achieved by Botswana have rewarded a minority of the population, new élite groups have emerged whose own economic interests now make the promised redistribution more difficult to achieve.

The recent favourable trends, therefore, may not be very significant, particularly in view of the apparent changes in the Government's salaries policy since 1976 which have reversed the earlier tendencies to reduce income differentials within the formal sector. The fundamental fact remains that, on account of the highly inegalitarian ownership of capital in the rural areas, the poorer households have not received much benefit from the agricultural recovery that has occurred since independence due to the very limited access to cattle of such households. The distribution of formal employment opportunities appears, if anything, to have exacerbated this bias against the poorest groups. Over the whole period the distribution of income in the rural areas has deteriorated, and unless specific action is taken to change access to or ownership of capital—and particularly of cattle—amongst the poorest 40 per cent of rural households, it is probable that inequality in rural income distribution will intensify in the future.

Notes

[1] All currency values in this chapter are expressed in Pula. This currency replaced the South African Rand in Botswana in 1976. Although introduced at the same par value as the Rand, the Pula was subsequently revalued, and in mid-1979 its value stood at US$ 1.20.

[2] Household needs, for purposes of calculating the PDL, were based upon what the Government believed to be the basic minimum consumption requirements of families of different age and sex compositions. These included food, clothing, housing, durable goods, personal care, and health, education, recreation and social items, based upon typical rural prices.

[3] Some outside observers have felt that Botswana's PDL is high when compared with those of other countries at similar levels of development. But a careful examination of the methods of calculation suggests that the allowances made for food, clothing and housing needs (which represent the major part of the index) were not over-generous. It should, however, be noted that small allowances for tobacco and alcohol were included, which are often excluded in other studies.

[4] This depends, of course, upon whether one regards minimum consumption requirements as a suitable basis upon which to weight different family members. Given fixed weights for different family members it is also possible to think of Income/PDL as an index of income per 10-year-old child—or per "average" family member.

[5] In this case the PDL for each household has been expressed as a ratio of the estimated minimum income needed to sustain an average rural male. (See Central Statistical Office: *The rural income distribution survey in Botswana, 1974-75* (RIDS) (1976), p. 213.) This is done purely for presentation reasons, as otherwise very small absolute values would be obtained for some of the estimated regression coefficients.

[6] See Central Statistical Office: *A social and economic survey in three peri-urban areas in Botswana* (1974).

[7] Some direct evidence for this comes from agricultural surveys conducted in 1974. The authors found that "only 74 per cent of households which at one time held trek oxen presently have them, and 26 per cent do not". These ratios are broadly comparable with those given in the text. See Ministry of Agriculture: *A study of constraints on agricultural development in the Republic of Botswana, 1974.*

[8] See Christopher Colclough and Stephen McCarthy: *The political economy of Botswana: A study of growth and distribution* (Oxford, Oxford University Press, 1980), table 5.5.

[9] Central Statistical Office: *A social and economic survey...*, op. cit.

[10] See, for example, Republic of Botswana: *Transitional plan for social and economic development, 1966,* where this strategy was clearly spelled out.

[11] Table 49 includes other rural expenditures in addition to those formally included in ARDP.

EQUITY AND GROWTH: AN UNRESOLVED CONFLICT IN ZAMBIAN RURAL DEVELOPMENT POLICY

6

Charles Elliott

Since Zambia's best-known product is copper, it is readily assumed that mining dominates the economic life of the country, leaving to agriculture the function of a labour reservoir and retirement home. The copper industry accounts for over 90 per cent of all exports, while agriculture accounts for less than 2 per cent. In years of high international prices, the mining industry provides two-thirds of the revenue of the central Government. Even in years of low copper prices, the industry contributes three times as much to GDP as does commercial agriculture. In a year of high copper prices and poor harvests that difference can be a factor of 30 (table 50). In terms of employment, however, the mining industry takes on a very different perspective. Employment in the industry grew slowly and erratically from 45,000 to 52,400 between 1965 and 1974.[1] A minimum estimate of those employed in agriculture in 1971 is 800,000[2]—over two-and-a-half times as many as are employed in the whole of the "modern" sector. While copper may provide foreign exchange and, on occasion, finance for development projects or redistributive services such as education and health care, the majority of Zambians depend for their immediate livelihood on agriculture.

It does not follow from the foregoing, however, that the links between the modern sector (of which mining is a major component) and agriculture are confined to the allocation to the rural sector of surpluses from mining on the one hand and, on the other, the provision of labour and food from the rural areas to the modern sector. It will be a constant, though often implicit, theme of this chapter that the mining sector, acting as a powerful wage leader,[3] plays a key role in perpetuating the urban-rural gaps that in many subtle ways increase the poverty of the rural poor. The relationship between high urban wages and rural poverty can thus be seen to be *dynamic* and *systemic*, being incorporated in a set of processes that result in the impoverishment of the poor. Some such processes are already well known: the rural-urban terms of trade; the allocation of scarce resources, such as investment, current expen-

Table 50. Gross domestic product of type of economic activity in current prices, 1965-76
(in millions of current Kwacha)

Economic activity	Former SNA						Current SNA					
	1965	1966	1967	1968	1969	1970	1970	1971	1972	1973	1974[4]	1975[4]	1976[4]
Agriculture, forestry, fishing	97.4	106.8	109.9	114.1	118.5	120.7	136.1	154.0	171.7	186.3	191.0	195.5	247.0
Commercial sector	18.3	22.3	22.5	21.7	23.5	26.1	41.5	52.7	66.3	67.1	59.0	53.5	81.0
Subsistence sector	79.1	84.5	87.5	92.4	95.0	94.6	94.6	101.3	105.4	119.2	132.0	142.0	166.0
Mining and quarrying	291.8	380.3	380.5	412.7	639.3	460.1	467.7	303.7	326.1	544.7	635.0	156.0	204.5
Manufacturing	48.0	69.0	86.1	105.8	113.9	127.4	127.5	142.6	182.3	196.8	249.5	279.0	320.0
Construction	40.9	54.8	56.9	62.3	67.5	71.5	91.4	99.1	102.6	107.3	120.3	158.0	166.0
Transport, communications and storage	32.8	33.3	50.0	48.4	44.1	42.5	49.9	62.7	63.4	70.4	76.4	82.6	91.5
Services	185.7	191.7	258.7	299.7	302.4	367.1	390.7	425.6	470.3	507.9	616.4	664.9	746.0
Community, social and personal services[1]	64.0	70.1	93.7	102.1	108.2	125.8	150.5	178.4	195.1	201.0	230.2	260.6	303.5
Electricity, gas, water	5.4	7.4	8.4	12.6	14.2	15.5	15.5	18.2	25.7	27.0	32.7	34.5	38.0
Other services[2]	116.3	114.2	156.6	185.0	180.0	225.8	224.7	229.0	249.5	279.9	353.5	369.8	404.5
Other[3]	14.5	12.3	15.0	19.0	27.8	27.7	15.7	16.3	18.3	2.6	15.4	26.0	18.0
Gross domestic product	711.1	848.2	957.1	1 062.0	1 313.5	1 216.9	1 279.0	1 204.0	1 334.7	1 616.0	1 904.0	1 562.0	1 793.0

[1] Includes public administration, defence, education, health, recreational and personal services. [2] Includes finance, real estate, commerce and trade. [3] Import duties less imputed bank service charges. [4] Provisional.
Source: Central Statistical Office: National accounts and input-output tables, 1971 (Aug. 1975); and idem: Monthly Digest of Statistics (Mar. 1977).

diture and scarce skilled manpower; and the flow of migrants from the rural to the urban areas have all attracted recent comment.[4] In suggesting others, we shall need to differentiate sharply between those persons who are sufficiently incorporated (or incorporable) in the market economy to be affected by such processes; and those insulated from them—whether by distance, or by the very scarcity of their own economic resources.

Before we turn to a more detailed examination of these processes, however, it will help to put them in a proper perspective if an indication is given of the scope of income disparities among four roughly drawn groups:

(a) the vast majority of farmers and subsistence producers who are, by any standards, exceedingly poor;

(b) their much smaller (though rapidly growing) urban counterparts—the so-called informal sector of unemployed and underemployed, and grossgroups;

(c) the prosperous "modern" farmers; and

(d) those employed in full-time jobs in the formal sector.

As a rough order of magnitude, the above four-part division may be associated with total (that is, cash and non-cash) incomes as indicated in table 51.[5]

Although some of the relevant data are not available, it is clear that the size and income levels of the above four groups have been growing at very different rates. Formal sector employment, for instance, grew at a rate of 3.6 per cent per annum between 1965 and 1974, though the deceleration since 1973 may trim that figure to under 3 per cent per annum for the 15 years from 1965 to 1980. Although this rate of growth compares favourably with those of many other African countries, it has been wholly insufficient to meet the rising demand for urban jobs in the formal sector on the part of migrants from the rural areas, as well as from school-leavers already living in the cities or towns, and from the growing number of women and more elderly people who would, in the 1950s and 1960s, have "naturally" returned to their villages. Although it is very difficult to arrive at a precise figure, it is likely that this gap between demand for, and supply of, urban jobs has been rising very rapidly since 1966—possibly three times as fast as urban employment in the formal sector.[6]

In so far as this high rate of growth of the urban poor was fuelled by migrants from the rural areas, it is itself a comment upon urban-rural income disparities. Yet the group of "prosperous farmers" (a concept that is admittedly extremely fragile when put to any analytical use) seems to have been growing steadily since independence. Estimates of their numbers vary hugely: from fewer than 1,000 when assessed according to the very rigorous definition of the agricultural Production Sample Survey, to between 15 and 20 per cent of rural households if one takes the likely number of rural households farming with the intention of selling sizeable proportions of their

Table 51. Economic groups and their incomes: a rough guide

Group	Number	Percentage of population	Average income per head (Kwacha)	Percentage of total income
Poor farmers and subsistence producers	850 000	70	< 40[1]	25
"Informal" employees and urban unemployed	300 000	25	< 120[2]	27
Prosperous farmers	20 000	1	> 424[3]	6
"Formal" employees	50 000	4	1 150[4]	42

[1] Calculated from table 3. [2] K600 is the median income of households in all low-cost housing areas: but it is not the case that all low-cost housing area residents are employed in the informal sector or are unemployed. See National Housing Authority: *Survey of site and service area (Lusaka, 1973).* [3] Assumes average sale of 400 bags of maize and autoconsumption of 350 bags per household of maize: average household size/sex/adults/equivalents. This is clearly an underestimate since it would imply that nearly 90 per cent of marketed maize originated in either the European or the state farm sector (i.e. highly capital intensive) or in the "poor farmer" sector. [4] *Monthly Digest of Statistics* (Nov. 1978). Note that this is a "genuine" per head figure. It overstates the disparity with the others, since in few urban households will all members of the household be earning. If we assume that the average urban household of formal employees has 1.25 employees, and that household size averages 5, a more strictly comparable figure of K287 is obtained.

total output. Whatever the definitional niceties, indirect evidence (sales of fertiliser, production of maize, reports of extension staff) all points towards the very rapid development of a still small group of prosperous small farmers concentrating on maize production, and able to increase real income per household member despite adverse rural/urban terms of trade[7] by both extending acreage and increasing yields.

The remaining group of this crude quadripartite division—the rural poor—constitutes the main focus of this study. We shall adduce evidence which suggests not only that this group has become relatively poorer, but also that at least significant sections of it have become so absolutely. The natural result has been a flight from the land in general, and from the agriculturally backward areas most of all. Though in some such areas special circumstances have been operating—such as, for example, the Lumpa risings in North-Western Province—it is clear from analysis of the most recent population census that in huge tracts of Zambia (some of it, admittedly, sparsely populated already) the rural population has been falling, despite a national population growth rate of 3 per cent per annum.

The problem of rural-urban "gaps" thus constitutes one pole of discussion in Zambian rural development. It is given proper urgency by a simple projection of table 50, for this suggests that by the end of the century the imbalances in the socio-economic structure of Zambia will have become uncontainable, at least within the context of "normal" processes of the kind of democracy Zambia has enjoyed since 1964. But the "gaps" are not the only pole of discussion. As will be seen in greater detail below, the failure of the sector to produce sufficient food for the urban population and, to a somewhat lesser degree, agricultural exports to reduce dependence on copper, has implied, on the one hand, heavy *macro-economic* costs for a country in which two key constraints on economic growth have been transport capacity (continuously) and foreign exchange (intermittently since 1969); and, on the other, *political* costs for a country whose leaders have set a high priority on reducing economic links with Zimbabwe (until Zimbabwe's own independence) and South Africa, the traditional suppliers of Zambia's food deficits. The implication of the foregoing is that the technical and political aspects of rural development in Zambia transcend a merely routine interest: it is widely appreciated that the economic and political future of Zambia will be determined in the rural areas.

Before these themes are further pursued, however, it is important to attempt to quantify the scale and nature of rural poverty in Zambia. Various efforts have been made to do so and these will be very briefly reviewed here. Baldwin's was the first attempt to derive *national* income distribution data;[8] deducing a Gini coefficient of 0.48, his was a trail-blazing attempt to show how impoverished the mass of rural Africans were. In statistical terms, however, his analysis was largely invalidated by the fact that he was obliged to rely for data concerning rural income distribution on the work of Morgan Rees and Howard describing the Sala of the Mumbwa District. Later work has shown that Mumbwa is a relatively high-income area and, although it is true that the Sala shared in this general prosperity to only a limited extent, it is clearly highly unsatisfactory to take a distribution from one limited and special case and apply it to a sector which includes the "improved" farmers of Southern Province as well as the *chitemene* cultivators of the north and the cattle-keeping Lozi in the west. No other attempt was made to produce national or rural income distribution figures until the mid-1970s, when further estimates became available, notably that of James Fry,[9] as well as others based on the Central Statistical Office's Pilot Household Budget Survey of 1972-73.[10] These estimates suffer from serious problems concerning the collection and assessment of the relevant data. If we eschew statistical exactness, however, there are a number of tentative conclusion that emerge from them. First, rural income is less unequally distributed than urban income: this follows from the small size of the "prosperous farmers" group in relation to subsistence or quasi-subsistence farmers. Second, it is likely that rural incomes have become more unequal in the course of time as a result of the emergence of the "prosperous farmer" group. If the Agricultural Census is right in maintaining that 54 per cent of rural households sell no produce, then even though the value (either farm-gate or local market) of their autoconsumption may have risen sharply in money terms (and possibly to a small degree in real terms), it is probable that the concentration of income will have risen. Third, it is clearly true (as in any society where subsistence income is significant) that rural net cash incomes are more unequally distributed than rural total incomes. It follows from the previous point that the disparity between these distributions has probably increased over time.

Turning to local, as opposed to national, studies, we encounter some problems of comparability. Different authors have used different concepts of income, conducted surveys in different years and, most important of all, have been working in wholly different areas of the country, where the natural resource base, the economic infrastructure and the agricultural systems in use can vary greatly. The limitations of data in table 52 should therefore need no further emphasis. The only general conclusion to emerge from them is that, based on the shares of total income accruing to the poorest 25 per cent of households, the more advanced areas of the country show greater inequality than the less advanced. That is *a fortiori* true in the case of cash incomes. To pursue the statistics further is unhelpful. Two basic facts are clear: there is a greater concentration of exceedingly poor people in the areas still largely untouched by "modern" agricultural methods and systems than in those areas where the new technologies are being applied; and, in Zambia as a whole, between one-half and two-thirds of rural households have incomes that are so far below official minimum wages that malnutrition and seasonal famine are a constant fear and a too frequent reality. Thus Latham's work on the 1969 Nutritional Status Survey, based on 1,550 children, showed 24 per cent with mild Protein/Calorie Malnutrition (PCM). Bliss, working in Luapula Province (where the alleged high consumption of fish, actually disproved by Bliss, might suggest above average protein in the diet) found that 24.5 per cent of children were below 80 per cent of Nigerian anthropometric standards, indicating, in her view, moderate to severe PCM. If those showing indications of mild PCM are included, well over half of all children (55.1 per cent) are affected.[11]

EQUITY AND GROWTH

In ecological terms Zambia can be roughly divided between the Central, Southern and Eastern Provinces, together with some of the Copperbelt and North-Western Provinces which are mainly maize-growing areas. Northern and Western Provinces are primarily millet-producing areas, while cassava production is common in North-Western and Luapula Provinces. Cattle-keeping is confined to the tsetse-free areas and is particularly important in the Zambesi flood plain of Western Province. In terms of land tenure, the few—between 400 and 500 but, in production terms, very important—European producers have a modified form of freehold. The other large-scale farms are either state production units or subsidiaries of the mining companies which operate on state land. For the rest, the vast majority of land is held in various forms of communal tenure, though there is some evidence of improvements in communally held land being saleable as one operator succeeds another. While absentee landownership has become increasingly common over the past 15 years as urban élites invest in land along the railway (usually buying farms formerly owned by Europeans), the land is seldom operated by tenants in the normal sense but more frequently, by managers. These managers are usually kinsmen of the owners in question, and the level

Table 52. Percentage shares of household income accruing to poorest 25 and 50 per cent of the population

Province	Concept	Date	Average K	Share of 25%	Share of 50%
Eastern[1]	Total net farm income	1973	132	14.9	35.3
Central[2]	Gross revenue from crop sales	1977	803	2.2	11.1
Central[3]	Net farm cash income	1977-78	273	6.6	23.9
Central, Southern,[4] Eastern	"Crop and livestock income" (net cash?)	1975-76	219	neg.	5.0
Northern[2]	Gross revenue from crop sales	1977	97	(11.5)	(29.8)
Luapula, Northern,[4] North-Western	"Crop and livestock income" (net cash?)	1975-76	28	neg.	neg.

Sources. [1] R. H. Harvey: *Some determinants of the agricultural productivity of rural households: Report of a survey in Kalichero District, Eastern Province* (Chipata, MRD, 1973). [2] C. R. Joseph: *Report on a random pilot survey of traditional farmers in the Central and Northern Provinces of Zambia* (Lusaka, MIA, Sep. 1977). [3] CIMMYT: *Demonstrations of an interdisciplinary approach to planning adaptive agricultural research programmes*, Report III (Nairobi, 1978). [4] A. Marter and D. Honeybone: *The economic resources of rural households and the distribution of agricultural development* (UNZA/RDSB, 1976; mimeographed).

of management expertise is not normally high. To all intents and purposes, therefore, there is not a "land tenure problem" in Zambia and, since land is in relatively abundant supply, the size of holdings tends to be determined more by the size of the labour unit than by availability of capital for land purchase.

In many respects, however, the problems of Zambia's agricultural sector are not unusual. Being a relatively highly urbanised and relatively wealthy country, the demand for food (and particularly higher-quality foods) has outstripped her ability to produce them, with the result that 40 per cent of all foodstuffs is now imported—despite untapped agricultural resources which include plentiful, if somewhat infertile, land,[12] major unexploited water resources, a relatively benign and predictable climate and rapidly improving rural communications. Indeed, by comparison with some countries facing the same basic structural transformation (rapid urbanisation and rising urban incomes) the increase in output of Zambian agriculture is impressive. To triple maize output, double production of sugar cane and groundnuts, and establish sunflower as a major commercial crop in six years (see table 53) is no mean achievement, even when seen in the rather dimmer light of stagnant tobacco production, falling milk, beef and pig sales and erratic supplies of horticultural products.

Nor is there anything unusual about the fact that, rather than becoming easier to solve, the structural problems of the agricultural sector in general and of the rural poor in particular have become more difficult. The myths of the 1960s—that rural development follows some linear path of progress to a

Table 53. Marketed production of some selected crops, 1971-78

Year ending 30 April	Maize ('000 bags)	Tobacco		Sugar cane ('000 tonnes)	Seed cotton ('000 kg)	Sunflower ('000 bags)	Shelled groundnut ('000 bags)
		Virginia flue-cured ('000 kg)	Burley ('000 kg)				
1971	2 791	4 805	255	322	5 446	–	45.0
1972	1 388	6 248	389	331	12 675	–	84.7
1973	4 137	5 544	385	397	8 349	–	81.4
1974	6 367	6 222	471	488	5 225	20	40.2
1975	4 290	6 201	430	570	2 173	70	45.3
1976	6 491	6 466	502	768	2 602	129	81.2
1977	8 334	6 115	212	780	3 885	322	118.3
1978	7 734	5 588	312	691	8 928	266	93.3

– = magnitude nil.
Source. *Monthly Digest of Statistics*, Nov. 1978.

high-wage, high-technology modern agriculture—were never very convincing, and are now wholly discredited. It is a more common experience, and one which Zambia certainly shares, that those structural problems become more deeply embedded, more resistant to manipulation and more pernicious in their effects as "development" (or at least some interpretations of it) proceeds. Differentially increasing output tends to be accompanied by worsening income distribution, increased social and economic stratification, the generation of political pressures that reflect the interest groups of the more vocal and more powerful, at the same time as urban/rural relations tend to introduce new forms of "urban bias" to the detriment of the rural areas in general, and the poorest in those areas in particular. To that extent, the scale and severity of rural poverty and inequality in Zambia are not unusual. Although these themes are played with local variations throughout Africa, they are the cacophonous accompaniment of rural life throughout the continent.

The unusual feature of Zambia (indeed, the feature which makes it a mandatory area of study) is the unresolved tension at the heart of official decision-making between two normative approaches to rural development strategy. It is important to sketch these at the outset because no adequate appreciation of Zambian rural development policy and its effects on the lives of the rural poor is possible without a proper sensitivity to this tension. To represent it as a tension between two clearly defined and mutually exclusive polar opposites is, undoubtedly, to simplify: if the reality were as stark as that, the tension might possibly have been resolved in the 15 years since independence. In fact, in the minds of Zambian decision-makers, ideas that may be inconsistent on academic grounds can coexist either because both are needed to satisfy different groups, or simply because by the nature of decision-making under pressure they are never presented as coherent, though mutually exclusive, bodies of thought.

Aware, therefore, that in doing so we do violence to the perceived reality of the policy-makers themselves, we shall state the two views as polar opposites. The first of these, which is also the easier to identify and describe, is a technocratic view of rural development that is concerned with the need to maximise agricultural output and to ensure the efficient use of scarce resources, more particularly public resources. Such a view puts rural development in a macro-economic context and is therefore concerned to ensure that agriculture fulfils given structural roles within an over-all strategy of economic development. Dominated by variants of neo-classical economics, this approach is supported not only by aid donors and multilateral agencies but also, though sometimes ambivalently, by a number of senior Zambian politicians and civil servants.

The second approach is much more concerned with the ideological issues that surround the political economy of agriculture. This is not to imply that those who adopt this approach have a settled solution to those issues. This they certainly do not have; indeed, rather than conflict between the ideologists and the technocrats, one of the fundamental difficulties of Zambian rural development has been the confusion within the ideological group and therefore its inability to reach any consistent and enduring accommodation with the technocratic group. The ideological group is centred on the idea of Zambian humanism, which has been more successful in raising issues than in solving them. Thus, in the name of humanism, various forms of co-operative have been favoured, allegedly on the grounds that such a system recreates a traditional form of Zambian village culture: yet both the ambiguities of different forms of co-operation and their relationship with traditional village forms of co-operation have never been adequately explored and expounded.[13] Almost simultaneously, the ideal of the individual peasant family farm has been reasserted with no apparent awareness of the potential conflict between co-operative forms of organisation and the ideals of individual enterprise.[14] Although state farms and state production units have received relatively less public support from the ideological group, they have not been actively opposed, with the result that three quite different forms of agrarian organisation have been in both the public and the bureaucratic consciousness simultaneously. Given the constraints on the capacity of the bureaucracy (and supremely in the area of rural development), it is no wonder that resources have been switched from one to the other in a confusing series of policy shifts, which have left not only officials at provincial and district level lost and demoralised, but also farmers themselves increasingly sceptical regarding the utility of such policies.

The difficulties, however, go deeper still, for the ideological group has never taken adequate account of the realities of local political processes, with the result that *ex cathedra* judgements about desirable forms of agricultural organisation have been turned into forms of political patronage in the hands of local politicians. This is a theme to which reference will be made again with respect to credit, but it applies to most resource transfers from the centre to village and district level. It achieved perhaps its quintessential expression in the land-clearing bonus which, although designed to stimulate opening of new land for co-operative farming, in the long term benefited only the local officials of the United National Independence Party (UNIP) and their clients.[15] This episode illustrates precisely a further dimension of the dilemma in which the ideological group finds itself. Fired by a (probably romantic) view of village life in the past and its potential for the future, this group is uneasy with the way in which a small number of emergent and small-scale farmers, principally along the railway, can quickly increase their income levels and come to dominate agricultural institutions designed to serve the rural community as a whole.[16] At the same time, however, the ideological group is obliged to recognise that more co-operative, equity-oriented approaches to rural development have not only proved exceedingly difficult to implement but have also too often failed to increase production. The group thus sees at the heart of its thinking a fundamental conflict between the technical requirements of a rapid growth in output and its ideologico-political requirements that such output must avoid polarising rural society into rich successful farmers on the hand and excluded subsistence producers on the other. It is not the case that the ideological group ignores or denies many of the concerns of the efficiency group but rather that it has so far found it impossible to integrate those concerns with its own in a way which is both technically and politically possible to implement.

The central concern of this chapter is less with the playing-out of the conflict which exists within the ideological group and between that group and the technocrats, than with the effect that the confusion at the heart of rural development policy has had on the rural poor: for one of the themes of this chapter is that the way in which the various agencies of rural development have coped with this confusion has made at least some elements of the rural poor more, rather than less, vulnerable to the fundamental pressures to which they are subject.

RURAL POVERTY

Although the roots of rural poverty in Zambia are to be found far back in history, three factors of more recent origin have put at least some poor rural households under intense pressure. In this section, these sources of pressure will be reviewed and it will be argued that poor rural households can be differentiated both in terms of the degree to which they are subject to these pressures and in their capacity to respond to them. The factors at work are ecological, economic and social. They operate both separately and in combination reinforcing each other, so that in the extreme case a rural household may be subject to the operation of all three in a multiplicative rather than an aggregative sense.

The mechanisms of the ecological factors are well known, namely the effect of population pressure on a shifting cultivation system of agriculture.[17]

Although this is by no means a new phenomenon in Zambia (it is well documented, for instance, in the Eastern and Southern Provinces in the early 1950s),[18] the manner in which the population pressure is generated is relatively new. As is clear from table 54, rural areas as a whole are not subject to high rates of growth of population, with the result that the classic analysis of falling land/labour ratios is hardly applicable. More significant is the increasing nodality of the rural population noticed by Jackman in her analysis of the 1969 Population Census.[19] This nodality, arising from the perceived advantages of operating a shifting cultivation system within walking distance of a range of social and economic services, clusters the population, so that although there may be no evidence of declining land/labour ratios at provincial level, in the operation of the agricultural system itself there is clear evidence of declining fertility. This may take the form of extended periods of cultivation; of returning to land that has not fully been regenerated; or even of maintaining two gardens, one within easy reach of the node (subject to declining fertility) and the other further away where yields may be higher. Joseph reports evidence of this in Northern Province.[20]

Faced with a situation of increased competition for *local* land (despite abundant land at some distance from the node), the household is likely to experience two forms of deprivation: declining yield on cultivated land; and increased competition for, and therefore increased scarcity of, gathered products. Both of these imply declining labour productivity and increased exposure to seasonal scarcity.[21]

The second source of pressure on poor rural households is economic, arising from increased household demand for cash, partly as a result of the deep penetration of the cash economy in general—to which reference is made below—but particularly on account of the high rate of urban inflation and the associated fall in the real income, however tiny, of the household.[22] There are very few households that have absolutely no cash income.[23] Even the village beer economy, which plays such a central role in the social functioning of the village and the economic survival of women, has become partially commercialised, a trend that is much accelerated by the replacement of village brewed beer by bottled beer imported from the urban breweries. Although primary schools are non-fee-paying, uniforms have to be bought and most schools have some informal mechanism for raising cash for additional facilities. In even the poorest household, therefore, some access to cash is required: that cash requirement is increased in money terms as prices rise. Since clothing prices haven risen by a factor of three and food prices by a factor of two-and-a-half since 1969, economic pressures on poor households are clear.[24]

The third source of pressure on poor households stems from a wide range of social penetrations into traditional village life. This is not a new phenomenon: there is a danger of too easy an antithesis between a romanticised view of stable village culture of a pre-European era and a broken and disintegrating pseudo-culture of today. Village culture has always been in the process of response to new challenges and stimuli, and in some respects has shown

Table 54. Annual average population growth rate, 1963-74

Province	Average annual growth rate (%)		
	1963-69	1969-74	1963-74
Central	5.9	5.2	5.6
Copperbelt	7.0	5.1	6.1
Eastern	1.0	2.3	1.6
Luapula	−1.0	−1.0	−1.0
Northern	−0.6	1.4	0.4
North-Western	1.6	0.8	1.2
Southern	1.0	1.5	1.2
Western	2.1	2.3	2.2
Total rural	0.5	1.0	0.73
Total Zambia	2.5	2.9	2.7

Source. M. Jackman: *Recent population movements in Zambia: Some aspects of the 1969 census*, Zambian Papers, No. 8 (University of Zambia, Institute for African Studies, 1973), p. 15.

remarkable adaptability and endurance. None the less, it is true that the pressures on both the village as a unit and the household as a component of that unit are perhaps stronger now than they have ever been, at least with respect to the economic and agronomic demands those pressures make upon the household. While the strength of these pressures obviously varies from region to region, they tend to operate both to increase the economic *demands* of the household (for example, through rising aspirations and/or expectations, and through increasing inequalities in income and expenditure patterns in the village and the district) at the same time as they reduce the household's *capacity* to respond to those demands through weakening authority patterns within the household, greater instability in family structure (as a result, for instance, of migration) and a decreasing acceptance of a common responsibility for the economic welfare of the village as a whole, and of its more vulnerable inhabitants in particular.

Under certain circumstances, this asymmetry between a changing pattern of demand on the household and the capacity of the household to meet such demand within traditional patterns of production and exchange may be thought to be reconcilable through the adoption of new technologies. For example, the hiring of a tractor to do the heavy ploughing after the harvest or at the onset of the new rains may simultaneously compensate the household for the migration of some of its workforce, who are most likely to be young males, and promise—at least seemingly—a higher net income through the extension of acreage which will enable it to satisfy some of the aspirations it has acquired on the example of small-scale commercial farmers in the neighbourhood.

Accordingly, when these three sources of pressure—the ecological, the economic and the social—act in a way that is mutually reinforcing, the household is likely to be projected out of its traditional agricultural system towards the adoption of new agricultural systems, either in whole or in part. It is at this point that the household becomes especially vulnerable, for it surrenders a degree of self-reliance and independence and becomes dependent upon centrally provided inputs and services. This vulnerability is heightened by the fact that the technology it is seeking to adopt is to a degree indivisible. Table 55 shows the increase in yields associated with different components of maize-growing technology and underscores the fact that the marginal returns to any individual component of that technology are fairly low, while the marginal returns on the whole package are high. From the point of view of the individual household, the implication is that once it is on the threshold of growing maize commercially, particularly in a situation of declining fertility or scarce labour, there is a very strong premium on adopting the whole package of the technology; in other words, that there is a high penalty for failure to incorporate any single component. Three elements are therefore critical: the farmer's understanding of the technology; his ability to implement it; and the availability of inputs implied in that implementation.

These are questions that are taken up later in this chapter: in concluding this section it is important to emphasise that not all households are under such pressures that they feel it necessary to respond either by increasing their sales of maize or by changing their productive technology. The most universal form of pressure is undoubtedly that of the rising price of bought necessities; the rest are far from universal in their incidence or undifferentiated in their vigour.[25] Moreover, one must avoid a mechanistic account of these forces: there is a degree of choice in the way and in the extent to which households respond to perceived changes in their circumstances. While a proper regard for the resource base of the household (for example, its ability to respond to these pressures) is essential, it should not be allowed to occlude a realisation that households are differentiated with respect to their desire or motivation to respond. If the head of the household is in semi-retirement;[26] if the household is closely integrated into the affairs of the village, so that its principal occupation is with village society and politics rather than with increasing its economic surplus; more particularly if the household has, *or expects to have,* alternative sources of income, it might either ignore or resist these pressures and continue to organise its farming exclusively round its own subsistence.[27] (This does not necessarily imply, of course, that it sells no surplus: it does, however, imply that it does not plan to produce a surplus.) In the following section we shall endeavour to differentiate between, on the one hand, poor households who are attempting to change their productive technology in response to the pressures outlined above and, on the other, those farmers that either cannot or do not wish to do so.

POVERTY AND POLICY

The rural poor comprise several categories which overlap in such a way as to make sharp distinctions a distortion of reality. At one end of the spectrum

Table 55. Some technical systems for maize production and their associated productivity

	Villagers [1]		Farmers [1]	
	V_1	V_2	F_1	F_2
Attributes				
Planting method	By hand	By hand	Behind plough	Planter
Plant spacing	Broadcast	Rows without spacing	Rows without spacing	Rows with spacing
Type of seed	Local	Local	Improved open-pollinated or hybrid	Improved open-pollinated or hybrid
Fertiliser used [2]	No (100)	No (97)	Sometimes (16)	Mostly (60)
Number of weedings [2]	1-2 (83)	1-2 (100)	1-3 (83)	2-3 (80)
Average performances				
Acres per cultivator	2.05	3.24	19.98	31.84
Labour hours per acre grown	350	396	116	85
Yield per acre grown (lb.)	520	883	935	911
Yield per labour hour (lb.)	1.49	2.23	8.08	10.77
Production per cultivator (lb.)	1 066	2 860	18 680	29 000

[1] The villager/farmer distribution is primarily one of intention: farmers intend to produce sufficient maize to enable them to sell more than half of total output. [2] Guides to fertiliser use and number of weedings are only indicators of practice. The figures in parentheses show the percentage of cultivators actually covered by the general guides given.
Source. Universities of Nottingham and Zambia Agricultural Labour Productivity Investigation: *Report No. 3* (Lusaka and Sutton Bonington, 1970; mimeographed).

is the farmer who has the resources, motivation and, to a degree at least, managerial skill to grow a range of "modern" crops but who finds that his attempts to do so are continuously frustrated by poor prices, inadequate delivery systems, corrupt distribution of credit, ineffective marketing, erroneous extension advice and distortions in the rural labour market. Clearly, there is a sense in which his poverty is directly attributable to mistaken policies and ineffective administration.

At the other end of the spectrum, however, is the remote subsistence farmer who has neither the resources, motivation nor management expertise to produce any surplus in a planned and systematic way and who is therefore impervious to pricing policy, market or delivery systems, the vagaries of credit and even, since he neither hires labour nor offers himself for hire, the rural labour market. Since his overriding concern is to produce food for his family's needs, the most significant determinant of variation in his ability to do so is more likely to be the weather, the physical fitness (and, of course, age) of his wives and children, and the availability of new fully regenerated land on which he can practise his *chitemene* system of farming. While there may well be senses in which his environment is indirectly affected by the development (albeit of a long-term nature) of government policy, the very fact of his physical inaccessibility (which may well be a symbol of a political and even epistemological inaccessibility) ensures that for him marginal adjustments in rural development policy as currently conceived in Zambia are wholly irrelevant.

These two descriptions are, of course hypothetical types: they mark each end of the spectrum, and the bulk of the rural poor fall somewhere along that spectrum. It is, however, part of the argument of this chapter that the most acute forms of rural poverty, and the largest number of the rural poor, tend towards the latter end of that spectrum. From this it follows that current debates concerning the *form* of rural development policy—for example, intensive development zones, or rural reconstruction centres, or multipurpose co-operatives—may be relevant on one plane of political discussion but have *absolutely no relevance whatsoever* as constructive vehicles for the elimination of rural poverty in Zambia. In the same way, the stock remedies of donor missions (border prices, fertiliser subsidies, improved product stores) may have meaning for one type of rural poverty (and incidentally one type of urban poverty) but are almost totally irrelevant in the case of the second type.

For ease of discussion we shall call the first type of rural poverty identified above "stratum 1 poverty", and the poverty of isolation of the subsistence producer "stratum 2 poverty". The present discussion will be confined to these two strata, but it must be emphasised that they are no more than hypothetical types.

Stratum 1 households are those which are willing and able to produce a marketable surplus, since they have the motivation, the resources and access to supplies of imputs and markets. Yet they are simultaneously on the fringes of the commercial sector, operating on a small scale a technology with which they are not wholly familiar and from a resource base that is sufficiently small to leave them extremely vulnerable to changes in relative prices, crop failure, or failures of the input delivery system. However, because they have the potential for rapid increases in output, they are one of the most important target groups of agricultural policy. In this section, therefore, we shall describe four areas of government policy that are designed to assist this group of farmers to progress smoothly along the spectrum from quasi-subsistence to fully commercialised production.[28] It will be a fundamental theme of our treatment that as a result of confusion in policy-making, combined with defective implementation, the process of transition has been made more difficult for the majority of such farmers, with the result that many now find themselves caught in a kind of agricultural poverty trap from which it is going to be exceedingly difficult to rescue them.

To place what follows into a national context, table 56 shows the proportion of capital and current expenditure that has been devoted to agriculture in the period between 1965 and 1977. These data need to be read against those of table 51, which showed that 75 per cent of the population, receiving only 25 per cent of total personal incomes, are in the "poor and subsistence" farmers category. Although a high proportion of health and education expenditures were made in the rural areas, it is not the case that the Government has typically devoted substantial resources to increasing the productive potential of the rural sector. Comparisons with the "transport, communications and power" sector in the capital table, and with both that sector and "general services" (army, police, and above all, government administration) in the recurrent table, are particularly revealing. Apart from these global allocations, however, we need to assess the effect of the various programmes thus financed.

Let us begin with credit. The inglorious history of the Credit Organisation of Zambia (COZ) has already been written.[29] In 1970 its functions passed to the Agricultural Finance Company (AFC), established as a statutory credit agency and a subsidiary of the Rural Development Corporation. As such, it has the obligation of corporate profitability and is therefore extremely conservative in its lending to stratum 1 households. In 1974-75, 96 per cent of all agricultural loans (from the AFC and the commercial banks) went to the three agriculturally developed provinces, and there is little doubt that even in those provinces the vast majority of credit went to small-scale commercial producers rather than to stratum 1 households.[30] Whatever the commercial pressure on the AFC and the commercial banks, however, it is the brokerage relationships in the village that are perceived by the households concerned as the major problem. In other words, even if current policy were not slanted so heavily towards the small-scale (and large-scale) commercial producer, the structure of political control at the village level would tend to discriminate against stratum 1 (and, *a fortiori,* stratum 2) households for, in general, it is not these households who are effective manipulators of political power at the village level: they therefore have little of political substance to offer and can thus expect no rewards from their political patrons at district or provincial level.[31]

One small example will serve to demonstrate the point. In 1974 Ward Development Committees (WDC) were promised powers to determine distribution of AFC credit to farmers within the ward. To the efficiency school of thought, this arrangement, itself an unhappy compromise between the two world views analysed above, promised to dissolve quickly into a repetition of the COZ disaster. Pressure was therefore brought to bear on the Cabinet and the decision was reversed. This provoked an angry response from many Ward Development Committees throughout the country, and a few were allowed to exercise the powers in question. They were, however, instructed by the AFC to favour farmers with all the characteristics of emergent or small-scale commercial farmers, with particular emphasis on a willingness to hire labour.

Zambia

Table 56. Functional classification of central government current and capital expenditure

Sector	1965-66	1966-67	1968	1969	1970	1971	1972	1973	1974	1975	1976	1977
Capital												
Agriculture and rural development	9.6	9.1	5.2	13.5	19.9	20.4	26.7	17.0	10.6	10.9	10.4	16.6
Land, mining and natural resources	3.5	6.3	6.5	7.2	7.8	4.8	2.5	2.4	2.7	3.3	3.6	4.6
Transport, communication and power	24.0	37.0	50.8	42.7	36.4	42.7	43.5	47.3	47.2	52.2	53.5	44.6
Trade and industry	0.8	7.1	6.4	8.0	6.8	1.2	1.0	1.7	9.8	5.3	0.5	5.0
Education	25.0	12.2	7.0	6.2	7.3	11.0	11.9	14.4	12.7	9.2	12.7	11.8
Health	1.6	2.6	2.5	4.2	4.8	5.6	4.2	2.5	2.4	2.3	4.8	3.8
Other social services	–	2.0	1.6	0.1	0.1	0.1	0.1	0.1	0.1	0.3	0.3	0.4
General services	35.3	23.6	19.9	18.1	16.9	14.2	10.1	14.7	14.5	16.5	14.2	13.2
Recurrent												
Agriculture and rural development	4.5	5.7	5.9	13.0	10.3	14.7	11.5	12.7	11.2	14.3	13.4	13.3
Land, mining and natural resources	5.1	4.0	3.2	1.9	1.7	1.9	1.8	1.6	2.5	1.5	2.0	1.7
Transport, communication and power	9.7	7.3	12.6	13.3	16.0	13.4	10.9	9.0	9.2	7.9	7.6	7.5
Trade and industry	0.4	0.3	0.2	1.7	1.8	0.3	0.7	0.2	0.2	2.7	0.3	0.3
Education	16.0	15.7	15.6	17.7	17.1	16.4	18.5	17.7	17.1	13.8	16.5	16.4
Health	7.3	6.2	7.1	7.7	7.9	7.4	7.8	7.6	7.8	6.4	7.7	8.4
Other social services	1.2	1.9	2.1	0.7	0.7	0.4	0.5	0.4	0.8	0.4	0.7	0.6
General services	41.1	35.8	41.5	36.5	37.3	39.2	41.9	36.5	40.4	43.6	41.9	37.3

– = not available.
Source. Government of Zambia, Financial reports and estimates.

In other words, the Ward Development Committees were put under pressure to follow an efficiency algorithm in the distribution of credit which was inevitably to discriminate against stratum 1 and 2 households. Ward councillors unanimously agreed that only the "best" farmers should get loans, and a study of the Kasama WDC revealed that that committee has consistently followed the AFC's guidelines, discriminating against stratum 1 households. Significantly, all the WDC councillors were at least emergent farmers; one revealed precisely the characteristic mixture of self-interest, concern for efficiency and disregard of the long-term implications when he justified the committee's decisions with the view that "people like us know the work".[32]

In this way the efficiency algorithm of the centre is reinforced by the mixture of motives at the local level. The inevitable result is that the stratum 1 farmer is excluded from the possible benefits, unless he can become a local power-broker and use political leverage to break through the built-in biases. Clearly, only a small number are able to do this. It is thus not surprising that an agricultural assistant in a predominantly subsistence area near Kabwe wrote lugubriously to the District Agricultural Officer in these terms: "So many forms have been sent but none was approved in Chitanda's area. This carries no weight with the agricultural extension staff in the area. Farmers have the suspicion of not sending their forms to Kabwe by the agricultural staff."[33]

Let us now turn to extension. How far is it the case that the extension effort is biased against stratum 1 households at the same time as they are made more dependent upon it? That dependence is worth emphasis since it might be assumed that, given that the cash crops most likely to be produced by stratum 1 households are maize and, much less frequently, cotton, the role of the extension service might be minor. But as has already been stressed, stratum 1 households are likely to be either on the fringes of the transition from shifting to settled cultivation, or relatively newly settled. In these circumstances, the maintenance of soil fertility soon becomes a major problem, except in the very exceptional cases where stratum 1 households have settled on first-class land. This implies an increasing technical dependence upon fertiliser, stimulated by substantial fertiliser subsidies in the period between 1971 and 1978. Even in a relatively simple regime like maize, however, returns to fertiliser depend critically upon the application of the right fertiliser, at the right time, in the right quantity. If a farmer does not appreciate the difference between basal fertiliser and top dressing, his expenditure on fertiliser—which in the case of stratum 1 households may constitute as much as 25 per cent of his total annual income—is totally wasted. It is here that the extension service has a major role to play in familiarising stratum 1 farmers with the relatively simple techniques involved. By contrast, all the research that we have been able to review suggests that such assistance is very unlikely to be forthcoming.

For this there are three major reasons. The extension service has traditionally operated with an efficiency algorithm, reflected in instructions to extension officers to identify the most promising farmers and concentrate wholly upon them. Now, it is perfectly true that the extension service does not always follow this instruction, since field officers evidently find it difficult to identify the most progressive farmers in their districts. Even in selecting farmers to attend farmer training courses, the extension services exclude stratum 2 households fairly rigorously, but do not evidently discriminate between stratum 1 farmers and genuinely emergent or small-scale commercial producers.[34] We thus have the paradoxical situation of an extension service that is directed to discriminate against stratum 1 households but in fact is so disorganised[35] that the impact of its operations is largely random, except for the exclusion of the most obviously poverty-stricken rural families, the stratum 2 households.

To this extent a crude efficiency-bias theory cannot be applied: rather, two more fundamental objections have to be made. The first is that the resources put into the extension service by the central government authorities are wholly inadequate to the task that the extension service is set, irrespective, of whether that task is couched purely in terms of output maximisation or whether it includes also a concern for help to the neediest. Although the over-all ratio of technical and professional staff in the Department of Agriculture to farmers is one of the highest in Africa at 1:350, this figure is seriously misleading since the great majority of field staff are poorly trained demonstrators, the supervision of whose work is rendered even less effective by the chronic failure to fill vacancies in the rest of the establishment. Typically, the most experienced and able technical and professional staff are found in Lusaka:[36] a posting off the railway is regarded as a public censure and a private disaster. Under these circumstances the *quality* of the extension service is quite disporportionate to its quantity. One simple example can illustrate the point. In a survey in the Kalichero District of Eastern Province, Harvey found that cotton growers were very dependent on the extension service for advice on spraying, an operation that crucially affected yields. Not only were half the growers not visited at all, but, more revealingly, such confusion surrounded the distribution of, and instructions on, spraying equipment that those who were visited were in fact no better off than those who were not. The sprays were packed in quantities applicable to one-fifth of a hectare of cotton, but each grower, irrespective of his acreage, was given one pack. Neither the demonstrators nor the staff of Eastern Province's Co-operative Marketing Association (EPCMA) were aware of this confusion. As a result labour returns per hour were less than 10 ngwee; 10 per cent of Harvey's sample actually made a cash loss and the number of cotton growers was falling.[37]

The third major objection is that, given the concentration of skills in the centre, the low quality and relative immobility of the field staff and the increasing dependence of poor farmers upon the extension service, the resultant difficulties have been compounded not only by a regressive spatial distribution of the extension staff (following directly from the efficiency algorithm with which agricultural planning staff at the centre operate), but also by the tendency to concentrate extension services in *project* areas, whether those be settlement schemes, Intensive Development Zones or Rural Reconstruction Centres. By definition, these project areas exclude virtually all stratum 2 and the great majority of stratum 1 households. The criteria for selection includes adequate supplies of family labour; existing adoption of improved practices; experience of a modern crop; and even, in extreme cases, some cash savings. Thus, an internal evaluation of the IDZ programme, completed in May 1978, concluded that "the predominant group among the primary beneficiaries of this programme consist of small-scale farmers, who have just emerged from the subsistence farming level. The farmers at lower levels have benefited from the IDZ-Agricultural Production Services only to a smaller extent."[38]

Let us now turn to marketing. Is there evidence that stratum 1 households, by definition becoming more dependent upon marketing structures for both inputs and produce, are discriminated against in terms of the structure of marketing? This will be considered first in relation only to the structure of marketing: a discussion of pricing policy will be held over to a later section. To stratum 1 households three transactions are crucial: purchases of hybrid seed and fertiliser for the maize crop; sales of hybrid maize; and purchases of clothes and very small amounts of manufactured foods from local stores.[39] These will be examined in turn.

In terms of purchases of inputs, there is little evidence of discrimination against stratum 1 households: indeed, the rapid development, since 1969, of the National Agricultural Marketing (NAM) Board and co-operative marketing union depots, has been impressive. Although the sparsely populated and relatively less developed areas are inevitably less well served than the more prosperous areas, there is little doubt that in the *development of the marketing structure* the equity school of thought has triumphed.

It should not be assumed that the relatively equitable distribution of marketing depots has gone unchallenged or, indeed, that it will remain unchallenged. There are those, including the World Bank, who regard the undoubted difficulties of the National Agricultural Marketing Board and the Regional Co-operative Marketing Unions as either fundamentally insoluble (because the real value of the crop will not support so lavish a marketing infrastructure) or as the consequence of a misallocation of scarce public resources and executive manpower. The view is therefore put forward that the NAM Board depots and buying stations should be closed down and the whole of the primary end of the marketing chain should be passed over to private entrepreneurs. In this way, spatial comparative advantage would quickly be reflected in prices received by farmers, and inefficiencies in the marketing structure would be competed out of existence. In other words, there is, at the centre, a continuing struggle between efficiency and equity criteria which has been resolved in favour of the latter, perhaps only temporarily and certainly in the face of bitter hostility from influential donors.

When we turn from intention and strategy to effect and implementation, the result is rather different. Just as the rising dependence of stratum 1 households on the extension service is misplaced because it cannot support the weight of that dependence, so we find the same with respect to marketing, for although close investigation does not support the many rumours and complaints about the evacuation of crops,[40] the same cannot be said concerning supplies of inputs. There are too many well attested cases of inputs arriving at the local distribution depot either late or not at all, so that the farmer is either deprived of them entirely or likely to get a greatly reduced benefit from their application.[41] More importantly, however, the uncertainty and waste of scarce labour time involved in frequent visits to the depot are particularly stressful for the stratum 1 households, given the extremely high opportunity cost of labour time in the period immediately before and after the onset of the rains.

Let us now summarise the argument with respect to services: namely, that while pressures analysed in the preceding section have made stratum 1 farmers *more* dependent upon these services, the services for their part have responded to that dependence in different ways. The credit service has become almost exclusively concerned with the efficiency algorithm; the extension service has become dominated by that algorithm but is actually incapable of implementing it; and the marketing service has been deeply affected by the equity algorithm.[42] With respect to the extension service and the marketing service, the discrimination against stratum 1 families stems not from over-all policy, or from a crude kind of strategy aimed at the reinforcement of success, but rather from poor implementation. This is partially—but only partially—explicable in terms of shortage of resources made available to these services; it is much more explicable in terms of generally low levels of administrative capacity, lack of political and administrative discipline, and a confusion at the centre between form and substance.

With this summary in view, we turn now to the no less vexed but more public discussion on pricing policy in Zambia. This discussion has traditionally revolved around two poles. The first is the differential productivity between advanced (basically European) farmers on the one hand, and emergent and small-scale producers on the other. Given this differential productivity, the effect of raising prices is to increase the surplus of the high-productivity farmers. Particularly in the post-independence period, this differential effect posed an acute political problem. Second, and more obviously, producers' prices are consumers' costs. In Zambia this familiar conflict is given a particular twist by the fact that for a number of reasons, some of which are purely geological in origin, production costs of the Zambian copper industry have been increasing rapidly, with the result that by 1972 it had some of the highest-cost mines in the world. This meant that, in a period of falling copper prices, mines had to be subsidised from other (extremely narrowly based) sources of revenue, or taken out of production altogether.[43] If this is the logic behind a marked and persistent reluctance to raise consumer prices (and thereby make a further rise in production costs almost inevitable), it has traditionally been reinforced by strong political pressure in the urban areas and on the Copperbelt in particular. Just how strong those political pressures are, the United National Independence Party (UNIP) learnt through a very sharp lesson in 1966 and, in a rather different form, again in 1971 and 1974.

Two illustrations must serve to indicate the vigour of these pressures and therefore to explain the history of rural/urban terms of trade since 1964. Let us take, first of all, the subsidisation of imported wheat flour, a commodity consumed only by upper-income groups in preference to maize meal. When the Cabinet took a decision to reduce the subsidy in November 1974, there was such an outcry from the Copperbelt in general, and the Zambian mineworkers in particular, that within weeks the Government was forced to reverse its decision and continue to shoulder the subsidy despite the fact that the level of subsidies to agriculture in general was already running at 25 per cent over the generous budget. The second example concerns sugar, a commodity more widely consumed than wheat flour, but still more heavily in demand along the railway than in the rural areas. In 1979 the producers were demanding a 12 ngwee increase in price to meet projected losses of K12 million. The respective application went to the main Board of Indeco politicians representing urban constituencies, who resisted the rise, while the technocrats insisted that the company could not continue to finance losses on such a scale. In the end a revealing compromise was struck: despite the fact that central government revenues were so low that the Ministry of Finance was delaying release of funds to the NAM Board to pay farmers for their maize, the Cabinet was persuaded to rescind the excise tax on sugar, itself conveniently valued at 12 ngwee per pound. In this way the company was given an increase in its effective price without the urban consumer having to pay. The real costs, needless to say, were paid by those deprived of the services and/or jobs thus forfeited by the Ministry of Finance.

In the face of pressures of this intensity, it is no surprise to find that the evidence of declining rural/urban terms of trade in general, and in particular for stratum 1 households producing limited quantities of maize, is overwhelming.[44] Table 57 shows the relative prices of maize and the clothing component of the low-income cost-of-living index. For a number of reasons, the movement of the terms of trade against stratum 1 households is almost certainly understated. The table is based on urban prices where price control was much more effective than in the rural areas, and it ignores the high and rising premium of rural prices over urban prices. None the less, the table shows a decline in the terms of trade of stratum 1 farmers (thus measured) of nearly 50 per cent between 1970 and 1978. Given the much stronger prices of groundnuts (both confectionery and oil) and cotton, as well as the difference

Table 57. Index of price of clothing in terms of maize, 1970-78

Year	Index	Year	Index
1970	100	1975	120
1971	105	1976	103
1972	102	1977	129
1973	113	1978	149
1974	122		

in consumption patterns, it is unlikely that small-scale commercial farmers saw their terms of trade fall as severely.

There is, however, a more important point. Not only have farmers in general been taxed heavily through a maize price that is under import parity to subsidise urban consumers,[45] but it is also the case that, through the operation of the fertiliser subsidy and the pattern of fertiliser use, the incidence of that tax has been highly regressive so that farmers who use relatively little fertiliser have been effectively taxed more heavily than commercial farmers who use much more (tables 58 and 59).

If all the farmers are taxed and non-fertiliser users are taxed more than average, is it also the case that farmers living off the railway are taxed more than those living along it? In other words, is there a spatially regressive effect as well as an input regressive effect? It is in this context that the introduction of a universal maize price is most important, for if the farmers off the railway receive the same price as farmers with easier access to the urban markets, it follows that if the domestic producer price is less than import parity and if the cost of production and delivery to the consumer is higher off the railway than it is along it, the implicit tax paid by the distant farmer is less than the implicit tax paid by the farmer along the railway. In other words, the universal price, so bitterly assailed by the efficiency school, has the effect of offsetting in a crude and imperfect way the regressive character of the fertiliser subsidy.

For that reason, as well as in response to crude political pressure, the maintenance of a universal price has become the touchstone of agricultural pricing policy in Zambia. It is almost the classic issue which divides the equity school from the efficiency school. Further, since the universal price can be discussed in isolation from the general consumer price, it does not tend to divide the urban from the rural politicians. Although it may well be true that the cost of the operation of the universal price was two-thirds of the value of all marketed maize in 1975-76, the equity school is unmoved by the alleged efficiency loss associated with these costs, for to yield the point to the efficiency school would have a high political cost in the rural areas as a whole. To implement comparative cost pricing—namely, to abandon the universal price—would involve alienating either the powerful urban lobby (by raising

Table 58. Fertiliser subsidies

Year	Per bag of maize (Kwacha)	As percentage of producer prices
1973-74	1.15	27
1974-75	1.40	28
1975-76	2.18	35
1976-77	1.60	25
1977-78	1.86	27
1978-79	2.76	31

Source. World Bank.

Table 59. Comparative prices of principal crops, 1975-76 and 1978-79

Kwacha per unit	1975-76			1978-79		
	World price	Zambia price	World price / Zambia price %	World price	Zambia price	World price / Zambia price %
Maize	83.6	70.0	119.4	89.1	100	89
Groundnuts (confect.)	511.4	312.5	163.6	469	400	117.3
Cotton	332.2	300	110.7	496.2	460	107.9

Sources. D. J. Dodge: *Agricultural policy and performance in Zambia* (University of California, Berkeley Institute of International Studies, 1977); and World Bank.

prices to those with high comparative advantage and leaving unchanged the prices to those without such advantage), or losing the support of representatives of the more distant producers (by pegging the along-the-railway price and reducing the off-railway price). Neither the Cabinet nor the Central Committee of UNIP relishes such a struggle, particularly given the wider implications of a shortage of maize that could follow the choice of the second option.

If we inquire how pricing policy compares with the three services analysed above in terms of bias against stratum 1 households it will be found that it is not precisely analogous to any of them, for it represents an almost pure confusion of objectives. The deliberate (and at times savage) tax paid by the rural sector could hardly be justified on grounds of equity, either as between different rural groups or as between all rural groups and urban interests. Similarly, the total neglect of comparative advantage, however attractive to the equity school, is highly inefficient on the criteria used by central and sectoral planners. The effect of this confusion is to leave stratum 1 households vulnerable, since the changes most likely to be yielded to the efficiency school are the eventual abandonment of the universal price and a further large reduction in the fertiliser subsidy. Only if these are accompanied by import parity pricing (and therefore large increases in urban food prices) can poor households escape serious falls in income. As we have already seen, however, the balance of political forces makes that improbable.

To summarise our argument, as households are projected into this stratum in consequence of the pressure outlined earlier, they become increasingly dependent upon government services for the operation of the technologies they are obliged to adopt. These services are not *uniformly* biased against them either in intent or in fact; but the lack of clarity of objective at the centre, the interplay of central and local patronage systems, and weak implementation leave stratum 1 households in an exposed position where some (the fittest?) may survive adequately, but a great many do not. The current extreme resource scarcity in Zambia has put much leverage in the hands of the efficiency school and especially of its international exponents. If the use of this leverage is to institute comparative cost pricing, transfer marketing to private traders, concentrate extension and credit on "progressive" farmers and remove subsidies on agricultural inputs, Zambia will certainly have a less confused rural development policy. She will also have an even greater number of rural poor.

LIMITS OF POLICY

Almost by definition, stratum 2 households are not affected by the confusions and shortcomings of government policy. They do not seek to produce a significant surplus for sale. That does not exclude the possibility of *occasional* sales, whether of crops or livestock, but it does exclude the organisation of the farming enterprise in such a way as (in intention at least) will consistently produce a cash income. There are two explanations of this behaviour. The first may be couched in terms of motivation. There are those who, viewing the risk/reward ratio and the likely returns per hour of exceedingly hard work, do not think it worth while to seek cash in that way. Given that rewards may be as low as 6 ngwee per hour,[46] this is not an irrational or perverse view of reality. On the contrary, it is entirely rational: unless the farmer can be confident of securing much higher yields without greatly increasing his labour input, he may well decide to wait for, or look for, alternative forms of cash generation, or to refrain from entering the cash economy as far as he is able. For small households, with relatively limited demands for cash (clothes and expenses related to education are typically the two major sources of expense for stratum 2 households) this may have represented a viable strategy in the past. It was made the more viable if there was the hope, which may in the end have been largely unfulfilled, of alternative sources of cash: remittances, occasional farm or non-farm employment and, particularly important for women, the proceeds from village brewing. We have already seen that this strategy has come under increasing strain as a result of economic pressures on the household.

The second explanation of stratum 2 poverty follows from Marter and Honeybone's demonstration that the labour resources of stratum 2 households are extremely limited.[47] Particularly in a less than fully mechanised maize system, labour supply is of crucial significance in determining both the area cultivated and yields. In a survey of 683 households in seven areas of Zambia, they therefore classified households on the basis of labour supply. The first group, in their terminology "group 1", consisted of households with specific disadvantages, usually a shortage of labour resources (due to migration, illness or family breakdown). This group roughly corresponds to our stratum 2 group: it should be contrasted with their second type "group 2", defined as "a nuclear family which normally produces a surplus and thus is able to sell crops, but also engages in other activities for additional income".[48] Of their sample, 41 per cent fell in group 1 and 36 per cent in group 2. Table 60 makes possible a comparison of labour supply in all four of their groups. The first thing to note is the relatively small average household size of group 1 farmers: there are rather less than 1.5 fully productive members of the household, which implies a dependency ratio of 1.52—more than double that of "group 4", consisting of effectively emergent commercial farmers. Second, the sex ratio in group 1 households is such as to suggest a high incidence of females as heads of households, or indeed households in which all members are women; again, by contrast, group 4 households have a sex ratio of over 100 per cent. Third, there is a tendency for the adults in group 1 households to be older than in any of the other groups. Since they lack the cash resources to substitute draught or tractor power for human labour, group 1 households are caught in a trap of inadequate labour, high dependency ratios and total absence of labour-saving aids. (Indeed, Marter and Honeybone found that some groups of households do not even own their own hoes and axes.)

It is, however, somewhat facile to assume that this shortage of labour makes shortage of food inevitable. That it makes the earning of *cash* difficult is certainly suggested by table 61, but that does not necessarily imply a chronic shortage of food, since even 1.5 adults per household should be enough to produce adequate food for a small household under traditional systems. This ceases to be true when maize, a labour-intensive and time-sensitive crop, displaces traditional crops that are less time sensitive and less labour demanding, as the favoured staple in stratum 2 households. The competition for labour thus generated is exacerbated by the fact that only a very small proportion of the land occupied by stratum 2 households is sufficiently fertile to grow *successive* crops of maize. This implies either declining yields from the second year onwards, or further demand for labour as new land is cleared and brought into production. Shortage of food can occur when a household misjudges the trade-off between the greater palatability of maize and the lower labour demand of cassava.

Zambia

Table 60. Composition of average households in each group

Household members	Group				
	1	2	3	4	All
Children under 8 years	0.94	1.51	1.84	2.58	1.42
Children 8-14 years	0.66	1.34	1.60	2.66	1.22
Men 15-55 years	0.53	1.19	1.35	2.25	1.04
Women 15-55 years	0.89	1.46	1.52	2.22	1.30
Men and women over 55 years	0.56	0.45	0.47	0.46	0.50
Total (household size)	3.58	5.95	6.78	10.17	5.47
Sex ratio [1]	60	82	89	101	80

[1] Males per 100 females aged 15-55 years.
Source. Marter and Honeybone: *The economic resources of rural households...*, op. cit.

Table 61. Crop production and sale by households

Crop	Percentage of household *growing* in group					Percentage of households *selling* in group				
	1	2	3	4	All	1	2	3	4	All
Maize	69	83	94	98	80	15	45	64	78	39
Groundnuts	26	48	39	49	38	5	11	18	29	11
Millet	10	21	10	8	14	1	1	1	3	1
Cassava	46	45	35	12	41	4	4	3	3	4
Sorghum	17	13	5	3	12	5	3	3	2	4
Cotton [1]	2	2	17	22	6	2	1	15	20	5
Sunflower [1]	3	8	19	54	11	1	3	13	49	8

[1] Cotton and sunflower are grown as cash crops but not all households are in a position to sell, either because of a complete failure of the crop or because it was being grown for the first time at the time of the survey and the harvest position was uncertain.
Source. Marter and Honeybone: *The economic resources of rural households...*, op. cit.

Maize and cassava, however, only supply starch: nutritionally there is as much interest in the minor vegetable and nut crops as in starch crops. Here again there is the same pattern of labour competition. Many (though certainly not all) of the minor vegetables need attention at the same time as maize, particularly for planting and weeding. Traditionally, this competition has been mediated by a division of labour in the household according to sex. Although there are many variants of precise division of labour,[49] the division was adequate to ensure reasonable security of food supplies. Clearly, however, such a division of labour breaks down in small households and becomes inoperable when the household becomes dependent upon female labour. It is then physically impossible for the household to undertake both normal agricultural activities, the quasi-investment activities associated with clearing new land and the domestic activities involved in cooking, the collection of firewood and the drawing of water. Even if maize and groundnuts are abandoned entirely as being too labour-intensive, the long-term food supply of the household is at risk.

If the Honeybone-Marter figures may be taken as at least roughly indicative of the size of the stratum 2 population, broadly consistent as it is with the 1970-71 agricultural census which showed 53 per cent of traditional farms not producing a surplus, it must be asked how it has come about that the great majority of these are so seriously short of labour in general, and young adult male labour in particular. What has produced the extreme poverty associated with labour shortage? The answer is not hard to find. Table 54 gave, inter alia, demographic change by province for the period 1963 to 1969. It is immediately clear that the less accessible, more ecologically disadvantaged provinces have suffered substantial net emigration. However, there is some evidence that much of this migration, perhaps particularly that from Northern and Luapula Provinces, was made up of women joining their husbands on the Copperbelt. To a lesser degree, the same may be true of some of the migration from Eastern Province to Lusaka. Under the colonial and federation regimes, while the settlement of women with husbands in the urban areas was not prohibited, it received little encouragement, and the normal pattern was for the man of the household to migrate first, to be joined later by his wife and children. After independence this waiting period seems to have become much shorter, so that whole families are more likely to migrate now with the very minimum of time-lag between the husband and the rest of the family. However, a more detailed examination of the 1969 census suggests two patterns of migration flow that are relevant to our purpose. The first is that in the flow of migrants *to* the rural areas (whether from other rural areas or, as is more likely, from urban areas) women are over-represented. Thus 54.4 per cent of immigrants to Serenje were female, while only 45.6 per cent were male. Second, although young women tend to migrate from the rural areas, the proportion tends to drop sharply after the age of 25, while that for men declines much more steadily to the age of about 50.[50]

Further, there is some evidence that dependants, both young children and more elderly people, are over-represented in migrants *to* rural areas. Given the tendency for older people to be under-represented in migrants to towns, this further reinforces the high dependency. We can illustrate the effects of migration on household structure from the 1969 census. Table 62 shows that in Luapula, a province of which the population declined by about 5 per cent between 1963 and 1969, over one-third of all households were headed by females and nearly one-third of households comprised fewer than three people. If we take the district which suffered the highest rate of outmigration, both those figures rise to 40 per cent.

It might be thought that in so far as stratum 2 households are those who suffer most directly from migration, they are also those who will benefit most directly from remittances from family members in employment in urban

Zambia

Table 62. African households in Luapula Province classified by sex of head, number of wives present (male heads) and size of household

Male head	Total number of persons	Total number of households	Size of household (persons)									
			1	2	3	4	5	6	7	8	9	10+
No wife	29 231	12 300	5 637	2 626	1 677	980	561	347	198	143	29	102
1 wife	201 957	42 533	0	6 744	7 714	7 510	6 603	5 408	3 610	2 807	901	1 233
2 wives	5 142	629	0	0	17	46	62	70	96	100	45	193
3 wives	517	48	0	0	0	0	0	2	3	10	4	29
4 wives	78	3	0	0	0	0	0	0	0	0	0	3
5+ wives	7	1	0	0	0	0	0	0	1	0	0	0
Total	236 932	55 514	5 637	9 373	9 408	8 536	7 226	5 827	3 908	3 060	979	1 560
Female head	98 184	28 569	5 562	5 689	5 363	4 359	3 183	1 967	1 133	806	212	295
Total	335 116	84 083	11 199	15 062	14 771	12 895	10 409	7 794	5 041	3 866	1 191	1 855

Source. Final report of the Census of Population and Housing, 1969 (Nov. 1973), Vol. II (d).

areas. By contrast with some other African countries, most notably Kenya, the flow of remittances from the urban areas appears to have been meagre in the immediate post-independence period.[51] No recent survey of rural incomes has identified a substantial remittance flow.[52] One major reason is clear: the function of remittances is largely to preserve the remitter's status, rights and reciprocated obligations to the village. This is only important to him in so far as he expects to return to the village and resume his position there. While it is true that many older men and women do still retire to the village, it becomes a less common expectation as the nature of urbanisation changes.

Almost by definition, stratum 2 households have little opportunity for cash employment, either as agricultural labourers elsewhere or in non-farm employment. There is, however, an important geographical distinction to be made. Stratum 2 poverty is not confined to the agriculturally backward provinces—for example, Northern, Luapula, North-Western and Western Provinces. It is also to be found, though in less concentrated form and in smaller numbers, in the more advanced provinces: indeed, according to the agricultural census of 1970, the proportion of non-surplus producing households in Central Province was only marginally lower than that in Luapula (44 and 46 per cent respectively). However, the households in the more developed parts of the country have very much greater opportunities for alternative employment, both farm and non-farm, than do those in the less accessible, less developed, less agriculturally differentiated north and west. Certainly, there is evidence of the increasing employment of hired labour at least to meet peak demands in Southern and Central Provinces, and possibly in Eastern Province too.[53] Although wage rates can be very low, they are not necessarily so. For instance, in the Gwembe Valley, where the demand for farm employment greatly exceeds the supply of jobs, wages were reported in 1977 to be 55 ngwee per day, compared with a rate in Mumbwa of more than double that figure, in addition to free transport and food. The significance of these wage rates lies in their implication that in less than three days the stratum 2 household could earn as much cash from selling its labour as it could from selling its crops.[54]

Further, there is clearly an important distinction between, on the one hand, those for whom the only employment opportunities come at a time that conflicts with their own peak labour demand for food production and, on the other, those who have possibilities of off-peak employment. Observation suggests that in the former case security of family food supply is given paramount importance, and that only if the demand for cash is felt to be extremely pressing will cash employment be preferred over the cultivation of the basic staples. In the latter case, when there is no such competition, wage employment may be taken relatively without risk.[55] In general, competition is the more common case, because areas in which there is substantial local demand for peak period labour are co-terminous with areas in which maize is grown as a preferred staple.

The basic argument of this section, then, is that stratum 2 households are in an economic and geographical position of such a kind that typical strategies of rural development have very little to offer them. They are neither equipped nor motivated to take high risks; they have very little spare capacity to produce a marketable surplus, and, with the resources at their command, the return to their labour is so low that even if they could physically handle a larger acreage, they have little incentive to do so. Given, then, that they are almost wholly untouched by the normal variables of rural development policy, are they wholly beyond help? Our analysis has suggested that there are a number of lines of possible action. The first relates to migration. It is clearly a gross over-simplification to believe that "narrowing the gaps" will reduce migration and improve household structure and therefore the labour resources of stratum 2 households, especially since the average age of such households and the dependency ratio may well be made worse by *reverse* migration. It may be nearer the truth to state that greater opportunities for wage employment in the rural areas, both on farms and elsewhere, would help retain more young males. Striking evidence of the effect this could have came from Maud Mutemba's study of the Kabwe Rural District (Central Province), where she detected a significant difference between Zambian and Zimbabwe farms.[56] The former were starved of labour; the latter relatively prosperous since the young men would not leave the farm because they knew there was no hope of their getting a job in the towns against competition from young Zambians.[57] How wage employment could be provided particularly in the remoter areas, it is more difficult to say. Rural capital works are well known as such a device—as, indeed, are their associated problems. More controversial is the promotion of cash crops and therefore labour peak employment in the remoter areas. This was one of the implicit objectives of the universal price. However, we have already seen that *peak period* employment is less attractive since it competes with labour demand on the family farm. The extension is thus implied of price incentives to non-competitive crops grown in the remote areas, and of these cassava appears the most attractive.

This raises a last point. Cassava is a much neglected crop not only in Zambia but in other parts of the world. We have seen that its displacement by maize has far-reaching effects on stratum 2 families. The implication is that a programme of genetic improvement that increased palatability and yield and of research on processing that economised on labour time would, in the longer run, have a profound effect on the cropping patterns of the rural poor. If such work also made cassava acceptable to urban consumers and for stock-rearing, a source of cash income could be added to a more secure food crop.

From this it is clear that although stratum 2 households do not, in general, benefit from the standard array of rural development policies, it is not the case that they are inherently inaccessible. Rather they are presumed to be inaccessible by those who operate a conventional efficiency algorithm; and the "success" of the equity school in Zambia has been confined to areas—such as universal maize pricing and the structure of marketing—that are not significantly relevant to stratum 2 households. A determined long-term attack on the poverty of stratum 2 would need to extend that range to encompass areas that would be certain to generate intense opposition from the ranks of the bureaucracy, the party and international donor institutions.

CONCLUSION

The principal conclusion which emerges from the above discussion is that the condition of the majority of rural dwellers in Zambia and the vast majority of the poorest section of the rural population is unaffected by so-called rural development activities. Indeed, we have suggested that in so far as those policies have any impact at all, they tend to reinforce both the absolute and relative poverty of most of the rural poor.

Why should this be? It is suggested that part of the answer lies in the terms within which the debate about rural development policy is set in Zambia. It is important to end, however, by emphasising that the confusion that surrounds the formulation of such policy in Zambia, while going a long way to explain the impact of policy on stratum 1 households, throws considerably less light on its impact on stratum 2 households. Even if any one policy had been enunciated with great clarity and implemented with great vigour (a condition that would have made Zambia almost unique in sub-Saharan Africa), the penetration of stratum 2 households by the agents of central government policy would have been at best modest; for the truth is that the panoply of institutional forms, which central planners and administrators regard as the instruments of rural development policy (co-operatives, farm training centres, the extension service, credit, improved marketing facilities, improved infrastructure) are, as it were, beyond the bounds of stratum 2 households. Indeed, those agencies themselves may become the agents of impoverishment rather than development by introducing new elements of disequilibrium in the social, economic and ecological environment of stratum 2 households. This is, admittedly, in no way unique to Zambia. Although the problems may be intensified in Zambia through the effect of migration on family labour supply conditions in stratum 2 households, the fundamental problem—that of fashioning policy instruments capable of revolutionising the productive capacity and commercial possibilities of the rural poor—is universal. It is this observation that gives rise to the more extreme forms of agrarian pessimism that see as the only ultimate solution the absorption of all the rural poor into non-agricultural activities in the towns. As a fact, this may still be a long way off in Zambia: as an objective, it may well offer better prospects than any of the range of policies that have been discussed in this chapter.

Notes

[1] African employment only.

[2] Excluding unpaid family and "other" workers.

[3] See J. B. Knight: "Wages and Zambia's economic development", in C. Elliott (ed.): *Constraints on the economic development of Zambia* (Nairobi, 1971), pp. 97-102.

[4] See J. Fry: *Employment and income distribution in the African economy* (London, 1979); and ILO/JASPA: *Narrowing the gaps: Planning for basic needs and productive employment in Zambia* (Addis Ababa, Jan. 1977).

[5] Compare ILO/JASPA, op. cit., p. 41, where a different approach is taken, but which seems broadly consistent with the table.

[6] This estimate is based on the labour participation rates assumed in the Second National Development Plan, and formal sector employment as revealed by the Employment Inquiries. There can be little doubt that the explosive growth of the "informal/unemployed" category took place after 1969, when the growth of formal sector employment slowed and yet the income gap between urban and rural households was perceived, by rural households, to be very large.

[7] See below, p.176.

[8] R. E. Baldwin: *Economic development and export growth: A study of Northern Rhodesia, 1920-1960* (Berkeley, University of California Press, 1966), pp. 44-48.

[9] Fry, op. cit.

[10] See for instance, ILO/JASPA, op. cit.

[11] C. R. Bliss: *Xerophthalmia and food intake frequency of children in the Luapula Province of Zambia*, Unpublished Master of Nutritional Science thesis (Cornell University, 1974), p. 86.

[12] Over most of Zambia, land is not a constraint in the normal sense. Apart from some localised signs of population pressure (e.g. on lake margins) there is no crude shortage of uncleared land. There is therefore no "landless labourer" class in Zambia, in contrast to Kenya. Size of holdings is a function of labour supply, since it is almost literally true that a farmer can have as much land as he can handle. For statistical demonstrations, see Universities of Nottingham and Zambia Agricultural Labour Productivity Investigation: *Report No. 3* (Lusaka, and Sutton Bonington, 1970); and R. H. Harvey: *Some determinants of the agricultural productivity of rural households: Report on a survey in Kalichero District, Eastern Province* (Chipata, MRD, 1973). For a socio-anthropological account, see N. Long: *Society and the individual* (Manchester University Press, 1968). But see below for the effect of local shortages.

[13] See S. Quick: *Bureaucracy and rural socialism: The Zambian experience*, Unpublished PhD thesis (Stanford University, May 1975), pp. 162 ff.

[14] Gavin Kitching points out that it is a hallmark of populism (in the classical *narodnik* sense) that this potential conflict is never seen. See D. Mitrany: *Marx against the peasant* (North Carolina, 1951), pp. 49-51.

[15] See M. Bratton: *Peasant and party state in Zambia: A study of political organisation and resource distribution in Kasama District*, Unpublished Ph. D. thesis (Brandeis University, 1977), p. 427.

[16] This is a recurrent theme in many of President Kaunda's speeches, especially in the period 1966 to 1969.

[17] W. Allen: *The African husbandman* (Edinburgh, 1963).

[18] See, for instance, *Land holding and land usage among the Plateau Tonga of the Mazabuka District*, Rhodes-Livingstone Papers, No. 14 (1948); M. J. S. W. Priestly and P. Greening: *The Ngoni Land Utilisation Survey* (Lusaka, Government Printer, 1955); and H. A. M. Maclean: *Resettlement problems in the Eastern Province of Northern Rhodesia* (Lusaka, Ministry of Agriculture, 1963; mimeographed).

[19] M. Jackman: *Recent population movements in Zambia: Some aspects of the 1969 Census*, Zambian Papers, No. 8 (University of Zambia, Institute for African Studies, 1973), p. 15.

[20] C. R. Joseph: *Report on a random pilot survey of traditional farmers in the Central and Northern Provinces of Zambia* (Lusaka, MLA, Sep. 1977), p. 6.

[21] Bliss found in Luapula that there was great spatial variation in the number of items included in the diet—from 14 in Kabole District to only one in Kalasa. Bliss, op. cit., p. 114.

[22] We return to rural-urban terms of trade below.

²³ Bliss found that only 7 per cent of her sample spent no money at all on food; 44 per cent spent up to K2 per week. Bliss, op. cit., p. 118.

²⁴ The price rises quoted are for urban low-income groups. There is no satisfactory index of prices for poor rural households. Even Maimbo's data refer to relatively affluent small farmers. However, the poorest farmers in his sample spent between one-quarter and one-half of their cash income on clothing and footwear and over 20 per cent on food, beer and beverages. See F. J. M. Maimbo: *An analysis of the pattern of income and expenditure of a sample of farm families of Chiefdom Hamaundu of the Choma District of the Southern Province of Zambia* (University of Zambia, Rural Development Studies Bureau, 1975), p. 34. The significance of the fact that Maimbo's survey was in *Southern* Province should not escape the reader.

²⁵ For instance, there is an obvious difference in the impact of land scarcity in Southern Province which is leading to deterioration of the range (though not of arable land) and in the Northern Province where variants of *chitemene* are widespread.

²⁶ Note, for instance, the volume of urban-rural migration and the age structure of these returnees. See Jackman, op. cit., p. 48.

²⁷ Marter and Honeybone seek to discount motivation as an important variable. While their own emphasis on resource endowment of the household is important and helpful, their strictures on those who have sought to explain differential farming strategies by motivation seem exaggerated. They ignore the fact, for instance, that in the UNZALPI survey they criticise it was shown that "villagers" have a higher input of labour per acre than "farmers". This is a finding inconsistent with the Marter-Honeybone hypothesis that farming strategy is dictated by labour availability. See Universities of Nottingham and Zambia Agricultural Labour Productivity Investigation, op. cit., p. 21, tables 11 and 12.

²⁸ That there is such a spectrum or ladder is itself an assumption that needs challenging. Elements of such a challenge are to be found in the next section.

²⁹ See, for instance, C. S. Lombard and A. H. C. Tweedie: *Agriculture in Zambia since independence* (Lusaka, Institute of African Studies, University of Zambia, 1974); and R. A. J. Roberts and C. Elliott: "Constraints in agriculture", in C. Elliott (ed.): *Constraints on the economic development of Zambia* (London and Nairobi, 1972), pp. 293-294.

³⁰ For instance, in Joseph's sample in Central Province, the *average* size of loan was over K2,000.

³¹ See Bratton, op. cit., pp. 296, 306-308.

³² ibid., p. 310; see also pp. 311-313 for further examples.

³³ M. Muntemba: *Rural underdevelopment in Zambia: Kabwe Rural District 1850-1970*, Unpublished Ph. D. thesis (Los Angeles, University of California, 1977), p. 346.

³⁴ A. Marter and D. Honeybone: *An evaluation study of Zambia's farm institutes and farmer training centres* (UNZA, Rural Development Studies Bureau, 1975); cf. Universities of Nottingham and Zambia..., op. cit.

³⁵ T. Maluza: *Crop demonstration: Its contribution to farmers' training and extension teaching in Zambia* (UNZA, School of Agricultural Sciences, 1975; mimeographed), p. 33.

³⁶ According to a World Bank report, non-salary recurrent budget cuts have so reduced mobility that only 20 per cent of an extension officer's time is typically spent in the field.

³⁷ Harvey, op. cit. Gross margin is reported as K 6.8 per hectare: 700 hours per hectare has been assumed as the labour input given that cultivation is by hand.

³⁸ *Evaluation of the IDZ programme of the Government of Zambia* (Lusaka, MLA, May 1978), p. 119. This is a bureaucratically cautious assessment of the real distribution of benefits: a more telling indication of the distribution of benefits is the fact that free wire is issued to owners of cattle who had cleared grazing—provided that they owned more than 20 cattle. This excludes all but the very advanced farmers in the less favoured provinces and roughly 70 per cent of traditional, emergent and small scale producers in Central, Southern and Eastern Provinces.

³⁹ There is increasing evidence that a much higher proportion of total output is sold through informal channels than was believed in the past. Traditional crops and vegetables have always been sold in village markets, but there seems to have developed over the past ten years a much livelier unofficial market in products for which the National Agricultural Marketing Board and the three regional co-operative marketing units, as their agents, have a monopoly.

⁴⁰ Stephen Carr of the World Bank investigated these complaints in Eastern Province as systematically as possible. In general he found them unfounded. Personal communication, May 1979.

⁴¹ Hence Joseph on the use of fertiliser in Northern Province: "Usage is more dependent upon availability than cost or choice" (Joseph, op. cit.); cf. Harvey, op. cit., who reported a high degree of dissatisfaction with both input and output marketing in Eastern Province, a fact that needs to be set against Carr's finding above.

⁴² There is now built into the marketing structures, particularly in the leadership of the NAM Board, a determination to cling to their self image as a service industry, since this enables them to justify drawing subsidies from the central Government, irrespective of the quality of the service they offer.

⁴³ In theory, devaluation is another option, but in political terms it is unattractive since it raises both urban living costs and costs of copper production.

⁴⁴ The work of Fry and Maimbo was concentrated in Southern Province: it is none the less revealing in this context. See F. Maimbo and J. Fry: "An investigation into the change in the terms of trade between the rural and urban sectors of Zambia", in *African Social Research*, Dec. 1971.

⁴⁵ See D. J. Dodge: *Agricultural policy and performance in Zambia* (University of California, Berkeley Institute of International Studies, 1977), pp. 5-9.

⁴⁶ Five bags per hectare at K6.30 per bag and 550 hours per hectare: no costs apart from labour; and the product delivered to the NAM Board depot.

⁴⁷ A. Marter and D. Honeybone: *The economic resources of rural households and the distribution of agricultural development* (UNZA/RDSB, 1976; mimeographed).

⁴⁸ ibid.

⁴⁹ In some villages the men traditionally did only the heavy work of clearing and preparing the mounds for *chitemene*; in others they also planted and tended the maize, while the women tended the vegetable gardens exclusively.

⁵⁰ Jackman, op. cit., pp. 45-49.

⁵¹ To some extent this is a technical problem: they are easily forgotten and/or concealed by respondents and they tend to be "lumpy" since cash and goods tend to be brought by visiting relatives rather than sent on a regular basis. Even allowing for this, however, it seems highly improbable that all but a small minority of stratum 2 households can expect significant sums from this source.

⁵² According to CSO sample surveys in 1974 and 1976, remittances accounted for less than 2 per cent of cash income in the rural sector.

⁵³ Joseph, op. cit., Agroprogress GmbH: *Development of cotton areas in Central and Southern Provinces: Report to the Government of the Republic of Zambia, 1978*, Vol. II.

⁵⁴ Based on household crop income of K6 for group 1 farmers as reported by Honeybone and Marter. If we go up to group 2, it would take roughly three weeks to earn as much from labour as from household crop sales; but note that group 2 farmers intend to produce a surplus for sale in a way that group 1 farmers do not.

⁵⁵ At the margin under the *chitemene* system there may again be conflict between clearing new land and thus improving the quantity and security of crops in the succeeding years and taking employment for immediate cash income.

⁵⁶ That is, farms of households whose head emigrated from Zimbabwe and still speaks a language from that country.

⁵⁷ Muntemba, op. cit.

OIL AND INEQUALITY IN RURAL NIGERIA[1]
Paul Collier

7

Nigeria is a country of central interest to the study of African rural development for three distinct reasons. First, it is by far the most important Black African economy. Although the size of the Nigerian population is uncertain, the rural population must be of the order of 60 million; this exceeds the combined rural populations of Ghana, Kenya, Mozambique, Tanzania, Uganda, Zambia and Zimbabwe.[2] Second, African commercial agriculture has a far longer history in Nigeria than in most other parts of the continent. On account of the absence of European settlers, Nigerian smallholders were being encouraged to grow cash crops at a time when the settlers of eastern and southern Africa were attempting to confine African integration into the market to the sale of labour. This fact renders meaningful the study of long-term trends in rural concentration. Evidence will be here presented of trends over half a century, whereas in most other African economies trends can be identified only for periods of ten to 20 years. Third, alone in Black Africa, Nigeria has large known reserves of oil. The development of the oilfields has coincided with the very substantial increases in oil prices, so that a massive and abrupt change has taken place in the economy. The resulting transfusion of foreign exchange into the revenues of the Nigerian Government may be seen as the fulfilment of the dreams of those who look to foreign aid as a major vehicle of poverty redressal. Thus, in studying the impact of oil upon rural poverty in Nigeria we are not only analysing an event which directly affected the largest rural population in Africa; we are also testing a strategy of rural development which is being pursued in more modest proportions across the continent.

The first section of this chapter is devoted to a review of macro-economic and sectoral indicators of performance, which tentatively suggest that the rural community in aggregate may have benefited only modestly from the oil bonanza. This inference is investigated more closely in the second section, in which a large number of village studies are analysed in order to identify trends in the level and distribution of rural incomes. In the third section an attempt is made to explain these trends with reference to Nigerian development strategy and, in particular, to the implications of the oil boom. Finally, we draw the conclusions from our research.

NATIONAL GROWTH AND RURAL STAGNATION

The development of the Nigerian economy may be broadly classified under three phases. During the colonial period Nigeria constituted a classic example of an economy based on agricultural exports. The interest of the colonial Government lay in encouraging smallholder production of export crops, thus supplying low-cost raw materials to the metropolitan Power and providing a market for its industrial output. After independence, the new Government followed a strategy of import-substituting industrialisation behind high trade barriers. By 1968 textiles, food processing, metal and plastic fabrication, and related industries, all enjoyed rates of effective protection in excess of 100 per cent.[3] Industrialisation was financed by export taxes through the operation of agricultural marketing boards.

The most recent phase of development, post-1968, is characterised by the reversion to export-induced growth generated by the oil boom. The contribution of oil was so substantial that by 1978 Nigerian GNP per head at US$ 560 was the second highest in Black Africa. Real GNP growth per head over the period 1960-78 was 3.6 per cent per annum, the third highest rate of growth in Africa, surpassed only by the very small economies of Togo and Lesotho. These macro-economic indicators suggest that Nigeria is probably the major African success story. Growth was so rapid that, unless income distribution changed powerfully, all groups in Nigerian society should have experienced substantial and perceptible improvements in living standards.

The first indication that this successful performance was not evenly distributed throughout the economy emerges when growth is disaggregated by sector (see table 63). The sectoral data in the table reveal that both during the import-substitution phase and during the oil boom, agriculture stagnated. Indeed, whilst the oil boom accelerated the growth of GDP, it coincided with an apparently accelerated decline of agriculture, though this observation must be qualified by the very poor quality of macro-level agricultural data available. Admittedly, even so rapid a sectoral re-allocation of economic activity as that observed in Nigeria need not necessarily entail an unequal distribution of the benefits of growth among households. However, despite rapid rural-urban migration the absolute rural population probably rose slightly even during the 1970s, so that rural agricultural income per head was at best stagnant in real terms. National consumption expenditure per head rose very rapidly during this period. For example, during the subperiod 1973-74 to 1978-79 national consumption per head rose in real terms by 38 per cent, whilst rural real agricultural income per head remained constant.[4] It

Table 63. Sectoral distribution of economic activity and performance

Sector	Annual real growth rates (%)		Composition of GDP		
	1960-70	1970-77	1960	1970	1977
Agriculture	−0.5	−1.5	54.0	37.1	22.7
Mining	31.7	8.1	5.2	30.1	37.1
Manufacturing	12.8	13.4	2.8	5.2	7.0
Public administration	4.8	24.6	1.3	3.3	7.7
Other	1.0	7.5	36.7	24.3	25.5

Source. World Bank: *World tables* (Washington, DC, 2nd. ed., 1980).

is, admittedly, possible that rural consumption per head grew by 38 per cent during this period even though income per head from agriculture remained constant. This could have come about either from rural dis-saving or from rising rural non-agricultural incomes. The former is neither sustainable nor desirable in the long term. The latter may indeed have occurred, for example, through the expansion of rural-based government employment in the form of posts for teachers and administrators. However, lacking the qualifications to gain access to such employment opportunities, the bulk of the rural population would have gained income only through the rural multiplier of those incomes. Given the stagnation of agricultural income, which must support a large component of non-agricultural rural expenditure, it seems unlikely that there could have been a sufficiently spectacular growth in rural non-agricultural income for rural consumption per head to grow at the national rate. Expressed in other terms, the macro data imply that agricultural output has been stagnant for a long period and that the rural population suffered relatively in the oil-financed consumption bonanza of the 1970s.

Whilst these inferences might merely reflect gross inadequacies in the underlying data, they are sufficiently disturbing to prompt two lines of analysis. First, if on average agricultural incomes were stagnating, what was happening to the distribution of agricultural income? Second, what were the economic processes by which rapid national growth might have widened the gaps between rural and urban consumption per head? It is to these two lines of analysis that we now turn.

RURAL INEQUALITY

Perhaps more important than over-all measures of the extent of and trend in rural inequality is the identification of inequalities between socio-economic groups. Typically in agricultural communities, access to land is the predominant determinant of income so that socio-economic status can be stratified by the mode of access to land in a hierarchy going from landlords, through smallholders, tenants and sharecroppers, down to the rural landless. In Nigeria such a classification is largely inappropriate because the country has an abundance of land suitable for cultivation. FAO estimates suggest that the current area of medium- and high-productivity soils is around 84 million acres (1 acre equals 0.405 hectare), of which only between 21 million and 29 million acres are cultivated. As a consequence there is no significant group of rural landless labourers in Nigeria. The majority of the rural population achieve access to land through the traditional system of land tenure, whereby an area of land will belong to a community which may exclude non-members. Membership is determined by family relationships through birth or marriage. Within the community a single custodian allocates land to its members, thus giving individuals the right to a parcel of land for so long as it is cultivated, at the end of which period the land reverts to the community. A survey by Famoriyo,[5] covering the whole of Nigeria, found that 65 per cent of rural households acquired land through inheritance and a further 12 per cent through communal allocation. However, this traditional form of tenure is giving way in the more commercialised areas of the country to systems of individual ownership, which in turn have introduced tenancy and sharecropping. Both the communal and individual ownership systems have in common the consequence that those from outside the locality do not have free access to land. This form of restricted access means that differences in living standards are likely to be generated by the enormous ecological variations between different parts of Nigeria which, in turn, are reflected in crop variation. Thus, in our analysis of rural inequality it is important to distinguish between farmers grouped according to type of crop. In particular, some farmers are engaged mainly in growing export crops, the most important being cocoa, palm oil, rubber and cotton. Other farmers, on the other hand, are solely engaged in growing food crops for sale primarily within Nigeria, notably maize, millet, sorghum, groundnuts, yams and cassava. In table 64 we present an estimate of the numbers of farm households engaged in the production of each crop.

Clearly, many households grow more than one crop; however, even on the limiting assumption that no household grows more than one export crop, the data in table 64 indicate that 64 per cent of households would have no participation in export-oriented agriculture. The actual proportion of households growing only domestic food crops is therefore likely to be rather higher than 64 per cent. There are three reasons why this distinction between food-producing and export-producing farmers is of importance. First, access to export-crop production is restricted by ecological conditions. Nigeria covers several ecological zones ranging from tropical rain forest to desert. To a considerable extent the ecological zones partition the country into distinct cropping areas. Barriers of culture, land ownership, information and distance inhibit the free movement of the rural population between ecological zones so that it is to be expected that ecological variations will be translated into

Table 64. Estimated number of farm households growing specified crops, 1963

Crop	Farm households ('000)
Food	
Sorghum	1 876
Millet	1 671
Maize	1 459
Groundnuts[1]	1 174
Yams	2 060
Cassava	1 366
Beans	1 214
Export	
Cocoa	343
Oil palm	1 000
Rubber	100
Cotton	252
All farm households	4 810

[1] Groundnuts were traditionally an export crop but are now a domestic food crop (see table 77 below).
Source. World Bank: *Agriculture sector survey, Nigeria* (Washington, DC, 1973), Annex 12, table 1.11.

significant inter-zonal income inequalities. These we aggregate for purposes of subsequent analysis into the export- and food-producing zones, whilst recognising that this simplification is inappropriate for those parts of Nigeria in which the zones overlap. Second, whilst food-producing farmers have until recently grown predominantly for subsistence, export producers have necessarily been integrated into the market economy for many years. Thus, by comparing the patterns of inequality of income and assets between food and export producers, we may draw inferences concerning the possible long-term consequences of commercialisation upon distribution. Third, we will argue that the consequences of the oil boom have been radically different for these two groups, owing to the impact of the oil revenues upon the exchange rate and aggregate demand. This implies that trends in both the level and the distribution of income would be divergent.

The foregoing suggests that descriptions of rural income distribution at the national level are of limited analytic value. In fact, the severe limitations of statistics at the national level means that they are also unobtainable. The only attempt at a national description of rural inequality is the rural economic survey showing land distribution in 1973 (see table 65). Since livestock is confined by climate and the tsetse fly to the pastoralists of the far north, land is indeed overwhelmingly the most important rural asset and thus might be thought a good proxy for income distribution. There are, however, two

Table 65. Land distribution, 1973

Size (hectares)	Households (%)	Area (%)
0-0.25	16	1
0.25-0.5	20	3
0.5-1.0	16	15
1.0-2.5	29	47
2.5-5.0	18	26
Over 5.0	1	8

Source. Rural economic survey 1973-74.

important qualifications to the results shown in table 65. First, no allowance is made for ecological differences; one hectare in a cocoa-growing area is likely to be far more valuable than several hectares of semi-arid land in a cereal-growing area. Second, the data in table 65 are regarded as so unreliable as to be worthless. While development economists are familiar with the limitations of rural data in general, the national statistics upon Nigerian land holdings appear to be peculiarly deficient. For example, the mean holding size implied by the distribution given in table 65 is 3.7 acres. In our subsequent analysis we shall present evidence from 53 village-level surveys, the data from which are considerably more reliable, and in scarcely a single village is the mean holding size as low as 3.7 acres. In fact, the average holding size from the village surveys is 7 acres. Admittedly, it is possible that the 53 villages, which form a comprehensive collection of the village studies undertaken in Nigeria, are all highly atypical; however, a more plausible inference is perhaps that the national data cannot be trusted.

In the absence of usable national statistics on rural income and asset distribution, we have turned to village-level studies usually conducted by social anthropologists. Since a single village study considered in isolation obviously makes it difficult to generalise, an attempt has been made to overcome this drawback by comparing all the village studies arranged into food- and export-producing groups.[6] An analysis is then made of similarities within groups, differences between groups and trends over time. This methodology is precarious since the studies cover different villages at different times and were conducted by different people. However, since the studies of individual villages can usually be taken to be far more accurate than nationally conducted surveys, the loss from not having a properly drawn random sample of villages is in part offset by having atypically reliable underlying data.

The first group of village studies focuses upon cocoa-producing farmers in the three Yoruba-speaking states of south-west Nigeria; Ondo, Oyo and Ogun. The second group of studies concerns the food-producing farmers of

Nigeria

Table 66. Estimates of mean household incomes (at 1960 prices) for cocoa farmers

Year	Source	Place and description		Sample size	1960 prices (N£)	
					Cocoa sales	Household incomes
1928-30	Forde	Owo	"Typical" cocoa farm	–	–	130
1951-52	Galletti	Cocoa belt	All	180	100	200
			Ondo ⎫ Ondo circle		100	
			Ife-Ilesa ⎭		100	
			Ibadan-Osun		70	
			Ogun		160	
1964	Upton	Alade	Near Akure, Ondo state	–	–	190
		Olugbo	Near Abeokuta, Ogun state			
1967-69	Essang	Ondo circle	Full-time farmers	100	100	–
			Whole sample	160	240	
1968-69	Oni	Cocoa belt	All	231	50	
			Ondo	67	100	
			Ibadan ⎫	48	30	
			Oyo ⎬ Ibadan-Osun	44	40	
			Abeokuta ⎫ Ogun	37	20	
			Ijebu ⎭	35	30	
1969-70	Oni	Cocoa belt	All	250	70	
			Ondo	69	170	
			Ibadan ⎫	52	50	
			Oyo ⎬ Ibadan Osun	46	50	
			Abeokuta ⎭	46	30	
			Ijebu	37	30	
1970	Olayemi	Cocoa belt	All	180	60	80
			Ondo circle		80	110
			Ibadan-Osun		50	70
			Ogun		40	50
1971-72	Essang	Ondo, Oyo and Ogun	Families with migrants	–	–	70
			Families without migrants			50
1972	Adeyokonnu	Isoya project area, Ife		61	–	60
1974	Clarke	Ifetedo	Community members	–	80	90
			Migrant farmers		40	40
1974	Osuntogun	Ogbomoso	Five villages	100	–	70

– = not applicable.
Sources. Derived from Gavin Williams: *Inequalities in rural Nigeria*, Paper prepared for the ILO (1980), table 5.1 Sources of individual studies in this table are given at the end of the chapter.

the Hausa-speaking northern states (Kano, Sokoto, Kaduna and Bauchi), the main crops being guinea corn, millet and groundnuts. Between them these areas cover some two-thirds of the rural population of Nigeria.

Let us consider, first of all, export-producing farmers represented by the Yoruba cocoa producers. Earlier in this chapter we documented, at the macro level, long-term agricultural stagnation, and our first concern is to check whether this corresponds to micro-level observations. In table 66 we present the findings of 11 sets of village-level studies spanning the period 1928 to

Table 67. Estimates of distribution of income among cocoa farmers

Year	Source	Place	Description	Sample size	% earned by household	
					Bottom 40%	Top 10%
1948-51	Galletti	Cocoa belt	Co-op sales		7	40
1951-52	Galletti	Cocoa belt	Cocoa sales	187	11	35
1951-52	Galletti	Cocoa belt	Net farm earnings	187	14	31
1951-52	Galletti	Cocoa belt	Net earnings	187	13	40
1967-69	Essang	Ondo circle	Cocoa sales	160	9	48
1967-69	Essang	Ondo circle	Net earnings	160	6	60

Sources. Derived from Williams, op. cit., table 8.1. Sources of individual studies in this table are given at the end of the chapter.

1974. Mean income is given at 1960 prices for cocoa sales and total household income, depending upon data availability. The table reveals that on both measures real incomes were considerably lower in the 1970s than in the period 1928 to 1964. Indeed, the most recent study, that of Osuntogun covering five villages in 1974, found real household incomes of only half the level reported as typical in the earliest study, that of Forde nearly 50 years earlier. Thus the micro-level data corroborates the inferences made from the macro-level data.

We now consider inequalities among cocoa farmers. Only two studies, those of Essang and Galletti, have measured income distribution among cocoa farmers, their results being summarised in table 67. Both studies find very considerable inequality. Whilst Essang's results need to be qualified since his sample selection perhaps exaggerated inequality, even the most favourable case, that of Galletti for net farm earnings, shows a differential of 9:1 between the poorest 40 per cent and the richest 10 per cent.

Essang's study shows that traders and professionals tend to have the largest holdings, that they are almost the only people with access to cheap loans from governments or commercial banks, that they have the best access to chemical sprays and extension services and that they are most likely to hold political office, itself a generator of resources. Clarke's study shows the differences in incomes between community members and migrant tenant farmers from beyond the cocoa belt. The average income of community members was estimated at more than double that of tenants (see table 66). Almost half the community members earned more than 250 Naira per annum, while only one tenant in 25 did so. The reason for this difference was that tenants had much smaller holdings than community members, to whom, moreover, they had to pay rents on such holdings.

The studies of Essang and Clarke both tend to the conclusion that in the long-commercialised environment of cocoa production the restricted access to land is indeed an important determinant of income inequality. In turn, since land is the major asset, its distribution reflects past inequalities in incomes. The seven studies of land distribution presented in table 68 all confirm the inference that land distribution is highly unequal. The differential between the top 10 per cent and the bottom 40 per cent ranges from 7:1 to 20:1.

Unfortunately, the paucity of observations precludes us from discerning any trends in inequality among cocoa farmers. To summarise the findings of tables 66, 67 and 68, the real incomes of these export producers have fallen substantially over a long period, and restricted access to land is an important determinant of inequality, the dimensions of which, both in terms of income and assets, are considerable.

We now turn to the Hausa food-producing farmers. In table 69 we summarise the results of 20 separate studies of mean incomes of food-producing households, spanning the period 1933 to 1978. No clear trend can be discerned from these data; however, in the most recent studies—those covering the period 1973-78—incomes are generally rather above those reported in the earliest studies undertaken between 1933 and 1939.

According to the results of five surveys which are summarised in table 70, inequality among food producers is not particularly severe. The differential between the top 10 per cent and the bottom 40 per cent is generally about 4:1. This is not to suggest, however, that the observed inequalities are socially random.

Hill's studies[7] of Batagarawa and Dorayi distinguish between three groups apart from the *masu sarauta* (office-holding aristocracy), according to their known ability to withstand the shock of a very poor harvest. The rich are able to make loans or gifts in such circumstances. The poor have to borrow grain and money to provide for their families: they are the ones who suffer. In the middle are to be found those who, while not rich, do not suffer. Hill shows that wealthy farmers are more likely to be older men with large households and sons in *gandu*: in the first place, they own and farm more land than others, per household, per working male and per resident; second, they can afford to manure their land more intensively, and are more likely to buy land and hire labour than smaller farmers; and, finally, they are more likely to engage in inter-village trade in grain and other commodities, to store grain for resale and to lend money. The poor who are most likely to be found among the old, as well as in households headed by young men, have smaller holdings which are less likely to be manured effectively. Heads of poor families usually engage in menial occupations which bring low returns, such as transporting manure, making corn-stalk beds or collecting firewood. They are likely to have to engage in wage labour. They may have to borrow money to buy grain in the rainy season, returning double the amount of grain at harvest. The worst-off are those who are unable to borrow; at best they may be lent a little land to farm, corn-stalks to make beds for sale, produce to sell or grain in exchange for labour later in the year. It is common for poor men to sell their farm manure and to sell farms or pledge them. Thus the poor are not only short of land, they also lack the means—cash, manure and family labour

Table 68. Estimates of distribution of land among cocoa-producing households

Year	Source	Place	Sample size	Mean acres	% used by households Bottom 40%	% used by households Top 10%
Cocoa land						
1950	Dept. of Agriculture	Cocoa belt	14 937	3.5	17	39
1951-52	Galletti	Oyo/Ogun	85	4.4	10	41
1969-70	Olayemi	Cocoa belt	180	6.7	14	32
1950	Dept. of Agriculture	Ondo				
		Ute	951	6.7	16	30
		Ondo	1 326	22.5	16	34
		Idoani	638	4.6	14	33
		Ekiti	880	1.7	12	36
		Owena	610	6.5	11	44
		Ife-Ilesa				
		Ondo road	671	6.2	10	37
		Osu	907	8.4	10	38
		Ibokun	513	6.2	8	39
		Ile-Igbo	195	4.3	8	40
		Ibadan				
		Gbongan	918	4.2	9	42
		Ajia	1 358	3.0	9	42
		Olojuoro road	923	3.8	10	40
		Egba (Ogun)				
		Ilaro	1 525	3.4	17	27
		Ota	3 822	2.1	15	31
All land						
1951-52	Galletti	Oyo/Ogun	81	22.6	10	40
1968	v. d. Driessen	Ife	1 000	5.9	17	28
1974	Patel	Badeku, Ibadan	254	7.6	14	34

Sources. Derived from Williams, op. cit., table 10.1. Sources of individual studies in this table are given at the end of the chapter.

Nigeria

Table 69. Mean incomes of food-producing farmers (at 1960 prices)

Year	Source	Place and description	Sample size	1960 prices (N£)	
				Farm earnings	Household incomes
		Cross Rivers state			
1952-53	Martin	Mpiokporo II and Usung Inyang	16		140
1970-71	Usoro	Itak Local Council	–		14
		Imo state			
1964	Upton	Uboma	70	100	130
1974-75	Lagemann	Owerri-Ebeiri		40	170
		Okwe	74	80	130
		Umuokile		60	120
		Bendel state			
1964	Upton	Ubuluku			160
		Akumazi } Near Asaba			120
		Onicha-Olona			
		Kwara state			
1933	Forde, citing Patterson	Shao, Ilorin province ("typical")			20
1964	Upton	Iloffa, near Ofa Ilorin province		120	120
	cited	Omu-Aran, Ode-Ore } Ilorin pro-			
1968?	RERU	Ipetu } vince		140	
		Kano, Sokoto, Kaduna, Bauchi, Niger states			
1934	Forde, citing Parsons	Kazaure Emirate, Kano ("typical")			40-60
1937	Forde, citing MacBride	Dawaki ta Kudu district, Kano ("typical")			35-50
1938	Forde	Environs of Kano City		50	90
1939	Forde, citing Shorter	Tarke village, Sokoto	15		20
1949/50	Smith	Zaria, a district, 4 towns and villages	49		70
		7 towns and villages, inc. above	90		60
1960	Luning	Katsina	384	60	70
		Kadandani	70	40	80
		Bugasawa and Ladanawa (hamlets)	75	60	70
		Bindawa	41	40	80
		Ilale and Makera (hamlets)	71	50	70
		Kasanki	57	50	70
		Birnin Kuka	70	20	50
1966-67	Norman	Zaria, Dan Mahawayi	42	80	100
		Hanwa	38	90	120
		Doka	42	100	110
1967-68	Goddard	Sokoto, Gidan Karma	38	110	140
		Kaura Kimba	31	100	110
		Takatuku	31	50	90
1968?	Cited RERU	Bauchi, Nasarawa		50	60
		Nabayi		50	50
		Bishi		40	60
1972-73	King	Kano, Dakwara		70	
		Gwarzo		50	
		Shonono		30	

Agrarian policies and rural poverty in Africa

Table 69. Mean incomes of food-producing farmers at 1960 prices *(concluded)*

Year	Source	Place and description	Sample size	1960 prices (N£)	
				Farm earnings	Household incomes
1974-75	Matlon	Bauchi, Ningi		140	–
		Gwaram		70	–
		Gar		40	–
		Kano, Rogo	32	40	60
		Barbeji	35	50	60
		Zoza	33	40	50
1978	Huizinga	Zaria, four villages, Giwa	103	60	–

– = not applicable.
Sources. Derived from Williams, op. cit., tables 5.2 and 5.3. Sources of individual studies in this table are given at the end of the chapter.

Table 70. Income distribution among food producers

Year	Source	Place and description		Sample size	% earned by households	
					Bottom 40%	Top 10%
Stratified by income per household						
1949-50	Smith	Zaria 4 places	Income/households	49		
		7 places		90		
1966-67	Norman	Zaria	Income/households	103	19	27
		Dan Mahawayi		42	16	34
		Hanwa		38	21	21
		Doka		42	22	21
1967-68	Goddard	Sokoto	Income/households			
		Gidan Karma		38	21	20
		Kaura Kimba		31	14	25
		Takatuku		31	24	21
Stratified by expenditure per household						
1970-71	Simmons	Zaria	Expenditure/households	120	14	39
		Dan Mahawayi		40	11	44
		Hanwa		37	22	30
		Doka		43	27	21
Stratified by income per resident						
1974-75	Matlon	S. Kano	Income/residents	100	21	21
		Rogo		32	20	21
		Barbeji		35	21	19
		Zoza		33	25	18
1974-75	Matlon	S. Kano	Income/households	100	28	19

Sources. Derived from Williams, op. cit., table 8.2. Sources of individual studies in this table are given at the end of the chapter.

—with which to farm land. They dispose of their few resources in order to meet their expenses, thus falling into a vicious cycle of impoverishment. The impoverishment of some is not unrelated to the enrichment of others. As Hill[8] points out, "profits from grain storage would be much lower if there were none too poor to farm", and wage labour would be less cheaply available.

Matlon argues that, although the rich farmers in his sample used more land than the poor, differences in farm size cannot account for differences in farm incomes. According to Matlon, poor farmers spend less time than their more successful counterparts on farm labour, including wage labour for others, for lower returns and probably for that very reason. While variations in productivity must result, in part, from differences in the way farmers manage their resources, they are also likely to result from differences in access to subsidised fertiliser and the capacity to grow high-value crops.

Most studies separate the *masu sarauta* and local office-holders from the rest of the population. Although they are no longer able to command free labour and appropriate resources in the manner which the colonial and the first civilian governments allowed, they are still significantly richer than most farmers, with better access to education for their children, and government resources and produce licences for themselves, and are still able to exact tolls on a variety of political and economic transactions. In Matlon's study, the households of the district head, village heads and *sarkin noma* ("head farmer") are much larger than others, farm far more land, and appropriate and allocate the resources which the Government supplies to the village, such as fertiliser, tractor-hiring services and even seed provided for famine relief.

If Matlon is correct in his conclusion that land holding is not the primary cause of income inequality, since land is virtually the only asset available to households, its distribution should reflect rather than determine income distribution. This view is particularly useful since data on the land distribution of food-growing households are more plentiful than those on income distribution. The summarised results of a large number of village studies, spanning the period 1930 to 1979, are presented in table 71. We may immediately conclude from these results that land is considerably less equally distributed than income—the opposite of what one would expect if income inequalities in a particular year reflected only transient deviations from a more equally distributed "permanent" income. Hence, the pattern of income inequality must be persistent.

The large number of observations in table 71 enables us to use statistical techniques to investigate whether there are systematic relationships between the extent of land concentration and population density, and whether there are trends over time in land concentration. It must be emphasised that the observations do not constitute a random sample and, hence, too much weight should not be attached to the results. In particular, the time trend might be highly inaccurate. Two regressions were run to explain respectively the share

Table 71. Land distribution among food-producing farmers

Year	Source	Place	Sample size	Mean acres	% land used by households Bottom 40%	Top 10%
c. 1930	Hill	Kano, Dawaki				
		Yargaya	–	7.2	14[1]	24[1]
		Runa	–	6.0	15[1]	28[1]
1949	Smith	Zaria 4 places	58	5.3	17	29
		7 places	109	4.5		
1960	Luning	N. Katsina				
		Kadandani	70	8.7	20[1]	25[1]
		Bindawa	41	6.1	20[1]	36[1]
		Ilale-Makera	71	7.9	20[1]	21[1]
1962	Kohlhatkar	Makarfi, Zaria	–	6.7	16[1]	30[1]
1963	Luning	Sokoto	220	6.3	22[1]	24[1]
1966	Norman	Zaria				
		Dan Mahawayi	103	10.1	13	33
		Hanwa	64	5.9	11	27
		Doka	153	8.7	14	29
1967	Hill	N. Katsina				
		Batagarawa	171	5.6	13	33
		Autawa-Makurdi	43	11.2	16[1]	32[1]
		Kankai	45	6.5	14[1]	26[1]
1968	Goddard	Sokoto			16	28
		Gidan Karma	103	11.4	17	27
		Kaura Kimba	170	3.0	12	27
		Takatuku	98	7.6	19	26
1970	Hill	Dorayi, Kano	544	2.2	11[1]	40[1]
1974	Huizinga[1]	Giwa, N. Zaria	103	7.4	13	32
1976	Kohnert	Nr. Bida	210	9.4	15	32
1979	Clough	Marmara S. Kat	118	7.4	11	40
1968	Goddard	Kaura Kimba				
		upland	170	3.0	12	27
	Sample 2:	upland	31	4.1	11	32
	Sample 2:	fadama	29	2.6	8	36
	Sample 2:	all	31	6.7	12	31
1973	King	Kano, co-operative	49	10.9	15[1]	27[1]
		Kano, non co-operative	53	7.4	11[1]	36[1]

– = not applicable.
[1] Estimated from distribution of holdings by land size group.
Sources. Derived from Williams, op. cit., table 10.2. Sources of individual studies in this table are given at the end of the chapter.

of land owned by the bottom 40 per cent and the top 10 per cent. The independent variables used were the mean size of holding, the year of the study, and regional dummy variables for Kano, Zaria, Katsina and Sokoto. The result of the regression explaining the share of the bottom 40 per cent was as follows:

$$S_{40} = 0.71M - 0.155T \quad r^2 = 0.53$$
$$(0.24) \quad -(0.054)$$

(standard errors in parentheses; regional dummy variables not reported)

S_{40} = share of land held by bottom 40 per cent (mean = 14.6)
M = mean holding size in village (in acres) (mean = 7.0)
T = time (in years)

Both time and mean holding size were statistically significant at the 1 per cent level. The negative sign on the time trend suggests that, controlling for mean holding size, the bottom 40 per cent have been losing land share. The coefficient implies that on average the share held by this group has been falling at about 1 per cent per annum. The positive sign on mean holding size implies that if population density increases, causing a fall in mean holding size, the share held by the bottom 40 per cent also declines. This must be qualified, however, for if household size changed over time, a change in mean holding size might not imply a change in population pressure. Converting the regression coefficients into elasticity from a 10 per cent fall in mean holding size would reduce the share of the bottom 40 per cent by 3.5 per cent. This is important because, with a national population growth of 2.5 per cent per annum, land pressure has been increasing. Such a consequence of population pressure might appear inconsistent with our earlier claim of land abundance. However, migrants from a low-income to a high-income area are not able to enjoy living standards comparable to those of the population which they seek to join. Clarke has pointed out that migrants had incomes only half those of the indigenous community because of their restricted access to land. Hence, to a considerable extent, population pressure in an area reduces mean holding size rather than being accommodated by out-migration. Since our time trend was estimated controlling for holding size, whilst in fact mean holding size is likely to be falling over time, the time trend coefficient understates the pace at which the bottom 40 per cent are estimated to have been losing their land share according to the regression.

We now repeat the previous regression with the dependent variable being the land share of the top 10 per cent. The results are presented below:

$$S_{10} = -0.76M + 0.24T \quad r^2 = 0.36$$
$$(0.38) \quad (0.09)$$

S_{10} = share of land held by top 10 per cent (mean = 30)

The r^2 is somewhat lower and, whilst time is still significant at the 1 per cent level, mean holding size is significant only at the 5 per cent level. The signs of the coefficients are the opposite of our previous regression, implying that the top 10 per cent have been gaining land share over time and that they also gain land share as population pressure increases. The value of the coefficients suggest that the effect of population pressure is a transfer from the bottom 40 per cent to the top 10 per cent, leaving the middle 50 per cent unaffected, whilst over time the top 10 per cent have been gaining land share more rapidly than the bottom 40 per cent have been losing it. However, these implications, which rest upon the comparison of coefficients each of which is subject to qualification, must be regarded as highly tentative.

To summarise our results concerning food-growing farmers, as represented by the Hausa, real incomes have probably risen over the course of the past 50 years, though without any steady trend. While income distribution is relatively equal, it is less so with respect to asset distribution. Finally, we have suggested that there may be a trend towards increasing concentration of assets which, if correct, would probably reflect a similar trend in income inequality. This last observation should be regarded as having the status of a hypothesis worth further research rather than as being an incontrovertible fact.

The contrast between food-producing and export-producing farmers is considerable. In the first place, they have been subject to opposite trends in real incomes. Whilst the earliest surveys (those of the 1930s) show export farmers to have had a mean household income some four times above that of food farmers, our most recent studies reveal no significant difference in mean incomes. Second, the two groups have had markedly different patterns of income inequality. The differential between the mean incomes of the richest 10 per cent and the poorest 40 per cent, which is 4:1 among food farmers, is in the range 7:1 to 20:1 among export farmers. Third, this corresponds to a less equal distribution of land among export farmers than that among food farmers, though the most recent observations show the extent of inequality to be close to that among export farmers.

Our observations both at the micro level within the rural economy and at the macro level discussed earlier have produced a series of puzzling phenomena. In an economy in which growth of GNP has been very rapid, agriculture has in aggregate stagnated. Within the rural community radically different patterns and trends have emerged between export and food farmers. Whilst we must repeat our caution about the underlying data, these events are not just a matter of a low percentage figure: they appear to be very substantial. They should therefore be translatable into identifiable macro-economic variables. It is to this that we now turn.

OIL AND RURAL INEQUALITY: MECHANISMS OF TRANSMISSION

The stagnation of Nigerian agriculture, and in particular the decline in export-based agriculture, during the phase of import-substituting industrialisation is of no surprise to development economists. The tendency of such a strategy to worsen the rural-urban terms of trade through a combination of export taxes and import barriers, and thus to reduce rural incomes, is well understood. What is surprising is that the transition from import substitution to oil exportation appears to have reinforced rather than reversed this dismal trend. A naïve expectation might have been that rural incomes would have risen sharply. First, rapid macro-economic growth might have "trickled down" to the rural poor. Second, oil money supposedly eliminated the need to finance industrialisation through the taxation of agriculture and promoted import liberalisation without balance-of-payments problems. It might, therefore, have reversed the earlier factors which had caused a deterioration in the rural-urban terms of trade. Third, the transformation of government finances provided the opportunity for the rural community in general, and the rural poor in particular, to become major beneficiaries of public expenditure. The failure of this naïve expectation to be borne out by Nigerian experience suggests that there must have been powerful transmission mechanisms from the oil boom offsetting these processes.

The value of exports of crude petroleum rose from N 74 million in 1968 to N 10,034 million in 1979. Government revenue rose from N 186 million in 1968 to N 6,265 million in 1977, the last year for which data are available. These two enormous changes, the inflow of foreign exchange and its channelling into public revenue, are the major economic manifestations of the oil boom. We will argue that each of these events has triggered mechanisms which have produced the consequences observed in the earlier sections of this chapter.

First, let us consider the oil boom as an enormous influx of foreign exchange. In the absence of measures to the contrary, this would tend to raise the real exchange rate. The *real* exchange rate differs from the exchange rate by correcting for different rates of inflation between countries. In table 72 we show the real exchange rate between Nigeria and the United States for the period 1968 to 1979. Column 1 shows the uncorrected exchange rate in terms of dollars per Naira. Columns 2 and 3 show consumer price series for the United States and Nigeria. Column 4 shows the real exchange rate, calculated as

$$\frac{\text{Nigerian price index}}{\text{United States price index}} \times \text{exchange rate}$$

and then re-based so that 1968 = 1,000.

The startling conclusion from table 72 is that the real exchange rate tripled in just 11 years (this does not, of course, allow for black market transactions). Now a change of this nature has predictable consequences for relative prices within Nigeria: a rise in the real exchange rate makes goods which are internationally traded cheaper in relation to goods which are only traded within Nigeria.

Whilst some industrial goods and all export crops are traded internationally and thus have their prices set on world markets, food (because of its low value/weight ratio) is scarcely traded internationally and thus largely has its price determined domestically. We predict, therefore, that the rise in the real exchange rate caused by the foreign exchange influx will have raised the price of food and other domestically traded goods relative to the price of internationally traded goods. This is confirmed by the data in table 73, which show that over the period 1968 to 1977 food prices rose by 46 per cent relative to the low-income consumer price index. Since the bundle of goods which

Table 72. The Nigerian real exchange rate, 1968-79

Year	Exchange rate[1] ($ per Naira)	Consumer price index[1] United States	Consumer price index[1] Nigeria	Real exchange rate (1968 = 1.000)
1968	1.40	89.6	79.9	1.000
1969	1.40	94.4	87.8	1.043
1970	1.40	100	100	1.122
1971	1.40	104.3	116.1	1.248
1972	1.52	107.7	119.3	1.354
1973	1.52	114.4	126.5	1.346
1974	1.62	126.8	142.1	1.455
1975	1.59	138.5	190.2	1.748
1976	1.59	146.5	231.8	2.016
1977	1.53	156.1	281.6	2.210
1978	1.54	167.9	350.2	2.578
1979	1.78	186.8	397.9[2]	3.038

[1] International Monetary Fund: *Financial Statistics*, various years. [2] Third quarter of 1979.

Table 73. Relative prices of food and non-food goods, 1968-77

Year	Consumer price index[1] All goods (1)	Consumer price index[1] Food (2)	Food price Relative to all goods[2] (3)	Food price Relative to non-food goods[3] (4)
1968	124	117	1.000	1.00
1969	137	138	1.067	1.14
1970	154	166	1.143	1.33
1971	175	204	1.236	1.62
1972	180	208	1.225	1.58
1973	186	211	1.202	1.50
1974	218	248	1.206	1.52
1975	286	354	1.312	1.91
1976	349	452	1.373	2.19
1977	427	589	1.462	2.72

[1] Federal Office of Statistics, Lagos Low-income consumer price index. 1960 = 100. [2] Column (2) divided by column (1) and rebased to 1968 = 1.000. [3] Column (3) data using the food weight 0.5 from Federal Office of Statistics expenditure surveys, as reported in World Bank: *Agriculture sector survey* (Washington, DC, 1973) Annex 12, table 7.3.

makes up the consumer price index itself contains a large weight for food, the relative price change between food and non-food must have been considerably greater than 46 per cent. The weight for food in the index is in fact 0.5 which enables us to derive in the last column of table 73 that food prices have risen by 170 per cent in relation to non-food prices in nine years.

A comparison of the right-hand columns of tables 72 and 73 shows that the relative price of food and non-food goods closely follows movements in the real exchange rate. We may safely conclude, therefore, that one consequence of the oil boom was a massive change in relative prices in favour of food producers and against export producers. Thus we have traced the opposing trends in the real incomes of food and export farmers observed above to an origin in the exchange rate consequences of the oil boom. However, the rise in the food price may also offer an explanation for the increasing concentration hypothesised among food farmers. It should be recalled that income and asset distribution was far less equal among the long-commercialised cocoa farmers than among the traditionally predominantly subsistence food farmers. While the sharp rise in the price of food has tended to commercialise food growing, however, the benefits of the price increase are confined to those farmers who can produce a regular net surplus of grain for the market. Our survey of inequalities among food producers suggests that it is generally the richer farmers who are in a position to make net sales of grain, so that the price increase would amplify existing inequalities. Hence, both the historical analogy with cocoa farmers and the evidence concerning which food farmers are most likely to benefit suggest that the oil boom, through its impact upon prices, has accentuated inequality among food farmers and has reduced the incomes of export farmers.

We now turn to the second manifestation of the oil boom, considering it, that is to say, as an enormous increase in government revenue. In the 1950s and 1960s the federal Government had secured much of its revenue through a high tax upon agricultural exports. An effect of this tax was to squeeze the incomes of export-producing farmers in the years prior to the oil boom. We thus observe in table 66 the incomes of cocoa producers falling well before the onset of the oil boom. Potentially the oil boom could have been of great benefit to export farmers since the Government no longer needed to raise revenue through taxation of export crops, and accordingly reduced the gap between the producer price and the world price. Unfortunately, however, as we have seen, the spectacular increase in the real exchange rate reduced the domestic purchasing power of a constant unit of world purchasing power to one-third of its initial level, thus swamping the more modest benefits experienced by export farmers by way of reduced taxation.

The enormous increase in government revenue has been broadly matched by the increase in government expenditure. The impact of this increase in expenditure upon rural living standards depends upon whether it was devoted to the provision of fresh resources for the rural community or whether it was used in such a way that it competed with the rural community for scarce resources.

Some components of government expenditure have had a national impact in which the rural community has shared, such as education and health. However, only a very small proportion of government expenditure has been allocated to promoting agriculture. In the Third National Development Plan (1975-80) only 5 per cent of planned expenditure was on agriculture. Morover, some of this expenditure—which is itemised in table 74—may tend to reduce rather than enhance the incomes of small farmers. (The major items not listed in table 74 are agro-industries, the river basin development authorities, research institutes, the National Centre for Agricultural Mechanisation, and rice, oil palm and rubber projects. It should be noted that the data refer to planned and not actual expenditures. World Bank and state government expenditure is not included.)

Four agricultural development strategies have been pursued as follows: investment in government plantations; irrigation projects for rice and wheat; extension services promoting high-yield seeds, chemicals and fertilisers; and, finally, planting projects for cocoa and palm oil. Clearly, the first of these (the use of oil revenue to establish state plantations) tends to compete with smallholder agriculture. Any benefits would be indirect through the relaxation of population pressure. However, even that expenditure directed towards smallholders has sometimes tended to accentuate rural inequality. Two large projects, the Kano River irrigation scheme and the Funtua extension services programme, may serve to illustrate this point.

Wallace[9] shows that farmers who exchanged rain-fed for irrigated land on the Kano River scheme are not equally in a position to benefit from this apparent windfall. Irrigated farming is expensive for farmers who have to pay for seeds, tractor use, fertiliser and water. It can be profitable, but it is also very risky. Wheat and tomato prices vary year by year and fluctuate over each season. Farmers must plant the crops dictated by the authorities; unless they have more than one plot they cannot diversify their risks by combining wheat and tomatoes. Nor can they rely on the supply of tractor services, fertiliser or water. Drivers plough the fields of powerful people first, and may have to be paid cash or lent land to get them to plough land promptly and thoroughly. Yields are much lower on fields planted late, as well as on fields situated far from the water supply. The scheme operates "inefficiently but to the benefit of scheme workers and the locally powerful farmers, and ... rich and powerful outsiders".[10] Irrigated farming means abandoning dry-season occupations and not growing guinea corn in the season. The poor cannot forgo their dry season earnings, and others are unwilling to do so, and therefore lease land to wealthier farmers. Farmers without married sons working with them are constrained by the high cost of labour. Traders and businessmen can employ wage labour for irrigated farming while pursuing their own occupations. Some urban aristocrats, officials and businessmen and irrigation staff are renting, buying and speculating in land on the scheme.

The scheme has created employment and an inflow of single workers for its construction and operation, thus increasing, in turn, the market for cooked food, rented housing, transport and building contracting. Income from supplying these goods and services may be used to pay for irrigated farming. Local traders can invest their profits from irrigated farming in selling provisions, building houses for rent, tailoring, transporting or money-lending,

Table 74. Federal expenditure on agriculture under the revised Third Plan (in N million)

Item	Expenditure
Irrigation	535
Nigerian Agricultural Bank	150
Large-scale food farms	132
Fertiliser procurement and distribution	100
Cash crop rehabilitation	100
World Bank projects	44
Strategic grain reserves	40
National accelerated food production programme and farm service centres	23
Roots and grain production companies	5

Source. T. Forrest: "Agricultural policies in Nigeria, 1900-1978", in J. Heyer, P. Roberts and G. Williams (eds.): *Rural development in tropical Africa* (London, Macmillan; New York, St. Martin's Press, 1981).

especially to construction workers. Foreign and Nigerian companies profit from lucrative consultancy, management and supply contracts.

Irrigated farming has deprived many farmers who are not involved in the scheme, and even some who are, of their livelihoods. It has increased the cost of farming to the disadvantage of the poor. Its effects are less clear with respect to those in the middle, where the corresponding greater costs and risks have to be balanced against the chance of higher returns. Those who have benefited most are in fact those who were better off in the first place. The scheme has conferred considerable benefits on some officials, aristocrats and contractors, enabling them to increase their influence over rural society.

As regards the Funtua project, the management concerned explicitly repudiated the approach offered by the Dutch Guided Change Team, which was to distribute small, and ideally equal, amounts of fertiliser to participating farmers. The Funtua project chose to work through existing patronage relations, assuming that this was necessary to get local co-operation. Contrary to this, Agbonifo[11] found from a study of a tomato-growing project conducted by Cadbury in Zaria that farmers resented being organised under the political leadership of village heads, who misappropriated project resources and used their control of them to increase their power over other farmers. When, as in 1978, insufficient fertiliser was available to meet demand, large-scale farmers and traders were able to appropriate a large share of the supply.

The Funtua project focused its extension activities on "progressive" farmers from whom the benefits would allegedly "trickle down". Progressive farmers are usually wealthy farmers who combine farming with trading and have the money to invest in land, labour and fertiliser to expand agricultural production. This class has long been established in Hausa society. The project's strategy accentuates an existing tendency towards differentiation among peasant producers.

A smaller class of "overnight" farmers, recruited from urban businessmen, civil servants, district heads and army officers, has been created by the liberal provision of subsidised bank loans to buy tractors and pay for labour and fertiliser. By 1978 the Funtua project had prepared farm management plans for 59 farmers with an average of 60 acres and had in preparation further plans for 167 farmers with an average of 140 acres. These are all much larger than the biggest holdings identified by the village surveys we have examined. Jackson[12] cites the example of a farmer in the Gusau Agricultural Development Project who, having expanded his farm from 200 acres after the inception of the project, subsequently expanded it still further to 555 acres when he acquired a tractor through bank credit, on the recommendation of the project's farm management unit. He was assigned an extension worker for his exclusive use at public expense.

Whilst government expenditure directly benefiting the poorer smallholders has, perhaps, been less than prodigal, total expenditure has risen enormously. It has directly financed the rapid growth of employment in public services and industry and, indirectly, has fuelled the growth of an informal sector geared to providing goods and services in response to the expanding demand from the urban community. The combined effect of these two developments has been to compete with smallholder households for the supply of young male labour. The resultant squeeze upon labour availability for smallholders has produced a predictable sharp increase in rural real wages, time series for cocoa- and food-producing areas being presented in table 75.

The rise in rural real wages, induced by labour shortages, has probably had a mixed effect upon rural income distribution. Those households from which labour has migrated to urban employment may either benefit or suffer. Some households will receive remittances from migrants and will benefit from reduced population pressure on the land cultivated. Others, however, will not benefit from remittances, either because migrants cannot afford to make them or because they choose not to do so: such households may then have lost a large part of their labour supply with no compensating advantage. In order to cultivate the family holding they may then have to hire labour at relatively high wages. Those households in a position to sell labour are not necessarily the poorest, for it will be recalled from the study by Matlon that poor farmers spent less time on wage labour than others.

Whilst the rise in rural real wages has redistributed rural income from labour-scarce to labour-abundant households, it cannot be interpreted as indicating a general increase in rural incomes. The reason for this is that the wage increase was induced not by a rise in the demand for rural labour, generated by increased agricultural output, but rather by a decrease in the supply of young male labour generated by increased urban demand.

The above explanation is consistent with over-all stagnation of agricultural output noted earlier in this chapter. Furthermore, the shortfall in agricultural output extends to both food- and export-producing farmers. Food

Table 75. Rural real daily wages for men
(in pence at 1960 prices)

Year	Source	Place and crop	Real wages Cocoa-producing area	Real wages Food-producing area
1914	Berry	Cocoa belt	37-73	
Pre-1920	Galletti	Cocoa belt	18-37	
1926	Berry	Cocoa belt	15-30	
1927-28	Berry	Cocoa belt	20-58	
By 1930	Galletti	Cocoa belt	28-37	
1935	Berry	Cocoa belt	12-48	
1938	Forde	Cocoa belt	22-44	
Pre-1939	Hill	Batagarawa, Katsina (food)		8
1940	Galletti	Cocoa belt	54-72	
1948	Pedler	Zaria (food)		11
1950	Galletti	Cocoa belt	33-47	
1950	Grove	Bindawa, Katsina (food)		30
1952	Galletti	Cocoa belt	37-52	
1952	Tiffen	Yamaltu, Gombe (food)		23-27
1957	Vigo	Ako, Gombe (food)		38-54
1963	Kohlhatkar	Ako, Gombe (food)		22
1963	Luning	Sokoto (food)		22-27
1964	Upton	Alade, Ondo (cocoa)	16	
1967	Tiffen	Gombe Emirate (food)		16-25
1967	Norman	Zaria Emirate (food)		22
1967	Hill	Batagarawa, Katsina (food)		21
1968	Hill	Batagarawa, Katsina (food)		17
1970	Olayemi	Ondo (cocoa)	40	
		Ibadan-Osun (cocoa)	37	
		Ogun (cocoa)	34	
1970	Oni	Cocoa belt	35	
1971	Berry	Araromi, Ibadan (cocoa)	17-43	
		Abanata, Ife (cocoa)	3-58	
		Orotedo and Omifon, Ondo (cocoa)	17-46	
		Ibadan, Ife, Ondo (cocoa)	9-116	
1971	Abaelu	Akure ⎱ Ondo circle (cocoa)	24-29	
		Akoko	25-46	
		Owo	23-40	
		Ife ⎰	12-29	
		Ibadan (cocoa)	58	
		Remo ⎱ Ogun (cocoa)	52-55	
		Ijebu ⎰	46-52	
1971		Western State cocoa belt	38-46	
1971	Hill	Dorayi, Kano (food)		45
1972	King	Kano, western (food)		44-55
1974	Clarke	Ifetedo, Ife (cocoa)	50-75	
1974	King	Kano, western (food)		26-51

Table 75. Rural real daily wages for men *(concluded)*

Year	Source	Place and crop	Real wages	
			Cocoa-producing area	Food-producing area
1974	Matlon	Rogo, Kano (food)		31
1975	King	Kano, western (food)		43-87
1976	Clough	Marmara, Katsina (food)		41-54
1978	Wallace	Chiromawa, Katsina (food)		55
1978	Clough	Marmara, Katsina (food)		52
1979	Clarke	Ifetedo, Ife (cocoa)	c.90-100	
1979	Clough	Marmara, Katsina (food)		c.65

Sources. Derived from Williams, op. cit., table 7. Sources for this table are given at the end of the chapter.

imports which averaged 46,000 tons in the mid-1960s had risen to 790,000 tons by 1977, the last year for which data are available. The virtual extinction of Nigeria's agricultural exports since the oil boom is documented in table 76.

CONCLUSION

The above analysis suggests that the benefits of the oil boom were unevenly distributed, and that the rural population in aggregate probably enjoyed a disproportionately small share of the total benefits. Within the rural community there were powerful and complex shifts in income distribution. Through the increase in the real exchange rate and the competition of public expenditure for smallholder resources, substantial sections of the rural community must have suffered both a relative and an absolute decline in living standards. In aggregate, food growers gained in relation to export farmers. Since the mean incomes of the latter were initially higher than those of the former, this reduced rural inequality. Among export-producing farmers real incomes fell sharply, plunging part of the rural community into poverty. Among food-producing farmers the main beneficiaries were those initially rich enough to make large net sales of food. Rural real wages rose, benefiting households well endowed with marketable labour at the expense of labour-scarce households. It appears to be unclear whether this improved or worsened rural income distribution. Migrants who acquired employment in urban areas gained from the oil boom; on the other hand, whether the rural households which they left gained or lost is itself a complex question which cannot be answered here.

This leads us both to a conclusion and to one remaining question. Our conclusion is that there can be no presumption that rapid growth at the

Table 76. Nigerian exports, 1965-78 [1]

Year	Cocoa	Palm oil	Palm kernels	Groundnuts	Groundnut oil
1965	259	152	421	553	84
1966	193	145	399	579	105
1967	248	17	162	546	61
1968	209	1.5	161	695	110
1969	174	10	179	524	98
1970	195	10	184	291	89
1971	272	20	241	138	41
1972	228	2	212	106	40
1973	214	–	138	199	111
1974	194	–	186	30	24
1975	175	11	171	–	0.3
1976	219	3	272	2	–
1977	168	–	186	0.8	–
1978	192	–	57	–	–

– = not applicable.

[1] The table does not include groundnut cake (2.3 per cent of export values in 1968), cocoa powder, butter and cake (2.9 per cent of exports in 1968).

Sources: G. K. Helleiner: *Peasant agriculture, government and economic growth in Nigeria* (Homewood, Ill., Richard D. Irwin, 1966), tables IV-A-7, 8-1; Central Bank of Nigeria (CBN): *Annual Reports*, 1960-79; and Williams, op. cit., table 1.

national level will automatically benefit the rural poor. In the Nigerian case which we have been considering it appears that even the massive income derived from the oil boom, which could have transformed rural living standards, has substantially been diverted to other beneficiaries.

This leads naturally to our question: given that the rural community did not benefit significantly from oil income, who did in fact benefit? A fully documented answer lies outside the scope of this essay, but some indication of the beneficiaries emerges from a recent study by a JASPA mission.[13] Comparing 1973-74 and 1978-79, the study estimates that mean rural household income fell slightly from N 733 to N 725 (at 1977-78 prices). Unskilled urban wage earners shared a similar experience, with real wages falling from N 780 to N 741. Yet average urban household income rose from N 2,520 to N 3,378. It is clear that unskilled wage earners were not responsible for this average rise.

Notes

[1] The author would like to thank Dharam Ghai, Samir Radwan, B. C. Rosewell and E. Chuta for their patient and helpful criticism. This chapter draws on material included in Gavin Williams: *Inequalities in rural Nigeria,* paper prepared for the ILO (1980). For details of data presented in tables 66-71 and 76 reference should be made to the paper. Particular acknowledgement is made for the following authors for material provided and for valuable comments: Tina Wallace, Jeremy Jackson, Bert Huizinga, Roger King, Tom Forrest, Julian Clarke and Paul Clough. Views and conclusions are the responsibility of the author alone.

Agrarian policies and rural poverty in Africa

[2] Rural population estimates derived from World Bank: *World Development Report, 1979* (Washington, DC, 1979), tables 17, pp. 158-159, and 20, pp. 164-165.

[3] idem: *The current economic position and prospects of Nigeria* (Washington, DC, 1971), Report No. AW-22a.

[4] Data are based on ILO/JASPA: *First things first: Meeting the basic needs of the people of Nigeria*, report to the Government of Nigeria by a JASPA basic needs mission (Addis Ababa, 1981), Annex G2, table 7. An implicit GDP deflator of 1.59 was used to convert the data to constant prices.

[5] S. Famoriyo: *Problems posed by land tenure in Nigerian agriculture* (Ibadan, NISER, 1977).

[6] A full account of the village studies is given in Williams, op. cit.

[7] P. Hill: *Rural Hausa: A village and a setting* (London, Cambridge University Press, 1972); idem: *Population, prosperity and poverty: Rural Kano 1900 and 1970* (London, Cambridge University Press, 1977); and idem: "The myth of the amorphous peasantry", in *Nigerian Journal of Economic and Social Studies,* Feb. 1968.

[8] idem: *Rural Hausa...,* op. cit., p. 198.

[9] T. Wallace: "The Kano River Project, Nigeria", in J. Heyer, P. Roberts and G. Williams: *Rural development in tropical Africa* (London, Macmillan; New York, St. Martin's Press, 1981).

[10] Unpublished material provided by R. Palmer-Jones.

[11] P. O. Agbonifo: "The introduction and impact of the tomato processing project in Nigeria", in *Samuru Agricultural Newsletter,* 18 Jan. 1976.

[12] J. Jackson: *The bank in Northern Nigeria,* research note (Norwich, 1979).

[13] ILO/JASPA: *First things first,* op. cit., Annex G2, table 13.

Sources for tables 66-71 and 76

J. N. Abaelu and H. L. Cook: *Wages of unskilled workers in agriculture and some characteristics of the farm labour market in the Western State of Nigeria* (Ile-Ife, 1975).

T. O. Adeyokonnu: "Rural poverty in Nigeria: A case study in Ife division", in *Poverty in Nigeria,* Proceedings of the 1975 Annual Conference of the Nigerian Economic Society (Ibadan, 1979).

S. S. Berry: *Cocoa, custom and socio-economic change in Western Nigeria* (Oxford, 1975), pp. 140, 147-148 and 180.

J. Clarke: *Agricultural production in a rural Yoruba town,* Ph.D. thesis (University of Ibadan, 1979).

S. M. Essang: *The distribution of earnings in the cocoa economy of Western Nigeria,* Ph.D. thesis (Michigan State University, 1970), pp. 43-44, or "The impact of the marketing board on the distribution of cocoa incomes in Nigeria", in *Nigerian Geographical Journal,* 1972, Vol. 15.

idem and A. F. Mabawonku: *Determinants and implications of rural-urban migration: A case study of selected communities in Western Nigeria,* African Rural Employment Paper No. 10 (East Lansing, 1974).

D. Forde and R. Scott: *The native economies of Nigeria* (London, 1948).

R. Galletti, K. D. S. Baldwin and I. O. Dina: *Nigerian cocoa farmers* (Oxford, 1956), pp. 148, 151, 211-214, 414, 443, 452, 458-459, 639-646 and 652-654.

A. D. Goddard, J. C. Fine and D. W. Norman: *A socio-economic survey of three villages in the Sokoto close-settled zone: 1, land and people; 3, input-output study;* Samaru Miscellaneous Papers, Nos. 37 and 65 (Zaria, 1971 and 1976), Vol. 2, tables.

M. M. Green: *Land tenure in an Ibo village* (London, 1947).

A. T. Grove: *Land and population in Katsina province with special reference to Bindawa village in Dan Yusufu district, 1952* (Kaduna, 1957), p. 21.

J. S. Harriss: "Some aspects of the economics of sixteen Ibo individuals", in *Africa,* 1944, No. 5.

P. Hill: *Rural Hausa: A village and a setting* (London, Cambridge University Press, 1972), pp. 105-108, 232-238.

idem: *Population, prosperity and poverty: Rural Kano 1900 and 1970* (London, Cambridge University Press, 1977), p. 122.

B. Huizinga: *An experiment in small farmer development administration among the Hausa of Nigeria*, draft (Zaria, 1979).

R. King: *Farmers co-operatives in Northern Nigeria* (Zaria and Reading, 1976), p. 28.

V. Y. Kohlhatkar: *Farm management survey report on Makarfi district* (Kaduna, 1967).

D. Kohnert: "Rural class differentiation in Nigeria—Theory and practice: A quantitative approach in the case of Nupeland", in *Afrika Spectrum*, 1979, No. 3.

J. Lagemann: *Traditional African farming systems in Eastern Nigeria* (Munich, 1977).

H. A. Luning: *An agro-economic survey in Katsina province* (Kaduna, 1963), p. 53.

idem: *Economic aspects of low-income farming* (Wageningen, 1967), p. 84.

A. Martin: *The oil palm economy of the Ibibio farmer* (Ibadan, 1956), p. 30.

P. J. Matlon: *The size distribution, structure and determinants of personal income among farmers in the North of Nigeria*, Ph.D. thesis (Cornell University, 1977), and *Income distribution among farmers in Northern Nigeria: Empirical results and policy implications*, African Rural Economy Paper No. 18 (1979).

D. W. Norman: *An economic survey of three villages in Zaria Province: 1, land and labour relationships; 2, input-output study*, Vol. 2, tables, Samaru Miscellaneous Papers 19 and 38 (Zaria, 1971).

J. K. Olayemi: "Peasant cocoa production in Western Nigeria: An economic analysis", Ph.D. thesis (McGill University, 1970), or idem: "Some economic characteristics of peasant agriculture in the cocoa belt of Nigeria", in *Bulletin of Rural Economics and Sociology*, 7 Mar. 1972.

S. A. Oni: *An economic analysis of the provincial and aggregate supply responses among Western Nigerian cocoa farmers*, Ph.D. thesis (University of Ibadan, 1971).

A. Osuntogun: "Poverty as an issue in rural development policy: A case study from the Western State of Nigeria", in *Poverty in Nigeria*, op. cit.

A. U. Patel, J. A. Ekpere, C. E. Williams and J. A. Akinwumi: "A socio-economic survey of a rural community in Western Nigeria", in *Journal of Rural Economics and Development*, 9 Jan. 1974, p. 27.

F. J. Pedler: "A study of income and expenditure in Northern Zaria", in *Africa*, Vol. 18, 1948, p. 267.

M. J. Purvis: *Report on a survey of the oil palm rehabilitation scheme in Eastern Nigeria*, Consortium for the Study of Nigerian Rural Development 10 (East Lansing, 1968).

Rural Economy Research Unit: *Farm income levels in Northern Nigeria* (Zaria, 1970; mimeographed).

M. G. Smith: *The economy of Hausa communities* (London, 1955), pp. 146, 150, 178-180, 226-227.

M. Upton: "Agriculture in Uboma", in H. A. Oluwasanmi, I. S. Dema et al.: *Uboma: A socio-economic and nutritional survey of a rural community in Nigeria*, World Land Use Survey Occasional Papers 6 (Bude, 1966).

idem: *Agriculture in south-western Nigeria* (Reading, 1967).

E. J. Usoro: "Producer prices and rural economic activity: A case study of two Itak villages in the South-Eastern State of Nigeria", in *Nigerian Journal of Economics and Social Studies*, 14 Feb. 1973, pp. 161-163.

H. van den Driessen: "Patterns of land holding and land distribution in the Ife division of Western Nigeria", in *Africa*, 1971, No. 1.

A. H. S. Vigo: *Survey of agricultural credit* (Kaduna, 1965).

T. Wallace: *Rural development through irrigation: Studies in a town on the Kano River Project* (Zaria, 1979).

STAGNATION AND INEQUALITY IN GHANA
Assefa Bequele

8

Ghana embarked upon its independence with a number of advantages, compared with most African and many developing countries.[1] To be sure, structurally it had in many ways a typically underdeveloped economy. Agriculture was the major source of income and wealth, contributing in 1955 approximately 50 per cent of GDP[2] and supporting a much larger proportion of the population. Its industrial base was extremely low and the country therefore depended on imports for most of its capital and many of its consumer goods. The economy was characterised by a technological dualism: traditional labour-intensive productive techniques coexisted with modern capital-intensive ones, especially in the industrial sector and, to a lesser extent, in agriculture.[3] It was an economy in which foreign trade constituted around 54 per cent of GDP.[4] A single crop—cocoa—contributed about 13 per cent of GDP and provided about three-fifths of the total export earnings.

But if Ghana's economy was structurally similar to that obtaining in many parts of the developing world, it also possessed certain features which gave it, to use Szereszewski's words, "a privileged place among the so-called underdeveloped or developing economies".[5] In the first place, its income per head in 1950 (equivalent to US$354 in 1974 dollars) was the highest in sub-Saharan Africa.[6] Second, although the density of population was higher than in many African countries, Ghana was endowed with an abundant supply of cultivable land which, given the absence of any labour constraint at that stage, provided possibilities for rapid increases in agricultural production. Third, Ghana had an aggressive and vigorous group of indigenous entrepreneurs in the cocoa sector who were capable of responding favourably to new economic opportunities.[7] Fourth, Ghana was endowed with a much larger stock of educated and skilled manpower than any other country in tropical Africa.[8] Fifth, it had, by the end of the 1950s, a large stock of capital[9] and foreign exchange, and was thus relatively unfettered by the constraints usually faced by developing countries.

Yet, 25 years later, the structure of its economy had changed for the worse and economic growth had ground to a halt. This performance is striking not only in the context of Ghana's own historical experience but also when viewed in relation to the experience of other developing countries. Furthermore, Ghana's society had also become more unequal over the same period. Given the over-all tendency towards stagnation, the rise in inequality must have meant a rise in absolute poverty as well, although no direct evidence on this is available. Here, then, is a society which, in spite of its considerable initial advantages, not only failed to sustain its momentum of growth but also experienced growing inequality and poverty.

The process underlying these developments forms the subject-matter of this chapter, which is organised as follows: in the first section we take a close look at the basic characteristics of the growth process in Ghana; next we review the evidence on trends in income inequality and poverty in rural areas; the third section draws together the separate observations, examines their internal consistency and formulates some explanatory hypotheses; and we present our conclusions in the last section.

CHARACTERISTICS OF THE GROWTH PROCESS IN GHANA

Structural changes in Ghana's economy over the past quarter of a century, though substantial, were fundamentally unhealthy. The figures given in table 77 are eloquent. Even though these data have been assembled from various sources and consequently show some minor inconsistencies, over-all tendencies are clear enough. The majority of the population continues to live in rural areas, and the bulk of the labour force continues to be engaged in agriculture. The share of agriculture in GDP has, however, declined sharply, and it should be noted that this is attributable, not to a growth of industries, but rather—as will become clear from further discussion below—to a sharp decline in agricultural production. While the share of industry in GDP has in fact remained roughly constant, the share of services has increased considerably. Perhaps the most striking change, however, is the disastrous decline, over the period 1970 to 1977, in the share of exports in GDP, due partly to a decline in production and partly to the smuggling across the border of cocoa, the crop that has always accounted for the bulk of Ghana's foreign exchange earnings. The change reflects a basic malaise in Ghana's economy, as we shall see below.

The growth process underlying these structural changes is clearly brought out in table 78. Once again the data have been assembled from diverse sources, and there are some minor inconsistencies; nevertheless, a consistent qualitative picture does emerge.

If one considers the period 1950-77, income per head would seem to have stagnated. This impression, however, is somewhat misleading. Positive growth was actually confined only to the period between 1950 and the early

Table 77. Ghana: some basic indicators

	1950	1955	1960	1970	1976	1977
GDP per head[1] (Cedis in 1968 prices)	182.4	186.8	219.8	222.4	196.2	.[3]
Percentage share in GDP[2] of:						
Agriculture, livestock, forestry and fishing	.	50.4	51.8 (40.8)	42.4 (46.9)	36.4	39.0*
Cocoa production	.	12.4	16.0	11.0	7.6	.
Industrial production	.	21.7	22.4 (18.6)	21.7 (17.8)	20.4	22.0*
Services	.	27.9	25.8 (40.6)	35.8 (35.3)	43.2	39.0*
Share of exports in GDP (%)	23.6	19.9	21.6	21.3	.	8.0*
Share of imports in GDP (%)	19.1	28.2	29.9	25.2	.	.
Percentage of population in rural areas[4]	.	.	77.0	.	68.0	.
Percentage of labour force:[4]						
in agriculture	.	.	64.0	.	.	54.0
in industry	.	.	14.0	.	.	19.0
in services	.	.	22.0	.	.	27.0

. = not available.

Sources. [1] Robert Szereszewski: "The performance of the economy", in Walter Birmingham, I. Neustadt and E. N. Omaboe (eds.): *A study of contemporary Ghana* (London, Allen and Unwin, 1966); Central Bureau of Statistics: *Economic Survey, 1969-71* (Accra, 1976); World Bank: *Ghana Economic Memorandum* (Washington, DC, 1979). [2] Figures in parentheses are from World Bank: *World tables, 1976*. Figures with asterisks are from idem: *World Development Report, 1979*. The remaining figures have been calculated from data provided in sources cited in reference[1] above. [3] In this column the figures with an asterisk are from idem: *World Development Report, 1979*. Others are computed from data provided in idem: *World Tables, 1976*. [4] idem: *World Development Report, 1979*.

Table 78. Annual growth rates [1]

No. indicator	1950-60	1955-60	1960-70	1970-76
1. GDP				
(a)	4.1	–	2.1	0.4 [4]
(b)	–	5.1	–	–
(c)	–	–	–	0.9
2. Agriculture [2]				
(a)	4.3	–	3.7	–0.7 [4]
(b)	–	5.7	–	–
(c)	–	–	–	–1.2
(d)	–	–	1.8	–
3. Food production	–	–	1.8	–0.1
4. Cocoa production				
(a)	–	9.0	–	–
(b)	–	–	–	–3.0
(c)	1.8	5.6	–0.02	–1.9
5. Industrial production				
(a)	4.6 [3]	–	6.7	0.8 [4]
(b)	–	6.3	–	–
(c)	–	–	–	0.4
6. Services				
(a)	–	–	–1.4	1.6 [4]
(b)	–	3.0	–	–
(c)	–	–	–	3.6
7. Population	2.2	–	2.4	3.0 [4]
8. GDP per head				
(a) (corresponding to 1 (a))	1.9	–	–0.3	–2.6 [4]
(b) (corresponding to 1 (b))	–	2.9	–	–
(c) (corresponding to 1 (c))	–	–	–	–2.1
9. Food production per head	–	–	–0.6	–3.1

– = not applicable.
[1] The growth rates, which are all in real terms, have been estimated by fitting semi-logarithmic regression equations to annual data. [2] Agriculture includes agriculture, livestock, fisheries and forestry. [3] Refers to the period 1970-77. [4] Refers to manufacturing only.

Sources. For 1 (a), 2 (a), 5 (a), 6 (a) and 7, World Bank: *World Tables, 1976*, and idem: *World Development Report, 1978 and 1979*; for 1 (b), 2 (b), 4 (a), 5 (b) and 6 (b), Szereszewski, op. cit.; for 1 (c), 2 (c), 4 (b), 5 (c) and 6 (c), *Economic Survey, 1969-71*, and World Bank: *Ghana economic memorandum*; for 2 (d) and 3, FAO: *The Fourth World Food Survey* (1977); for 4 (c), Gill and Duffus Group: *Cocoa Statistics* (London).

1960s. By about 1963 the economy had reached its peak in terms of income per head, and thereafter a process of steady decline became apparent. Thus growth rate per head steadily declined between 1950 and 1977. Indeed, over the period 1970-77, this was between −2.1 and −2.6, a growth performance which must be considered exceedingly poor even in terms of the experience of the developing world. Thus Ghana, which had entered the 1960s with the highest income per head in sub-Saharan Africa, was quickly superseded by other countries (see table 79). Although several factors were no doubt responsible for this state of affairs, the stagnation and decline of the agricultural sector and the consequent food and foreign exchange constraints appear to have been the most decisive. In order to understand this, it will be necessary to go into some detail.

Between 1955 and 1960 GDP increased at an annual rate of 5.1 per cent. At sectoral levels both agriculture and industry recorded comparable growth rates. Moreover, the high growth rate of agriculture, naturally enough, was largely due to a high growth rate in cocoa production. Between 1960 and 1970 GDP grew at a rate lower than that of the population.[10] At a disaggregated level both agricultural and food production grew at a rate lower than that of population, and cocoa production stagnated. Somewhat surprisingly, industrial production continued to grow at a relatively high rate. In the 1970s Ghana's economy appears to have been in the midst of a deep recession: agricultural, food and cocoa production declined absolutely, and growth of industrial production virtually ground to a halt; only the service sector would appear to have grown at all.

In short, in the 1960s industrialisation seems to have proceeded in the face of a near-stagnation in agriculture and a decline in food production per head. The serious consequences of this type of unbalanced growth became evident in the 1970s when Ghana's economy was set on a retrogressive course.

The considerable increase in cocoa and agricultural output in the late 1950s made possible the rapid growth in GDP, which extended into the early 1960s, in two ways: directly by contributing to output and employment, and indirectly as a source of foreign exchange and hence of investment. Ghana did not have a foreign exchange constraint in the 1950s. In 1960 its foreign exchange reserves were equivalent to 90 per cent of imports; however, this had dwindled to 22.0 per cent by 1965, and further to 11.5 per cent in 1970. The latest available estimate assesses it at 14.2 per cent in 1977 (see table 80). This almost continuous decline in Ghana's foreign exchange reserves was due not to increases in imports (in fact, imports declined in the 1960s and had a slow growth in the 1970s) but to the very slow growth and later decline in exports, which in turn was related to the decline in cocoa output.

The above observations are well supported by the data presented in table 81. The rate of growth of exports declined steadily from 3.2 per cent over the period 1950-60, to 0.1 per cent over 1960-70 and further to minus 1.9 per cent over 1970-77. The performance of exports, moreover, was clearly correlated to trends in cocoa exports, on the one hand, and to those in domestic investment, on the other. Indeed, there are reasons to believe that the serious foreign exchange constraint not only induced a deceleration in industrial growth but also hindered in some measure even the growth of the agricultural sector itself.[11]

The other factor that influenced the course of domestic investment was the sharply declining trend in food production per head. Such a decline must

Agrarian policies and rural poverty in Africa

Table 79. Income per head, Ghana and selected African countries, 1950 and 1975

Country	Income per head			
	In terms of 1974 US$		Ratio between Ghana and other countries	
	1950	1975	1950	1975
Ghana	354	427	1.00	1.00
Congo	303	460	1.17	0.93
Ivory Coast	283	460	1.25	0.93
Morocco	353	435	1.00	0.98
Zambia	310	495	1.14	0.86
Nigeria	150	287	2.36	1.48
Senegal	238	341	1.49	1.25
Africa	*170*	*308*	*2.08*	*1.38*

Source. Derived from David Morawetz: *Twenty-five years of economic development, 1950 to 1975* (Washington, DC, World Bank, 1977), table A1, pp. 77-78.

Table 80. Ghana: gross foreign exchange reserves as percentage of annual imports

Year	Percentage
1960	90.0
1970	11.5
1977	14.2

Sources. World Bank: *World Tables, 1976*; and idem: *World Development Report, 1979*.

Table 81. Average annual growth rates (in real terms)

Item	1950-60	1960-70	1970-77
Gross domestic investment [1]	8.9	−3.2	−8.6
Total imports [1]	8.9	−1.6	2.0
Total exports [1]	3.2	0.1	−1.9
Cocoa exports [2]	5.4*	−1.2	−4.1 (−3.8)

Sources. [1] World Bank: *World Tables, 1976*; idem: *World Development Report, 1979*. [2] Estimated on the basis of data provided in FAO: *Trade Yearbook*, various issues. The data refer to cocoa beans only. The figure with an asterisk refers to the period 1955-60. The figure in parentheses refers to cocoa beans, paste and cake.

have seriously restricted the growth of marketed surplus of food and thus the growth of investment. Surprisingly little effort was made to circumvent this problem through increased imports of food, as the data in tables 81 and 82 testify.

Thus the evident economic crisis of the 1970s had its roots in the agricultural crisis of the 1960s. The production of food and cocoa, the two basic sources of economic growth in Ghana, simply failed to increase, with disastrous consequences not only for over-all economic growth but also, as we shall shortly see, for income distribution and consequent aggravation of poverty.

POVERTY AND INEQUALITY

The discussion of levels and trends in incomes in Ghana is rendered difficult by the acute scarcity of distributional data relating either to a specific time or to a period of time. There are no poverty studies or income distribution data for the nation as a whole, or for individual rural and urban areas. We are therefore compelled to look at some rather sketchy data and, on this basis, draw some tentative conclusions.

The first set of data that may be considered relate to the distribution of money incomes among two important groups: wage and salary earners, and cocoa producers. These are presented in table 83. Some explanations and caveats are needed on the data relating to cocoa producers. The 1963-64 data are based on estimates by Beckman, who used the records of the Cocoa Marketing Board to determine the share of the various groups of cocoa farmers in total cocoa output; to this, Ewusi applied producer prices and arrived at the pattern of income distribution shown in the table.

These data suggest, quite unambiguously, a worsening trend in the distribution of money incomes among both groups, though the trend is sharper for cocoa producers than for wage and salary earners. However, they understate the increase in real inequality, the reason for this being that food prices increased much faster than other prices over the relevant period, as will shortly become clear. Under such conditions, the degree of worsening in the distribution of real incomes is necessarily higher than that implied by changes in the distribution of money incomes, since a rise in the relative price of food was more seriously felt by the poorer section of the population, who spend a high proportion of their incomes on food.

Unfortunately, no distributional data are available for a third important group, namely the food producers. Some conclusions, however, can be extrapolated from the data on land distribution among food producers, such analysis being facilitated by the fact that in Ghana cocoa production is almost entirely confined to the southern part of the country, while the cultivators in the north are basically food producers. Inequality in land distribution in the north can thus be taken as a proxy, albeit imperfect, for the inequality in

Table 82. Cocoa exports and food imports, 1959-77

Period	Cocoa exports as percentage of total merchandise exports	Food imports as percentage of total merchandise imports
1959-61	58.6	.
1962-64	.	16.6
1969-71	64.6 (65.4)	13.6
1975-77	59.2 (61.5)	9.8

. = not available.
Source. Calculated from data provided in FAO: *Trade Yearbook*, various issues. For cocoa exports, figures in parentheses refer to cocoa beans, paste and cake, and others refer to cocoa beans only.

Table 83. Distribution of money income in Ghana and other countries: various groups and years

Group	Year	Share of lowest 20%	Share of lowest 40%	Share of highest 20%
Wage and salary earners [1]	1956	11.5	24.9	36.6
	1962	10.9	23.6	39.2
	1965	10.7	23.2	40.2
	1968	10.1	22.1	41.9
Cocoa farmers:				
Share in cocoa earnings only	1963-64 [1]	2.4	8.0	56.4
Share in cocoa earnings only	1970 [2]	1.8	7.5	62.5
Share in total income	1970 [2]	5.4	14.8	50.8

[1] Adapted from Kodwo Ewusi: *Economic inequality in Ghana* (Accra, ISSER, 1977), pp. 20-74. [2] T. K. Buxton: *The distribution of cocoa income among the producers: Some policy implications* (University of Cape Coast, Centre for Development Studies, 1976).

income distribution among food producers. Before we examine the data, however, a few remarks concerning the evolution of the land system in Ghana are in order.

Notwithstanding minor differences between regions, traditionally the predominant agricultural system in Ghana was the bush fallow (or land rotation) system within the framework of communal land tenure.[12] Land belonged to the community or landownership group. However, individual members held definitely ascertainable rights, so that the individual who first cleared and cultivated a patch of land often established for himself the right to it.[13] Such an individual would have to cultivate the land on a more or less continuous basis in order to establish a permanent right. Nevertheless, even within this system the sale of land could still be effected and the individual could pledge its use to meet debt obligations, so long as he has the consent of the landowning group.

It was not difficult for "strangers" (those who did not belong to the landowning group) within this system to have access to land. The sole prerequisite was for them to have resided within the group for a period long enough to have established a name of good standing, after which they were granted land, freely or at a nominal fee, for use exactly like any other member of the community.

An integral aspect of this agricultural system was that of land rotation.[14] The low nutrient value below the level of the topsoil, and the importance of the natural protection afforded to it by forests or any other form of natural vegetation, permitted only a relatively low degree of agricultural exploitation. The farmer was thus obliged to leave some land fallow at any one time while clearing and cropping new or "restored" farms. This pattern of farming was facilitated especially in the early years by the "abundance of virgin forest land whose cost to the farmers moving into it was either zero, in the case of those having traditional rights to it, or very low in the case of 'strangers' having no such claims".[15]

However, the introduction of cocoa in the closing decades of the nineteenth century brought to greater prominence the important role of land and labour as sources of Ghanaian agricultural growth, while it simultaneously initiated major changes in the pattern of landownership and introduced new sources of social differentiation. The expansion of cocoa production was facilitated by the emergence of a stratum of vigorous rural entrepreneurs who reinvested the surplus from cocoa in new land and labour and continued to expand into areas where suitable land could be purchased or leased.[16] This led to widely differing types of holdings, from small family farms near existing villages to large plantations with numerous labourers deeper into the forest. As Beckman noted:

New relations were ... created over land and labour replacing a system dominated by family production for the individual household within a framework of communal land tenure. Private or semi-private property rights in land were extended throughout the cocoa areas. Land was bought and sold. Farms were mortgaged and auctioned. Farm labourers were hired even on quite small holdings, most commonly as sharecroppers on mature farms, receiving one-third of the crop, the so-called *abusa*. Wage labour was also used extensively, especially when establishing new farms. A growing number of farmers became tenants outside the areas where they could claim a share in communal lands. Rents were sometimes merely minor tributes but could rise to one-third of the crop or more.[17]

Moreover, the growth of commerce and trade as well as the associated increase in the demand for credit and transport further reinforced the forces that generated social differentiation.

Especially important for an understanding of the dynamics of social differentiation in the rural economy of Ghana was the impact of cocoa on capital accumulation and property rights, as well as that of regional inequality on the labour market and, indeed, on the generation of surpluses. The system of communal ownership of land was appropriate so long as production was directed toward self-consumption, and the relevant cash crop was of short gestation and involved little investment. The system, however, broke down when a cash crop was introduced which matured only after a long period of time. Cocoa was one such crop, and the predominant variety in Ghana took five to ten years to mature, and longer still before a cultivator could see a substantial yield. As the initial clearing as well as the successive operations of maintaining the farm required a great deal of labour, the long years of waiting therefore represented substantial labour investment on the farm. In the event, therefore, land began to acquire especial value, and property rights assumed especial significance. At the same time, labour assumed greater value since it was required to look after the farms of persons who, in addition to owning several farms, had other economic interests. This was, as Beckman called it, peasant capitalism "both in the important role of family labour as distinct from cash investment in capital formation, and in the landlord-tenant aspect of share-crop labour".[18]

Two further important elements that contributed to greater social differentiation were, on the one hand, inequality in access to land resulting from ecological factors and, on the other, inequality in regional development. Ghana is divided into two almost distinct ecological zones, the forest zone in the south and the savannah in the northern parts of the country. It is only in the southern forest zones that land is suitable for cocoa production: inequality in access to "appropriate" land—appropriate, that is to say, in market terms—was thus introduced. As the value of land in the south rose, it became increasingly difficult for those from the north to have access to it. On the other hand, increasing demand for labour made employment opportunities in the south more attractive than farm self-employment in the north. An active labour market was thus created which, by providing cheap labour, enabled a small proportion of farmers to establish a succession of new farms and expand their holdings, thereby leading to increased social differentiation in rural Ghana.

If the data on distribution of land can be relied upon, it appears that this process had over the years generated considerable inequality in wealth and income in the rural sector. This can be seen from table 84, which presents an "interpretation" of land distribution in Ghana, based on the 1970 agricultural census and on certain assumptions regarding distribution in each size class. The 1970 agricultural census provided data on the number of holdings in different size classes; it did not, however, present the total acreage in each size class. Table 84 was therefore constructed on the assumption that farm sizes are distributed normally within each class, and farms of over 50 acres were assumed to constitute the residual category.

As can be seen from the data in table 84, land distribution in Ghana is very unequal. It is also apparent that it is more unequal in the south (i.e. among cocoa producers) than in the north (i.e. among food producers). Confining our attention to the north, we observe that 67 per cent of the farming

Table 84. National and regional size distribution of holdings in Ghana, 1970[1]

Size of holdings (acres)	Holdings Number	%	Holdings (cumulative) (%)	Land within each group (%)	Cumulative land area (%)
National					
0-1.9	246 100	31	31	3.7	3.7
2.0-3.9	194 200	24	55	9.0	12.7
4.0-5.9	105 200	13	68	8.2	20.9
6.0-7.9	71 800	9	77	7.8	28.7
8.0-9.9	42 100	5	82	5.9	34.6
10.0-14.9	55 000	7	89	10.8	45.4
15.0-19.9	31 600	4	93	8.7	54.7
20.0-29.9	27 200	3	96	10.7	64.8
30.0-49.9	17 900	2	98	11.2	76.0
50 or more	14 100	2	100	24.0	100.0
Gini coefficient = 0.64129					
South					
0-1.9	217 700	35	35	4.0	4.0
2.0-3.9	137 100	22	57	7.9	11.9
4.0-5.9	71 600	11	68	6.9	18.8
6.0-7.9	48 200	8	76	6.5	25.3
8.0-9.9	32 300	5	81	5.6	30.9
10.0-14.9	42 500	7	88	10.3	41.2
15.0-19.9	23 900	4	92	8.1	49.3
20.0-29.9	22 600	4	96	11.0	60.3
30.0-49.9	16 400	2	98	12.7	73.0
50 or more	13 000	2	100	27.0	100.0
Gini coefficient = 0.66883					
North					
0-1.9	28 400	16	16	2.2	2.2
2.0-3.9	57 100	32	48	13.8	16.0
4.0-5.9	33 600	19	67	13.7	29.7
6.0-7.9	23 600	13	80	13.5	43.2
8.0-9.9	9 800	5	85	7.2	50.4
10.0-14.9	12 500	7	92	12.8	63.2
15.0-19.9	7 700	4	96	11.0	74.2
20.0-29.9	4 600	3	99	9.4	83.6
30.0-49.9	1 500	1.0	100	4.9	88.5
50 or more	1 100			11.5	100.0
Gini coefficient = 0.51774					

[1] The first three columns are derived from the *Report on Ghana Sample Census of Agriculture, 1970*. The rest are derived by assuming that farm sizes are normally distributed within each class, and farmers with average size of 50 acres or more are estimated as residual.

households operated, in 1970, less than 6 acres of land. Many of these households are likely to have been deficit farmers: in other words, their food production fell short of their consumption requirements. They were therefore likely to be net food purchasers.

The significance of this fact is that rising food prices are likely to have adversely affected this section of the food producers. Rising food prices, on the other hand, could not but benefit the surplus food producers operating large farms. The extent of such benefits would naturally vary directly with the volume of surplus. Thus, even in the absence of any change in land distribution, inequality of income among food producers must have increased. Unfortunately, it is not possible to discern the time trends in land distribution for lack of data. However, the logic of the situation rules out the possibility of land distribution becoming more equal: if it changed at all, it could only have been in the direction of greater inequality.[19]

The deficit farmers would of course be the ones who would seek wage employment in agriculture in order to supplement their incomes from farming. Indeed, according to the estimates presented in table 85, in 1970 about 31 per cent of the rural labour force depended primarily on wage employment for their livelihood in Ghana as a whole. Undoubtedly, there were others who depended on wage employment as a secondary source of income.

In this context, trends in agricultural wage rates would have a bearing on the standard of living of the deficit farmers. If money wage rates kept up with food prices, they would have been able to maintain their absolute standard of living, even though relative inequality among food producers would still increase. However, as the data in table 86 reveal, money wages actually increased at a much slower rate than food prices.

When we piece together the various observations, the conclusion seems inescapable that economic inequality in Ghana's rural areas increased quite significantly. This, combined with the fact that GDP per head itself declined, clearly implies a significant rise in rural poverty. Unfortunately, we cannot provide any direct evidence in support of this probability since the relevant data are simply not available.

TOWARDS A PERSPECTIVE ON GHANA'S ECONOMIC PROBLEMS

It is clear that Ghana's economy experienced a process of retrogression, engendering a growth of economic inequalities, in the 1970s. A careful analysis shows that both these tendencies originated from the same source, namely the decline of the agricultural sector. Indeed, such decline temporarily preceded the decline of the economy as a whole, and it is very probable that the root causes of the economic difficulties of the 1970s lay in the pattern of growth obtaining in the 1960s. As the foregoing analysis has indicated, two basic constraints underlie Ghana's economic malaise: a constraint on food and one on foreign exchange. It is possible to hypothesise, however, that the

Table 85. Employment in agriculture, 1960 and 1970 (thousands)[1]

Type of employment	1960		1970	
	No.	%	No.	%
Total employment	2 559.4	–	3 133.0	–
Employment in agriculture	1 581.3	100	1 790.7	100
Self-employed (peasant farmers)	1 059.6	67	1 238.2	69
(of which: cocoa farmers)	(315.3)	(20)	(386.6).	(22)
Agricultural employees	521.7	33	551.1	31
(Temporary)	–	–	(284.5)[2]	–
(Permanent)	–	–	(216.6)[2]	–
(Employees in large establishments)	–	–	(50.0)[3]	–
Total employment in cocoa production[4]	522.1	33	573.3 .	32
Total employment in forestry + fishing	79.3	5	75.2 .	4
Total employment in field crops + foodstuff production	979.8	62	1 142.2	64

– = not applicable.

Sources. These estimates are derived from various sources.

[1] From the 1960 and 1970 population censuses and cited in Ministry of Economic Planning, Manpower Division: *Data book on manpower resources in Ghana* (Sep. 1974). Vol. 1, tables A-52 and A-58; the definition of "the employed" according to the population census is very broad. It includes, among others, "all those who are in regular employment during the four weeks before Census Night...", "all those who worked for at least one day for pay or profit during the four weeks before Census Night", etc. [2] *Ghana Sample Census of Agriculture, 1970*. According to the census, a permanent labourer is one who has been employed on the holding for a total period of at least six months during the last year; a temporary labourer is one who has been employed for a period of at least one month but less than six months during the last year. [3] *Quarterly Digest of Statistics*, Mar. 1972. [4] These are estimated on the basis of Steel's data, which state that employment in cocoa and forestry and fishing was 20.4 per cent and 3.1 per cent respectively in 1960 and 18.3 per cent and 2.4 per cent respectively in 1970. See William F. Steel: *Analysis of changes in labour demand and supply in Ghana, 1960-1970* (Washington, DC, World Bank; mimeographed).

food constraint became binding in advance of the foreign exchange constraint, and that the former probably precipitated the latter. However, before the plausibility of such a hypothesis can be explored, a few other possible explanatory hypotheses need to be considered.

With reference to the emergence of a foreign exchange constraint in particular, it is necessary to examine the extent to which external factors may have been responsible for Ghana's dismal performance in the past two decades. It seems evident, however, that the usual problem deriving from unfavourable changes in the prices of exports from monoculture developing economies has not been encountered by Ghana. As the data in table 88 reveal, export prices of cocoa rose quite sharply in the 1970s, precisely the period when Ghana's exports of cocoa were declining. The foreign exchange constraint facing Ghana could not, therefore, be attributed to the vagaries of international markets. This conclusion is reinforced when one looks at the performance of other cocoa-producing countries. It is particularly instructive to contrast Ghana's performance with that of the Ivory Coast during the

Table 86. Indices of incomes and prices and relative changes, 1956-76

Item	1956	1957	1958	1959	1960	1961	1962	1963	1964	1965	1966	1967	1968	1969	1970	1971	1972	1973	1974	1975	1976
Money incomes																					
1. Wages in agriculture (male)	62	63	68	72	79	66	96	100	111	113	111	111	131	146	145	159	165	173	.	.	.
Consumer prices																					
2. Local foods	.	.	.	.	.	.	.	100	126	173	200	170	184	200	210	236	259	313	363	474	770
3. Other items	.	.	.	.	.	.	.	100	125	128	141	143	154	162	166	173	191	216	264	339	499
4. All items	79	79	79	82	82	87	96	100	120	151	171	157	170	182	189	206	227	265	315	409	639
Real incomes																					
5. GDP per head	82	.	.	.	96	97	98	100	100	98	92	92	90	93	97	102	99	98	100	93	96
6. Private consumption per head	81	.	.	.	100	105	97	100	94	92	87	94	87	93	100	102	97	.	.	.	.
7. Real wages in agriculture	78	79	86	88	96	76	100	100	99	75	65	71	77	80	77	77	73	65	.	.	.
8. Real minimum wages	100	100	108	104	122	115	104	100	83	66	58	69	68	64	61	56	68	59	64	.	.
9. Real earnings in non-agriculture	82	91	92	92	100	106	101	100	94	69	70	83	87	85	91	88	91	71	.	.	.
Relative changes																					
10. Relative prices of food[1]	.	.	.	.	.	.	.	100	101	135	141	118	120	124	127	136	136	145	137	140	155

. = not available.
[1] This was derived by dividing row 2 by row 3.
Sources: ILO: *Year Book of Labour Statistics* (Geneva), various issues; Central Bureau of Statistics: *Economic Survey, 1969-71* (Accra, 1976).

Table 87. Ghana and Ivory Coast: export volume indices (1973 = 100)

Exported product	1967		1974		1976	
	Ghana	Ivory Coast	Ghana	Ivory Coast	Ghana	Ivory Coast
All exports	80.2	63.6	77.6	124.6	51.1	125.9
Cocoa	89.3	73.4	83.5	143.4	48.8	136.3
Timber-logs	46.5	62.1	50.8	86.3	36.7	93.3

Source. World Bank: *Ghana economic memorandum*, op. cit., p. 18.

1970s. It is clear from table 87 that, although both were faced with a similar world market situation, their performances were vastly different. It can be argued, therefore, that the root cause of Ghana's disheartening performance with respect to production and export of cocoa lay in the poor domestic management of the economy, rather than in unfavourable trends in the world economy. Of critical importance were declining producer prices for cocoa, heavy taxation of the cocoa sector, lack of investment and, ultimately, the crisis of the food sector.

It can be observed straightaway that in Ghana farmers received a low price for cocoa. The price was low compared with *(a)* export prices in recent periods, and *(b)* producer prices in neighbouring cocoa-producing countries.

The movement of export and producer prices is summarised in table 88. As can be seen, the movement of producer prices has not always followed the movement of export prices over the period under consideration. During the 1950s and the 1960s producer prices remained largely unresponsive to export prices. In the 1970s export prices increased very sharply, but producer prices, while responding to export prices, rose at a far slower rate. Consequently, the gap between the two increased very considerably, and producers received only a small share of the benefits flowing from rising export prices.

Producer prices in Ghana in recent years were not only low in comparison with export prices; they were also low when compared with producer prices in neighbouring countries. As can be seen from table 89, up to 1967 producer prices were comparable in all the three countries: Ghana, Nigeria and the Ivory Coast. Over the period 1968-76, however, they rose much faster in both the Ivory Coast and Nigeria than in Ghana. In the 1970s, therefore, cocoa producers in Ghana were clearly receiving a significantly lower price than their counterparts in the other two countries.

In reality, the difference in cocoa prices between Ghana and her neighbours was much greater than is suggested by the figures in table 89. Ghana's currency is notoriously over-valued. Its foreign exchange value in the black market could be as low as one-eighth (or even less) of the official exchange

Table 88. Cocoa exports and producer prices, 1950-76
(£ per tonne)

Year	Export price	Producer price	Ratio of export to producer prices
1950	205.2	82.7	248
1951	280.9	128.6	218
1952	296.9	146.7	202
1953	283.1	128.4	220
1954	459.6	132.3	347
1955	297.2	140.9	211
1956	218.1	146.9	148
1957	243.3	138.4	176
1958	346.9	132.3	262
1959	280.9	117.6	239
1960	222.3	110.2	202
1961	177.1	110.2	161
1962	167.4	110.2	152
1963	204.9	108.4	189
1964	187.7	99.2	189
1965	138.4	90.9	152
1966	193.2	75.8	255
1967	238.0	85.0	280
1968	319.5	101.2	316
1969	415.5	113.7	365
1970	305.5	119.9	255
1971	232.4	119.9	194
1972	270.5	103.9	260
1973	585.4	135.4	432
1974	990.1	180.2	549
1975	722.7	226.3	319
1976	1 399.4	327.1	428
Average annual growth rates:			
1950-60	−0.16	0.93	−1.06
1960-70	6.83	−0.42	7.27
1970-76	33.70	18.84	12.49

Source. Gill and Duffus Group: *Cocoa Statistics* (London, Dec. 1978).

rate, and thus the producer prices of cocoa in Ghana, when expressed in sterling equivalents using the official exchange rate (as was done in constructing the figures presented in table 89), seriously overstate them. It is little wonder, then, that Ghana's cocoa producers found it profitable to smuggle their produce to neighbouring countries which offered better prices. It was, for example, estimated that in 1977 a minimum of 40,000 tons of cocoa were smuggled out of the country.[20] This amounted to well over 12 per cent of the

Table 89. Cocoa producer price: Ghana, Ivory Coast and Nigeria, 1957-76
(£ per tonne)

Year	Producer price (£) and index			Price differences as percentage of Ghana price	
	Ghana	Ivory Coast	Nigeria	Ivory Coast ÷ Ghana	Nigeria ÷ Ghana
1957	138.40	136.22	147.63	−1.6	+6.7
1958	132.28	158.88	147.63	+20.1	+6.7
1959	117.38	130.66	150.91	+11.3	+28.6
1960	110.23	137.97	157.47	+25.2	+42.9
1961	110.23	138.20	108.96	+25.3	−1.2
1962	110.23	101.74	100.06	−7.7	−9.2
1963	108.39	102.01	104.98	−5.9	−3.1
1964	99.21	102.29	111.54	+3.1	+12.4
1965	90.94	102.15	104.57	+12.3	+15.0
1966	75.78	85.97	72.17	+13.4	−4.8
1967	85.02	103.59	90.22	+21.8	+6.1
1968	101.23	118.09	100.60	+16.7	−6.2
1969	113.74	117.47	143.53	+3.2	+26.2
1970	119.97	123.80	174.15	+3.2	+45.2
1971	116.92	126.23	177.93	+8.0	+52.2
1972	102.62	134.71	177.98	+31.3	+73.4
1973	135.65	170.30	201.19	+25.5	+48.3
1974	181.00	223.99	313.35	+23.8	+23.8
1975	226.29	367.51	443.39	+62.4	+95.9
1976	327.13	410.06	547.30	+25.4	+67.3

Source. Gill and Duffus Group: *Cocoa Statistics* (London, Dec. 1979).

total cocoa production in 1976-77[21] and was believed to have entailed a foreign exchange loss to the country of US$ 100 million.[22]

However, although prices in neighbouring countries might be of relevance to those cocoa dealers with the capacity to sell directly or indirectly across the border, for a large number of producers the relevant parameter would be the real purchasing power of cocoa proceeds within the domestic market. The centrality of this element in explaining stagnation in Ghana's cocoa industry can hardly be overstressed. It has already been observed that food prices rose much faster than other prices in Ghana. And when we compare the growth of food prices with that of producer prices for cocoa, the results are startling. The sharp decline in the relative price of cocoa in terms of food, as shown by the data in table 90, must have acted as a strong incentive for cocoa farmers to shift increasingly to food production.

Indeed, there is some evidence to suggest that such a shift did occur. The 1970 and 1974 agricultural censuses provide incomplete but interesting insights into the changing pattern of agricultural production in Ghana. As can

Table 90. Trends in relative price of cocoa

Year	Price of local food	Producer price of cocoa	$\dfrac{\text{Col. 2}}{\text{Col. 1}} \times 100$
1963	100	100	100
1964	126	90	71
1965	173	83	48
1966	200	69	35
1967	170	77	45
1968	184	92	50
1969	200	104	52
1970	210	109	52
1971	236	109	46
1972	259	94	36
1973	313	123	39
1974	363	164	45
1975	474	206	43
1976	770	297	39
Growth rates:			
1963-70	9.37	2.10	
1970-76	22.27	18.86	
1963-76	12.56	7.99	

Sources. Central Bureau of Statistics: *Economic Survey, 1969-71* (Accra, 1976); World Bank: *Ghana economic memorandum*, op. cit.; Gill and Duffus Group: *Cocoa Statistics* (London, Dec. 1978); Kodwo Ewusi: *Economic inequality in Ghana* (Accra, ISSER, 1977).

be seen from table 91, although the number of holders between 1970 and 1974 increased by 6.5 per cent, the number of those engaged in the cultivation of the major food crops increased by a much higher proportion. The number of maize growers, for example, rose by as much as 48.8 per cent. The only cases where there was a decline or a slow growth in the number of holders were those of yam and guinea corn. In any event, the conclusion that emerges from the data in table 91 is that there was a significant response to price changes as indicated by the turning of farmers toward the production of food crops. And though table 91 does not include data on cocoa, another source estimated that about 50,000 hectares of land under cocoa were going out of production every year.[23]

The picture becomes clearer when one considers changes in acreage and production at regional levels. We have attempted here to disaggregate some of the available data, and the result strongly suggests asymmetrical development between the north and the south. As can be seen in table 92, although the total number of agricultural holders increased by 6.5 per cent between 1970 and 1974, their number in the north declined by 7.8 per cent while in the south it rose by 10.7 per cent. This fact, combined with our earlier observations concerning the relative increase in the number of holders producing

Table 91. Holders engaged in crop production (selected crops), 1970-74

Crop	1970		1974		1970-74 change (%)
	Number ('000s)	As percentage of all holders	Number ('000s)	As percentage of all holders	
Maize	413.8	51	615.6	71.8	48.8
Millet	146.0	18	154.4	18.0	5.8
Guinea corn	164.4	20	153.1	17.8	−7.0
Cassava	477.9	59	.	.	.
Yam	208.4	26	214.5	25.0	2.9
Cocoyam	298.5	37	393.6	45.8	31.9
Groundnuts	118.6	15	172.7	20.1	45.6
Plantain	333.7	41	363.3	42.4	8.9
All holders:					
Ghana	805.2	100	857.7	100.0	6.5

. = not available.
Sources. *Report on Ghana Sample Census of Agriculture, 1970; Current Agricultural Statistics, 1974*, various tables.

Table 92. Number of holders by region, 1970 and 1974

Region	1970		1974		1970-74 % change
	Total	%	Total	%	
Western	68 100	8.5	76 500	8.9	12.3
Central	81 100	10.1	88 100	10.3	8.6
Eastern	148 200	18.4	181 700	21.2	22.6
Volta	108 600	13.5	115 500	13.5	6.4
Ashanti	147 700	18.3	146 400	17.0	−0.9
Brong-Ahafo	71 600	8.9	83 700	9.8	16.9
Northern	61 200	7.6	59 800	6.9	−2.3
Upper	118 700	14.7	106 000	12.4	−10.3
Total Ghana	805 200	100.0	857 700	100.0	6.5
All southern	625 300	77.7	691 900	80.7	10.7
All northern	179 900	22.3	165 800	19.3	−7.8

Sources. Based on data from *Report on Ghana Sample Census of Agriculture, 1970* and *Current Agricultural Statistics, 1974*.

food crops and the decline in acreage under cocoa, strongly suggests a tendency on the part of farmers in the south to switch from cocoa to foodcrops (we may recall that cocoa is produced almost exclusively in the south). This is indeed confirmed by the data in table 93, indicating that acreage under foodcrops in the south increased by about 22,000 acres between 1970 and 1974. If, as noted above, 50,000 hectares were going out of cocoa production every year, then it must be concluded that only a small proportion of land

Table 93. Area under cultivation and holders engaged in cultivation of selected crops by region, 1970 and 1974
(number of holders in thousands; area in thousand acres)

Crop	South			North		
	1970	1974	% change	1970	1974	% change
Maize						
Holders	337.8	535.7	58.6	76	79.9	5.1
Acreage	887.0	805.4	−9.2	231.0	245.0	6.0
Millet						
Holders	.	0.9	.	146.0	153.5	5.1
Acreage	.	4.0	.	615.0	545.1	−11.4
Guinea corn						
Holders	4.6	4.9	6.5	159.8	148.2	−7.3
Acreage	12.0	15.8	31.7	588.0	518.3	−11.9
Cassava						
Holders	463.2	.	.	14.7	.	.
Acreage	848.0	876.7	3.4	18.0	83.0	361.1
Yam						
Holders	129.4	159.1	23.0	79.0	55.4	−29.9
Acreage	248.0	208.9	−15.8	178.0	119.0	−33.1
Plantain						
Holders	333.7	363.3	8.9	.	.	.
Acreage	711.1	846.6	19.1	.	.	.
Groundnuts						
Holders	29.2	46.3	58.6	89.4	126.4	41.4
Acreage	182.0	156.0	−14.3	60.0	116.0	93
Total acreage	2 033.1	2 055.5	+1.1	1 672.0	1 543.4	−7.8

. = not available.
Sources. *Report on Ghana Sample Census of Agriculture, 1970* and *Current Agricultural Statistics, 1974*, various tables.

going out of cocoa production was being diverted to food production, and the total acreage in the south was declining. At the same time, acreage under foodcrops in the north was declining: it fell by about 128,000 acres between 1970 and 1974. Thus the total acreage under foodcrops in the country as a whole also declined. Clearly, Ghana's agricultural crisis was a complex phenomenon, and we shall come back to this issue later. At this stage, we can reasonably conclude that the foreign exchange constraint, caused by a decline in production and exports of cocoa, originated in part from a food constraint generated by a continuous decline in food production per head since 1960.

The stagnation of Ghana's food sector thus had far-reaching consequences on living standards and long-term growth prospects. It may be pointed out that the over-all national price index (with 1963 = 100) had by 1977 risen to 1,729, in large measure due to the rapid rise in the local food price index, which soared to 2,678.[24] This enormous increase in prices, in combination with heavy taxation, brought about a staggering decline in the real producer price on cocoa, which further triggered a decline in cocoa production and exports. This in turn gave rise to a serious foreign exchange bottleneck which virtually brought the economy to a standstill. At the same time, such increase in prices over a fairly long period contributed to the continued erosion of real incomes and the impoverishment of a large proportion of the population in both urban and rural areas, thereby causing a contraction of the market frontier within which over-all growth and industrial expansion could take place.

The stagnation of agriculture in general and food production in particular adversely affected industrial expansion on the supply side as well. As Killick noted:

Industries processing local agricultural products, such as sugar, fruit canning and vegetable oils, encountered great difficulties in obtaining adequate and reliable supplies, and others which could have been based upon local agriculture...relied upon imported supplies—to the detriment of their production costs, their contribution to the balance of payments, and their spill-over effects on the rest of the economy. The incapacity of agriculture to keep pace with domestic demand for food maintained constant pressure for large allocations of foreign exchange for food, and such imports competed directly with foreign exchange for the needs of industry. Lastly, the inflation of food prices and its impact on the general price level helped to raise manufacturing production costs. This trend, coupled with the over-valued Cedi and the inelasticity of supplies of agricultural raw materials, effectively foreclosed the possibility of profitable exporting for most industries and kept them to the narrow confines of the domestic market.[25]

It was a domestic market, moreover, that was rapidly shrinking. A major though difficult task, therefore, is to explain this gap between food demand and supply which seemed to have widened especially towards the end of the period under review.

Until 1972, when the Acheampong Government initiated the "Operation Feed Yourself" programme, now considered to have been a failure,[26] Ghana's food sector was never the object of adequate public concern. Infrastructural facilities were directed mainly toward cocoa-producing areas and those among the rural population who had entrepreneurial inclinations were engaged mainly in cocoa-related activities. Producer prices for foodstuffs until the late 1960s were also low and unattractive. The lag in food production behind population growth was thus hardly surprising.

However, the late 1960s and the 1970s saw dramatic increases in producer prices for foodstuffs. Nevertheless, there continued to be a gap between food supply and demand and, in fact, food production and acreage declined in absolute terms. The process through which this occurred involved, as we have seen, abandonment of farms in the north, the main food-producing region of the country, the effect of which was only partially counterbalanced by the expansion of acreage under foodcrops in the south.

How could this be explained? While the switch from cocoa to food in the south is explicable in terms of changes in relative prices, it remains puzzling that the expansion of foodcrop acreage was not greater, since it appears that many cocoa farms were actually being abandoned. Likewise, it is difficult to understand why farms were being abandoned in such large numbers in the north. We have to admit that the available empirical materials do not enable us to resolve these puzzles in a satisfactory manner. The best we can do is to suggest some plausible hypotheses and thus delineate areas for future research.

Stagnation and decline in food production cannot be explained in terms of the usual official apologia to the effect that they were entirely attributable to the huge retail margin of speculators and to the inefficiency of the distribution system. This can be readily deduced from trends in producer, wholesale and retail prices. If the rapid increase in food prices was to be attributed to speculators, one would expect a more rapid increase in retail prices than increases in producer and wholesale prices. As can be seen from table 94, this did not happen. In almost all cases producer prices increased faster than retail prices and, in the majority, wholesale prices rose faster than retail prices. We therefore see that the alleged huge margin which is put forward as the major explanation for food shortage cannot be sustained. The most that can be said, as noted by Godfrey, is that "the speculators, such as they are, have been exploiting in existing situation of shortage, rather than causing the shortage".[27] We are therefore compelled to look more closely at the production side of the problem.

Ecologically, Ghana consists effectively of two different agronomic systems. The southern region generally has high rainfall in two seasons of the year, while the northern region, with a low rainfall in only one season and a probability of failure in about two in five years, is highly susceptible to occasional drought. The soil in the north is heavy, of low quality and intensively cultivated. Most of its productive valleys are afflicted with river blindness. In short, the region is ecologically disadvantaged, suffering as it does from "a steadily declining environment caused by increasing population, onchocerciasis and other health hazards, declining soil fertility, worsening soil erosion, sporadic lack of water, and the absence of proper infrastructure and services".[28]

If nature was unkind to the north, this was no less true of government policies. Development efforts, especially of an infrastructural nature, tended to be concentrated in the south. For instance, although the Upper Region had 10 per cent of the country's population, its share of the Ministry of Agriculture's budget between 1971 and 1974 was only 3 per cent.[29] In terms of social services also, the northern regions were clearly at the losing end in the competition for development resources. These disparities in natural endowments and agricultural potential, capped by unequal access to development re-

Table 94. Indices and changes in producer, wholesale and retail prices, selected commodities, 1973

Crop	Indices, 1973 (1967-69 = 100)		
	Producer	Wholesale	Retail
Maize	198	170	184
Millet	202	199	180
Guinea corn	182	217	132
Cassava	147	182	128
Yams	165	178	187
Plantains	239	257	200

Rural consumer index = 156.

Source. World Bank: *Fiscal and balance of payments aspects of Ghana's development*, 19 May 1975, various tables.

sources, could not but contribute to the inequalities in labour productivity and wage rates between the north and south.[30] More seriously still, they established a framework for powerful socio-economic forces that continuously disadvantaged the northern farmer in relation to his counterpart in the south who, being in "an ecologically more favourable area and in close proximity to markets [was] able to respond more rapidly and more effectively to incentives, such as higher prices, or changes in technology".[31]

Given this context, it is perhaps not surprising that food production grew at a slower rate than population in the 1960s. The negative growth observed in the 1970s, however, is another matter. An absolute decline in production in the face of a huge growth of prices is difficult to explain in terms of technical conditions of production alone. One could attempt to explain it in terms of a movement of farm population from the north to either the south or the cities, or to neighbouring countries.

If a drift to the south is to be a part of the explanation, we must consider the possibility that the change-over from cocoa to food in the south may have resulted in an increased demand for labour. With constant acreage, the change-over to food production would almost certainly have led to increased demand for labour in the south. This can be deduced from table 95. Labour requirement in cocoa production is highest at the time of planting and up to ten years later, after which point it declines substantially. On the whole, the labour requirement per hectare for all foodcrops shown here is significantly higher than that for cocoa. Unfortunately, we cannot assume a constant acreage; the evidence suggests a decline in total acreage in the south. No definitive conclusion concerning the changes in demand for labour in the south can therefore be drawn.

There is evidence, however, to suggest that the supply of labour in the south suddenly declined in the post-1969 period following the expulsion of alien migrant workers who were usually employed on cocoa farms. As a

Agrarian policies and rural poverty in Africa

Table 95. Labour input in man-days per hectare

Crop	Work-days	Crop	Work-days
Millet/sorghum		*Yam*	
Unimproved	80	Traditional	155
Improved	100	Improved	180
		Advanced	215
Maize			
Traditional	75	*Cocoa*	
Improved	90	Age-range (years)	
Advanced	110	0-10	60.62
		11-21	26.62
Groundnuts		22-32	17.95
Traditional	90	33-43	10.21
Improved	105	44-55	3.61
Advanced	125		

Sources. The data for the foodcrops are derived from estimates for the Upper Region. See World Bank: *Appraisal of the Upper Region agricultural development project, Ghana* (3 June 1976), Annex 2, various tables. The data on cocoa are derived from Richard A. Brecher and Ian C. Parker: "Cocoa, employment and capital in the Ghanaian economy: A theoretical and empirical analysis", in *Economic Bulletin of Ghana*, 1974, No. 3/4.

consequence, many southern farmers began to experience serious shortages of labour, and responded by attempting to recruit labourers from the north.[32] That they were only partially successful in this endeavour is demonstrated by the fact that total acreage in the south declined. Nevertheless, such a movement of labour could be responsible for abandonment of farms (and a consequent decline in acreage) in the north. It need not be assumed that the poorer peasants in the north abandoned their farms in order to work as labourers in the south; this would presuppose that the second option provided them with a higher standard of living—an unlikely occurrence in view of the soaring prices of food. The more likely situation is one in which the male members of the poorer peasant families took up assignments as labourers in the south, while the women stayed behind to look after the family plots. In this case, it would be the relatively large-scale farmers in the north who, faced with a labour shortage, would probably leave parts of their land uncultivated. Two associated implications are that some poorer peasant families in the north were able to improve their economic position, and that some relatively large-scale farmers may have suffered economic deterioration, in which case some modifications would be called for in some of our earlier observations on income inequalities. Lack of data prevents us from verifying any of these propositions satisfactorily, and they must remain as hypotheses for future research.

However if we tentatively accept them as valid, a consistent picture of Ghana's economic decline emerges. This would run as follows. The growth strategy pursued in the late 1950s and early 1960s was seriously unbalanced in the sense that little attention was paid to the growth of the agricultural sector in general, and of the food sector in particular. As a result, food production per head began to decline, while growth of incomes per head, given a positive income elasticity of demand for food, began to raise demand per head for food. The relative price of food in relation to cocoa was thus rising, making it increasingly profitable for cocoa farmers to change over to food production—a tendency accentuated by the fact that producer prices of cocoa did not keep pace with the international prices. Stagnation in cocoa production during the 1960s appears to be explicable in these terms.

Given this context, the Government's decision in 1969 to get rid of the "aliens" by creating a shortage of labour produced a real crisis in the cocoa sector. The cocoa farmers reacted partly by importing labour from the north, and partly by abandoning their farms. At the same time, migration of labour from the north created a labour shortage there and led to a decline in acreage. The over-all consequence was a decline in both cocoa and foodcrops acreages. Thus the production of both cocoa and foodcrops declined in absolute terms in the 1970s, and in this way both a food constraint and a foreign exchange constraint became binding in the 1970s, so that even the expansion of industries and services had to come to a halt.

At this point, a few observations concerning the neglect accorded to agriculture by the Government may be relevant. Notwithstanding the endless official pronouncements on the vital role of increased cocoa and food production in enhancing the growth of the economy and the well-being of Ghana's population, agriculture in practice played second fiddle to most other major policy considerations. As can be seen from the over-all taxation and expenditure policy, its place within the scheme of public policy rested mainly on its function as a source of revenue, both domestic and foreign. Even this important role was not correctly appreciated, for there were hardly any significant steps taken by the Government to provide the support required for sustaining agriculture's capacity to generate revenue.

As is clear from table 96, well over a quarter of total government revenues was derived from the cocoa sector. Expenditure on agriculture, on the other hand, was kept low, especially in the latter half of the 1960s and for most of the 1970s. Measured as a percentage of total government expenditure, agriculture's share was kept at less than 10 per cent and, as a proportion of total government revenues from cocoa, expenditure on agriculture was around 30 per cent in most years. Ghana's agriculture thus suffered a massive net extraction of resources. In the absence of compensating injections of imputs, this massive taxation of agricultural surplus undermined, as was only to be expected, the very foundation of the economy and the source of its surplus.

The inadequacy of governmental effort is also clear from other fragmentary evidence. The Government provided a variety of inputs, often at highly subsidised rates, yet in terms both of quantity and of distribution its performance was poor. For example, its cocoa seed production programme in 1976 could only produce an amount equal to the requirements of 20,400

Table 96. Government and the agricultural sector: some indicators (million Cedis)

Year	Government revenue	Revenue from cocoa	Government expenditure	Expenditure on agriculture
1959-60	114.3	30.2	175.8	13.8
1960-61	143.1	29.2	232.8	22.6
1961-62	191.5	33.4	284.4	22.4
1962-63	165.0	·	264.4	·
1963-64	293.4	39.2	377.4	·
1965	284.0	21.4	361.6	·
1966	115.7	9.5	160.1	·
1966-67	235.3	34.6	297.3	24.5
1967-68	293.1	77.6	361.8	25.1
1968-69	283.6	79.9	359.4	24.3
1969-70	360.3	124.7	439.3	27.1
1970-71	486.2	196.5	470.9	25.4
1971-72	421.9	122.4	547.0	31.9
1972-73	391.7	96.4	579.8	36.6
1973-74	583.0	173.0	779.7	41.3
1974-75	810.5	279.7	1 211.8	76.3
1975-76	869.8	179.8	1 604.6	101.1
1976-77	1 144.0	269.0	2 228.0	220.6

· = not available.
Sources. *Ghana Seven-Year Development Plan 1963/64-1969/70*; Central Bureau of Statistics: *Statistical Hand-Book, 1969*; World Bank: *Fiscal and balance of payments aspects of Ghana's development*; idem: *Current economic situation and prospects*, 25 Sep. 1972; idem: *Ghana agricultural sector review*; idem *Ghana economic memorandum*.

hectares, or slightly in excess of 1 per cent of the area under cocoa. To give another example, the total supply of insecticide against capside (a major problem in cocoa cultivation) was only sufficient to spray 150,000 hectares, or 9 per cent of the cocoa area.[33] The situation was even worse with regard to the supply of sprayers which, in 1975, were only available in sufficient quantity to spray about 50,000 hectares, or not more than 3 per cent of the total cocoa area.[34]

Input supplies and extension services to other crops, especially industrial crops, although relatively greater than those for cocoa, remained somewhat limited and in many cases mainly benefited a small number of large-scale farmers. Fertiliser had always been highly subsidised, yet the available supply was adequate for only 3 per cent of all areas under crops; moreover, the greater part of it went to the large-scale farmers. So also with credit: it is estimated that less than 10 per cent of the farmers received institutional credit, and that around 80 per cent of such credit went to medium- and large-scale farmers, while small-scale farmers depended largely on village lenders who charged interest rates of as high as 50 to 100 per cent, compared with the 8.5 to 12.5 per cent charged by banks.[35]

In short, therefore, the massive shift of resources from agriculture, mainly from cocoa, and the inadequate supply of inputs as well as the differential access to them could not but have resulted in an unequal rural society and an agriculture in a state of total disarray.

That the agricultural sector should bear the burden of development is familiar enough, and therefore hardly surprising. What is surprising, however, is the scale of this extraction of resources and the absence of any compensating measures to sustain, if not improve, the initial situation. One might justify this direction of public policy if it could be shown that these resources were efficiently utilised to generate new forces or sources of production such as new exports, industries, and the like. But this was not the case: the massive transfer of resources from cocoa was accompanied neither by new export items that could diversify Ghana's sources of foreign exchange and assure her a more solid foreign trade base, nor by an efficient industrialisation programme.[36]

CONCLUSION

Ghana's experience offers a number of unusual aspects. Starting with relatively high income per head at independence and with other economic and social advantages, the country went through a phase of stagnation and sharp decline in the 1960s and 1970s. At the same time there appears to have been an intensification of rural inequality and poverty. The available data and information are not adequate to provide a completely satisfactory explanation of changes in the level and structure of production. What seems highly likely is that the economic decline was triggered off by a stagnation of the agricultural sector, first in food and then in cocoa production. This in turn led to foreign exchange shortages, with negative repercussions on the entire economy. Wrong policies exercising a strong disincentive effect on agriculture, coupled with generally poor economic management, appear to have been at the root of the economic difficulties the country has faced over the past decade and a half. The sharp fluctuations in world cocoa prices added to the problems of economic management. These problems and difficulties have been greatly accentuated in the late 1970s and early 1980s by a general deterioration of the world economy. A reorientation of policies and more efficient economic management must be the first steps in the process of recovery and growth.

Notes

[1] For a detailed analysis of the Ghanaian economy in the 1960s, see Tony Killick: *Development economics in action: A study of economic policies in Ghana* (London, Heinemann, 1978). For an historical survey, see also Walter Birmingham, I. Neustadt and E. N. Omaboe (eds.): *A study of contemporary Ghana* (London, Allen and Unwin, 1966), Vol. I; and Bjorn Beckman: *Organising the farmers: Cocoa politics and national development in Ghana* (Uppsala, Scandinavian Institute of African Studies, 1976), especially Chs. I and II.

² Szereszewski: "The performance of the economy", in Birmingham et al., op. cit., table 2.11, p. 56.

³ Killick, op. cit., p. 3.

⁴ ibid., tables 2.7 and 2.9, pp. 50-53.

⁵ Szereszewski, op. cit., p. 41.

⁶ See David Morawetz: *Twenty-five years of economic development, 1950 to 1975* (Washington, DC, World Bank, 1977), table A1, p. 77.

⁷ Beckman, op. cit., p. 37.

⁸ Killick, op. cit., p. 4.

⁹ Szereszewski, in Birmingham et al., op. cit., pp. 205-209.

¹⁰ As a matter of fact, the figure quoted in table 78 probably understates the degree of decline in GDP per head. The rate of growth of GDP (2.1 per cent per annum) is based on an estimated 3.7 per cent rate of growth of agricultural production. Another source puts the rate of growth of agricultural production at 1.8 per cent per annum. This seems more plausible in view of the fact that cocoa production virtually stagnated (rate of growth −0.02 per cent per annum) and that food output grew at an annual rate of 1.8 per cent.

¹¹ For an elaboration of this point and the mechanisms by which the foreign exchange constraint impeded over-all growth, see Killick, op. cit.

¹² See G. Benneh: "Communal land tenure and the problem of transforming traditional agriculture in Ghana", in *Journal of Administration Overseas*, Jan. 1976, pp. 26-33; N. A. Ollenu: "Aspects of land tenure", in Birmingham et al., op. cit.

¹³ Benneh, op. cit., p. 27.

¹⁴ ibid., p. 28: Ollenu, op. cit., p. 254: Killick, in Birmingham et al., op. cit., pp. 217-218.

¹⁵ Birmingham et al., op. cit., p. 386.

¹⁶ See Beckman, op. cit., for a most interesting account of the process of social differentiation in Ghana. This section draws heavily on his Chapter II.

¹⁷ ibid., p. 37. Further, Ollenu describes the origin and nature of tenancy as follows: "The success of the cocoa and oil-palm industries led many of the forest landowners to prefer a system of land tenure based on crop sharing and called the *abusa* system, to an absolute sale of the land. Strangers who have not much money for the capital outlay usually accept the *abusa* tenancy. The *abusa* tenancy is a system under which a stranger with his own capital or labour, including the labour of members of his family, develops land belonging to another person either for a permanent crop like cocoa, oil-palm, coffee or rubber, or even a foodcrop, and gives one-third of the proceeds thereof to the owner of the land. In all other respects his position with respect to the land is as permanent as that of the purchaser of the possessory title. Where, for example, he is a cocoa *abusa* tenant, he can grow any amount of foodstuff on the land without having to give the landowner a share of it" (Ollenu, op. cit., p. 256).

¹⁸ Beckman, op. cit., p. 38.

¹⁹ Since rising food prices tend to impoverish the deficit (poor) farmers and benefit surplus (rich) farmers, land transfers, if they occur, are likely to be from the deficit to the surplus farmers.

²⁰ World Bank: *Ghana economic memorandum* (Washington, DC, 1979), p. 18.

²¹ idem: *Ghana Agricultural Sector Review*.

²² idem: *Ghana economic memorandum*, op. cit., p. 18.

²³ idem: *Ghana Agricultural Sector Review*, op. cit., Annex V, p. 14.

²⁴ Janet Girdner, Victor Olorunsola, Myrna Froning and Emanuel Hansen: "Ghana's agricultural food policy—Operation Feed Yourself", in *Food Policy*, Feb. 1980, p. 17.

²⁵ Killick, op. cit., pp. 205-206.

²⁶ Girdner et al., op. cit.

²⁷ E. M. Godfrey: "Labour surplus models and labour deficit economies: The West African case", in *Economic Development and Cultural Change*, Apr. 1969, p. 385. Godfrey bases his conclusions on Rowena Lawson's two surveys of the Ghanaian system of distribution: see Rowena Lawson: "Inflation in the consumer market in Ghana", in *Economic Bulletin of Ghana*, 1968, No. 1; and idem: "The distribution system in Ghana" in *Journal of Development Studies*, Jan. 1967.

[28] World Bank: *Appraisal of the Upper Region agricultural development project, Ghana* (3 June 1976), p. 5.

[29] ibid.

[30] See B. E. Rourke: *Wages and incomes of agricultural workers in Ghana* (Legon, ISSFR, 1971), pp. 59-63: Charles Elliott: *Patterns of poverty in the Third World: A study of social and economic stratification* (New York, Praeger, 1975), Chs. 3 and 5; Beckman, op. cit., pp. 38-39.

[31] Elliott, op. cit., p. 88.

[32] cf. J. Adomarko-Sarfoh: "The effects of expulsion of migrant workers on Ghana's economy, with particular reference to the cocoa industry", in Samir Amin (ed.): *Modern migrations in western Africa* (London, Oxford University Press, 1974). Also, for some relevant observations, see World Bank: *Ghana Agricultural Sector Review,* Vol. 1, p. 3; Beckman, op. cit., p. 219; N.O. Addo, "Employment and labour supply on Ghana's cocoa farms: The pre- and post-aliens' compliance order", in *Economic Bulletin of Ghana,* 1972, No. 2, p. 39; and J. C. Caldwell: "Migration and urbanisation", in Birmingham et al., op. cit., Vol. II, p. 138.

[33] Girdner et al., op. cit.

[34] ibid., p. 11.

[35] ibid., Annex VI, pp. 1-3.

[36] For critical appraisals of Ghana's industrialisation programme, see Killick, op. cit., and Morawetz, op. cit.

GROWTH AND DISTRIBUTION: THE CASE OF MOZAMBIQUE

9

R. K. Srivastava and I. Livingstone

Mozambique's experience with socio-economic development has been relatively short. Although independence from Portuguese colonial rule came on 25 June 1975 after a protracted war of liberation, Mozambique remained actively involved in the struggle for the liberation of Zimbabwe until early 1980. Thus, the period since independence can best be regarded as a "holding operation" during which more emphasis has been placed on socio-political mobilisation than on development as such. During this period the complex problems of economic reconstruction could only be dealt with ad hoc, under severe constraints on the availability of material and human resources. On the other hand, these formative years of Mozambique's history have, in a sense, been rather decisive in shaping the country's economic objectives: its development philosophy, as it has now emerged, has been strongly influenced by the interaction between the colonial patterns of economic exploitation and the ideological context of the liberation war.

The purpose of this chapter is to examine this experience and its implications for rural development, in particular its impact on the livelihood patterns in rural areas. The chapter begins with an analysis of the colonial economic structure and its effect on the rural sector. This is followed by a discussion of the changes since independence and the new rural institutions being created. The last section is devoted to a preliminary assessment of emerging trends in the rural strategy. Finally, a concluding section deals with the medium-term economic outlook for Mozambique.

THE COLONIAL ECONOMY

The setting

Although the Portuguese presence in Mozambique dates back to the early sixteenth century, the country was not effectively occupied until the 1885 Berlin Conference; however, it took another 50 years for Portuguese rule to become firmly established. With the abolition of slavery in 1869, a series of decrees forced the *libertos* to contract their labour to former owners. In addition, the practice of granting territorial concessions to foreign monopoly companies (the *prazo* system) in return for taxes and shares in favour of the Portuguese Government facilitated the dissemination of colonial economic practices throughout the countryside.[1] With the development of gold-mining in South Africa, Portugal also concluded a number of agreements with that country as well as with the former Federation of Rhodesia and Nyasaland for the "sale" of Mozambican labour after 1897, culminating in the Mozambique Convention of 1909. Under this Convention, South Africa agreed that 47.5 per cent of the Witwatersrand's cargo would be channelled through the port of Lourenço Marques (Maputo) on condition that mine labour could be freely recruited from Mozambican provinces below 22° S. Being the most heavily populated country in the region, Mozambique thus became the critical labour supply area in the formative years of the gold-mining industry. It was the centrepiece of the larger labour reserve from which controlled and cheap male labour could be drawn (see table 105). This arrangement was reinforced when South Africa began paying the deferred mine wages in gold through the Portuguese banking system, thus creating an enduring vested interest in labour trade.

The period 1902 to 1930 may therefore be characterised as one of steady expansion of the colonial economy into the rural sector and the establishment of labour use and production systems which have had a far-reaching effect on the future development of Mozambique; prior to that period, its speed had been slow because quicker returns could be obtained from the slave and coastal trade.

In the late 1940s and early 1950s Mozambique's role as a supplier of raw materials to Portugal increased in importance. As a consequence, the peasantry was compelled to cultivate specific crops, notably cotton. Price policies disadvantageous to the producer were adopted. On the other hand, Portuguese consumer and capital goods had assured markets in the country; import and tariff policies were geared to this purpose.

There is evidence to indicate that systematic policy decisions were taken to coerce the peasantry into entering the cash economy through sale of labour. The "head tax" which the peasants were obliged to pay in cash was one of the main sources of revenue for the colonial budget. The rural worker, in order to pay this tax, had three alternatives: he could either take low-wage work on the plantations, or sell his primary produce at low prices determined by the State, or was forced to join the migratory wage-labour.

This period was also significant for the development of transport links through Mozambique to service the needs of South Africa, especially the Transvaal, the Rhodesias, Nyasaland and the Congo. The infrastructure—roads, railways and ports—was developed mainly to meet the requirements of neighbouring countries rather than the domestic requirements of Mozambique itself. As these facilities developed, a number of service sector activities had to be established, including banking, trade and commerce, and import-export undertakings. As a result, there was a proliferation of expatriate trading companies and foreign firms. A modern social service system—education, health, housing—developed in the larger port towns of Lourenço Marques, Beira and to a lesser extent in Quelimane, Moçambique and Pemba, as well as in the provincial towns in the interior where plantation companies were located. The colonial Government adopted a protective tariff and foreign exchange system in order to ensure the near monopoly of imported goods, and to take the benefits of foreign trade out of the domestic economy.

Some agro-processing activity was undertaken in relation to plantation products soon after the Second World War. However, it was in the late 1950s and early 1960s, when sugar refining and the processing of cashew became important to the economy, that industrial activity began to grow. With the expansion of the three largest sugar plantations and the establishment of two new sugar companies, sugar refining became the leading industrial activity. Cashew processing became the second important subsector when decortication and canning facilities were established. Other agro-processing activities included cotton, sisal and tea. These were followed by tobacco, soap and beer manufacture and, later, by textile, clothing, cement and petroleum refining. However, the industrial sector remained small as it was linked to the plantation system and to the limited demand from the urban "modern" sector.

The economy of Mozambique thus developed to a remarkable extent as a subservient system to cater to the needs of the metropolitan Power and the trading companies. Its dependence on foreign trade, transit trade and urban services gave it an unbalanced structure unrelated to the needs of the indigenous population.

This lopsided development is reflected in the macroeconomic data (table 97), especially in the changes in the relative shares of agriculture and service sectors. Available statistics show an acceleration in the growth rate of GDP during the 1965-70 period, with corresponding increases in both private and government consumption and in exports and imports (table 98). It may, however, be noted that the external dependence of the economy is also reflected in the negative trade balance (table 99). The trade balance has historically shown a deficit with visible exports accounting for only approximately half the value of visible imports. Invisible items from transit trade and migrant wages, however, generally more than made up for this deficit. It is also noteworthy that proceeds from the sale of gold were never included in Mozambique's balance-of-payments statistics during the colonial period.

The economic data on Mozambique should, however, be interpreted with caution. There were several factors contributing to the unreliability of all growth and production statistics: in the first place, most of the available information was confined to the "modern" sector; second, commercial sector output was emphasised, while guesstimates of subsistence sector production were suited to particular interests; and finally, trade statistics were particu-

Table 97. Composition of gross domestic product, 1955-70 (at current factor cost) (percentages)

Sector	1955	1960	1965	1970
Agriculture	56.7	54.8	48.2	43.5
Manufacturing	6.3	7.7	10.2	5.9[1]
Other industry	0.9	1.3	2.2	6.0[1]
Services	36.1	36.2	39.4	44.6
Total	100.0	100.0	100.0	100.0

[1] Changes due to reclassification.
Source. World Bank: *World tables* (Washington, DC, 2nd ed., 1980).

Table 98. Selected economic indicators: annual average growth rates (market prices)

Indicator	1960-65	1965-70
Total GDP	2.3	8.3
GDP per head	0.2	5.9
Private consumption	2.4	7.2
Government consumption	4.2	10.8
Gross domestic investment	1.1	14.4
Export of goods and non-factor services	2.5	8.1
Imports	3.6	8.4
Population	2.1	2.3

Source. World Bank: *World tables* (Washington, DC, 2nd ed., 1980).

Table 99. Imports and exports in selected years (in million escudos)

Year	Exports	Imports	Balance
1968	4 420	6 740	−2 320
1969	4 081	7 491	−3 410
1970	4 499	9 363	−4 864
1971	4 213	8 773	−4 560
1972	4 768	8 912	−4 144
1973	5 541	11 415	−5 873
1974	7 560	11 741	−4 181
1975	5 050	10 472	−5 422

Source. (1968-70) *Africa south of the Sahara* (London, Europa Publications, 1973). (1972-75) Mozambique: *Indicadores Economicos*; Banco de Moçambique: *Boletin Conjuntura*, June 1977.

larly unreliable during the period 1965-75 because of the involvement of the colonial Government in trade with Rhodesia.

The rural sector

The rural subsistence sector would not seem to have derived any benefits from the post-Second World War growth; however, some components of the rural economy did participate in this growth. Among these were the following:

1. *Plantations.* This subsector comprised more than 2,000 large farms, of which over 400 were owned by foreign commercial enterprises, mostly non-Portuguese. Originating in the *prazo* system, these plantations specialised in the production of cash crops such as sugar, cotton, sisal, coconuts and tea, but derived their income largely from the extraction and subsequent sale of crops produced by traditional agriculture in which peasants were subjected to forced labour *(chibalo)* or forced production of specific crops.

2. *Large estates.* These estates, or *latifundios,* were formed through the granting of land concessions, though on a scale smaller than that of plantations, in the 1950s. In the course of three decades, several thousands of immigrants, mostly Portuguese, came to Mozambique and started farming in agricultural settlements and irrigation schemes patterned on Portuguese peasant communities.[2] The largest settlements were formed around irrigation projects at Guija on the Limpopo river (Gaza Province) and near Chimoio on the Revue river. The impact of *latifundios* on total production, however, remained small until many of them were modernised and converted into commercial farms producing mainly for the domestic market. They nevertheless extended the system of low-wage work and *chibalo* initiated by the plantations further into the countryside.

3. *Medium and small farms.* These farms, known as *colonos,* were mainly concentrated in the southern provinces and came to be established in response to the food needs of growing urban centres. They concentrated on the production of rice, wheat, potatoes, vegetables, fruits, meat and dairy products. Because of their size and reliance on family labour, these farms could offer better wages and regular work and thus could compete successfully for labour with South African mines. This did not prevent them, however, from using forced labour during the peak season.

4. *Merchants and middlemen.* These *cantineiros* constituted the main elements of the rural supply and distribution system. The marketable surplus of the peasantry was taken by them and, in turn, basic consumer products such as salt, sugar, soap, cloth and oil were sold through rural outlets. The *cantineiros* did not permit the development of indigenous marketing systems. They thus retained a complete monopoly on the terms of trade in dealing with the rural peasantry. This system worked closely with the banks, which ultimately also had a share in the appropriation of the peasants' surplus.

However, the benefits of growth and development did not trickle down to

Table 100. Farm size in the agricultural sector, 1970

Farm size (hectares)	Commercial sector			Traditional sector[1]		
	No. of units	%	Area ('000 hectares)	%	No. of units ('000)	%	Area ('000 hectares)	%
Under 0.5	–	–	–	–	306.1	18.6	92.1	3.7
0.5-5	141	3.0	0.4	–	1 296.9	78.6	2 068.3	82.9
5-10	108	2.3	0.7	–	37.9	2.3	244.2	9.8
10-20	183	3.9	2.4	0.1	6.8	0.4	88.9	3.6
20-50	1 733	37.4	61.3	2.4	–	–	–	–
50-100	290	6.2	19.0	0.8	–	–	–	–
100-500	1 397	30.2	297.6	12.0	–	–	–	–
500-1,000	285	6.1	189.5	7.6	–	–	–	–
1,000-2,500	270	5.8	402.8	16.1	–	–	–	–
Over 2,500	219	4.7	1 513.8	60.9	–	–	–	–
Total	4 626 (0.3%)	100.0	2 487.6 (49.9%)	100.0	1 647.7 (99.7%)	100.0	2 493.5 (50.1%)	100.0

– = not applicable.
[1] The composition of the traditional sector is broken down further in table 101.
Source: Estatísticas Agrícolas de Moçambique, 1970.

the largest group of the rural sector, the subsistence farmers. Despite the dominance of exploitative systems and the economic dualism which they fostered, the peasantry in the family farm sector *(sector familiar)* remained the principal producer throughout the colonial period. Although expatriate farmers constituted less than 1 per cent of the farming community, they controlled almost half of the farming area, including the greater part of the better-quality land. The other half was cultivated by over 1.6 million subsistence farmers producing mainly cassava, maize, sorghum, beans and oilseeds. The majority of these farmers also produced cash crops such as sugar-cane and the bulk of commercial cotton and rice. In addition, various uncultivated products were gathered by them, such as, for example, cashew nuts, coconuts, castor seeds and mafurra nuts for the market or for the owners of land concessions. Only the most rudimentary production techniques were used by the peasantry, the only tools used being the machete for jungle clearing and the hoe for cultivation. More advanced cultivation techniques using draught animals and ploughs were confined to limited groups—such as the Ngoni in Tete and the Thonga in the southern districts—who came to Mozambique with later Bantu immigration.

Tables 100, 101 and 102 show the pattern of land holdings and the share of the traditional and commercial sectors; the latter comprises the plantations, the *latifundios* and the medium and small commercial farms, most of which were "modern" only in name. The average size of the farm in the commercial sector was well over 500 hectares, as against only 1.5 hectares in the traditional sector. The former, of course, represents the wide variation in size between the large plantations and small commercial farms; yet the extreme concentration of land in this sector is evident. Much of this land was, however, not under cultivation. Indeed, the total area under cultivation remained, on the whole, small in relation to available arable land, as indicated in the land use table (table 102).

The pattern of labour exploitation in the commercial sector varied with crops. In the case of cotton, the policy to enforce cotton production in the traditional sector was adopted in 1926; zones were established in which each family farm had to cultivate 1 or 0.5 hectare and the produce was compulsorily taken over at a fixed price by the companies which had territorial concession. At the peak period, there were 700,000 registered cotton cultivators in the provinces of Cabo Delgado, Niassa, Nampula and Zambézia. Being a labour-intensive crop, cotton competed directly with subsistence food production, leading to the cultivation of low-labour/female-labour crops such as cassava for food.

Wherever such competition did not exist—for example, in the case of cashew nuts—conditions for production and delivery of produce were often laid down by zones and/or by households. Moreover, the colonial strategy of maintaining a cheap supply of labour as well as of cash crops was based on a series of laws and decrees known as *indigenato*.[3] Under one decree, subsistence farming was not even considered as valid work.[4]

Agrarian policies and rural poverty in Africa

Table 101. Farm size in the traditional sector, 1970

Farm size (hectares)	Proprietors		Area	
	No. of units	%	(hectares)	%
Under 0.5	306 077	18.6	92 166.5	3.7
0.5-1	412 245	25.0	307 714.9	12.3
1-2	540 608	32.8	781 297.3	31.3
2-3	232 871	14.0	651 169.3	22.5
3-4	75 313	4.6	259 531.1	10.4
4-5	35 850	2.2	158 574.2	6.4
5-10	37 925	2.3	244 249.4	9.8
10-20	6 813	0.4	88 852.2	3.6
Total	1 647 702	100.0	2 493 504.9	100.0

Source. Estatísticas Agrícolas de Moçambique, 1970.

Table 102. Agricultural production by sector, 1970

Crop	Commercial sector		Traditional sector	
	Quantity ('000 tons)	%	Quantity ('000 tons)	%
Crops produced mainly in traditional sector				
Maize	38	10	335	90
Beans	3	5	59	95
Sorghum	1	1	194	99
Oilseeds	4	7	57	93
Crops produced mainly in commercial sector				
Sugar-cane	2 497	97	74	3
Canape, sisal	33	100	–	–
Tobacco	3.5	80	0.9	20
Tea (green leaf)	68	100	–	–
Potato	39	100	–	–
Crops produced in both sectors				
Cotton (seed)	46	33	93	67
Rice	56	57	43	43
Copra	37	54	30	46
Uncultivated crops				
Cashew nuts	20	10	180	90

– = not applicable.
Source. Estatísticas Agrícolas de Moçambique, 1970.

Table 103. Land use, 1970

Type of land	Commercial sector		Traditional sector		Total	
	Area	%	Area	%	Area	%
Land in agricultural use	2 701	21	10 330	79	13 031	17
Land for crops	291	5	4 986	94	5 278	7
(cultivated)	(196)	(7)	(2 493)	(93)	(2 690)	(3)
(fallow)	(95)	(4)	(2 493)[1]	(96)	(2 588)	(3)
Land under permanent crops	190	15	1 099	85	1 289	2
Land under pasture	199	34	4 244	66	6 433	8

[1] Approximate.
Source: *Estatísticas Agrícolas de Moçambique, 1970.*

Regional differentiation of crops and labour supply gave rise to distinct "livelihood configurations" in the north, centre and south of the country. Northern Mozambique (Cabo Delgado, Niassa, Nampula) showed a concentration of family farms engaged in cash crop production (cotton, cashew nuts sisal, tobacco), oriented towards the export market; central Mozambique (Zambézia, Manica and Sofala, Tete) developed as an export-oriented plantation economy (tea, copra, sugar-cane, potato); and the southern provinces (Gaza, Inhambane, Maputo) produced food for urban consumption in settler farms and provided labour for the South African mines and plantations in Southern Rhodesia (see figures 1, 2, 3 and 4).

Estimates of the share of agricultural output from different sectors of the rural economy clearly indicate the impact of these livelihood configurations (table 104).

Employment and basic needs

The pattern of employment also shows the unusual predominance of wage employment in a highly rural and subsistence economy (table 105). The share of agriculture in wage employment was 38.16 per cent. In addition, a considerable number of Mozambicans worked in the agricultural sector of South Africa and Southern Rhodesia.[5] Agricultural wage labour, particularly in the southern provinces, largely consisted of those who were unable to go to the mines. The figures of employment in mining include migrant workers, only a small share being employed in the coal mines in Tete Province. However, the census figure of mining employment is, in all probability, an underestimate and does not include illegal migrants, whose number may be almost equal to the reported total. Close links were maintained between mine labour and the peasant economy. Mine labour recruitment showed a seasonal relationship to agriculture, and recruitment offices made monthly reports on the conditions of crops and weather and the value of cash crop sales in their areas.

Agrarian policies and rural poverty in Africa

Figure 1. Regional livelihood configurations in northern Mozambique

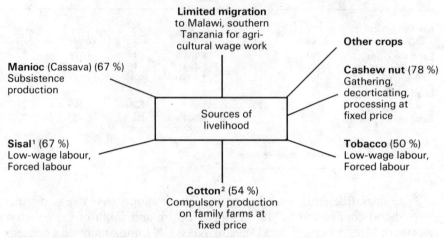

[1] Competes with food production for labour. [2] Withdraws labour from food production.
Figures within parentheses show regional share of total national production.

Figure 2. Regional livelihood configurations in central Mozambique

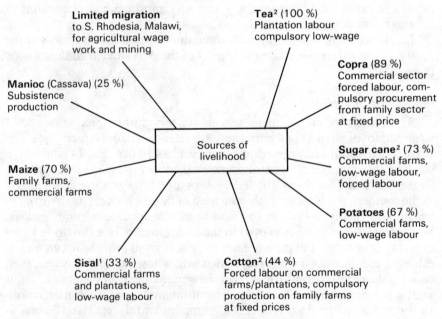

[1] Competes with food production for labour. [2] Withdraws labour from food production.
Figures within parentheses show regional share of total national production.

Mozambique

Figure 3. Regional livelihood configurations in southern Mozambique

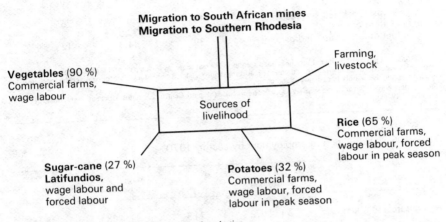

Figures within parentheses show share of total national production.

Figure 4. Over-all work opportunities

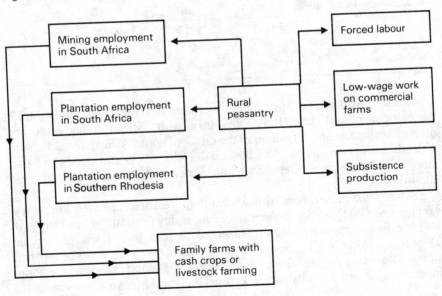

Table 104. Structure of agricultural production by region (percentages)

Type of production	South	Centre	North
Subsistence	49	52	60
Marketed output	51	48	40
of which			
(a) peasants' marketed surplus [1]	10	9	26
(b) plantation output	2	27.5	2
(c) output of *latifundios* and *colonos*	39	11.5	12

[1] The estimate of peasants' marketed surplus is not a surplus in the real sense; at least a part of it is attributable to the need for raising cash and a part to the low-price procurement of cash crops from registered cultivators. Note the relatively higher share of the northern region, due mainly to cashew-nut and cotton.

Source. Marc Wuyts: *Peasants and rural economy in Mozambique* (Maputo, Centro de Estudos Africanos, Universidade Eduardo Mondlane, 1978).

Table 105. Distribution of employment by sector, 1970

Sector	Total employment	%	Wage employment	%
Agriculture and allied	2 134 972	74.3	454 385	38.16
Mining	123 772	4.3	123 772	10.39
Manufacturing	155 996	5.4	155 996	13.10
Construction	81 469	2.8	81 469	6.84
Utilities	2 528	0.1	2 528	0.03
Trade	82 482	2.9	82 482	6.92
Transport	62 724	2.2	62 724	5.26
Services	229 880	8.0	229 880	19.30
Not adequately described	1 774	0.0	–	–
Total	2 875 597	100.0	1 190 707	100.0

– = not applicable.
Source. Mozambique: *Population census 1970; Anuario Estatístico, 1974.*

It is noteworthy that mine labour recruitment increased steadily during the colonial period (table 106). Obviously, while Mozambique could not determine the demand for labour in the mines of South Africa, it could ensure a steady supply on the basis of the relative cheapness and docility of that labour. The agreements between South Africa and Portugal ensured these conditions.

Before 1965 mine labour from Mozambique earned extremely low wages, ranging from 40 to 65 South African cents per day for underground work, of which a fixed proportion was withheld to be paid later for all miners on 18-month contracts.[6] The wages actually paid were expected to cover their immediate living expenses, the reproduction and maintenance of the population being left to the subsistence sector. Thus rural Mozambique continued

Table 106. Annual recruitment in gold-mining (thousands)

Year	Total recruitment	Recruitment from Mozambique	Percentage
1906	81.0	53.0	65.4
1916	219.0	83.4	38.1
1926	203.0	90.3	44.5
1936	318.0	88.4	27.8
1946	305.0	96.1	31.5
1956	334.0	102.9	30.8
1966	383.0	108.8	28.4
1976	331.0	44.1	13.3

Source. Quoted from J. Bardill, R. Southall and C. Perrings: *The state and labour migration in the South African political economy (with particular reference to gold mining)* (Geneva, ILO, 1977; mimeographed World Employment Programme research working paper; restricted).

to bear the infrastructural costs of supplying migrant labour to South Africa as well as the associated social costs of disruption in family life and household patterns. The women, children and older men who were left behind were compelled to look after farming and had difficulty in maintaining agricultural production and productivity levels; thus over the years traditional cultivation, irrigation and soil management practices fell into disuse in the southern provinces.

However, low wages in the mines were still attractive in relation to earnings from the subsistence sector or wages in *latifundios*. In the early 1960s subsistence production was hardly able to meet the food needs of the family in most provinces of the country. The highest earnings were from cotton: about 2,000 escudos annually per hectare, while cash wages in *latifundios* were legally fixed at 5 escudos per day. Mine wages were therefore about three to five times higher.[7] Thus, in the southern provinces, deferred wages resulted in considerable capital accumulation and investment in the rural sector both in livestock and in commercial farming. In the nexus of poverty reinforced by falling subsistence production, low wages in *latifundios* and forced labour, the socio-economic system encouraged migration as a way of life.

Relatively little information is available on income distribution. As has already been stated, the benefits of growth were hardly shared by the rural peasantry. Moreover, the mass of the rural population was also deprived of other basic needs. Malnutrition has always been a serious problem in Mozambique. The rural diet has been almost wholly based on "inferior foods", cassava and maize traditionally representing the main sources of energy and protein. The average availability of energy in 1964-66 was estimated at 2,130 calories and was reported to have fallen to 2,050 calories in 1969-71, against the estimated requirement of 2,300. Assessments of "undernourishment" (fewer than 1,500 calories per day) made by the FAO indicated that, in 1969-71, 34 per cent of the population could be classified as undernourished; this share increased to 36 per cent between 1972 and 1974.[8]

There was a conspicuous lack of social investment in the rural sector, particularly in education and health. In the early 1960s there were about 2,800 "rudimentary schools", almost all subsidised establishments of the Roman Catholic missions. These schools emphasised religious training and obedience to the Portuguese Government rather than academic instruction. This "indigenous" educational system was abolished in 1964 and replaced by ordinary schools, many of which continued to be run in mission stations. The new system required an entrance fee per child equal to several weeks of an adult's salary. In 1970-71 about one-third of children in the 7-12 age group were reported to be enrolled in 4,088 primary schools; secondary school enrolment was about 5 per cent of the children in the 13-18 age group.[9] Medical facilities were established mainly in response to the needs of expatriate settlements, while measures for the control of malaria and sleeping sickness were directed to making the area safe for settler farms and plantations.

Conclusion

The colonial economy of Mozambique exhibited a high degree of economic dualism, consisting as it did, on the one hand, of a distinct commercial sector (large farms, urban industry and services) and, on the other, of a peasant sector with no part in growth and development. This social structure has made the economy, and the rural peasantry in particular, extremely vulnerable. In the first place, the family farmer's opportunities were severely circumscribed, alternating as they did mainly between low-wage work and subsistence production (see figure 4 above). In the case of compulsorily grown crops, the risk of crop failure was borne entirely by the peasant farmer. Second, the absence of investment in subsistence farming led to falling levels of production and productivity, especially in areas of high external migration. The comparison in table 107 between average national output per hectare for selected crops in 1970 and the output achieved in Inhambane, one of the southern provinces and a major centre for mine labour recruitment, indicates the imbalances fostered by colonial practices. Third, the expatriate-controlled rural network of distribution and service systems allowed little opportunity for the development of local entrepreneurship, while at the same time imposing upon the peasants terms of trade that were adverse to their interests.

Among the factors in Mozambique's external dependence, the following four were especially significant: the pattern of foreign trade and composition of imports, as well as the exclusion of gold earnings from the economy, appear to have had an adverse influence on over-all domestic investment except in

Table 107. Agricultural productivity, 1970 (tonnes per hectare)

Crop	National average	Inhambane
Beans	0.35	0.10
Cassava	5.68	2.77
Cotton	0.30	0.11
Groundnuts	0.22	0.11
Maize	0.42	0.13
Millet	0.29	0.01
Rice	0.91	0.43

Source. Quoted in Ruth First et al.: *The Mozambican miner* (Maputo, Centro de Estudos Africanos, Universidade Eduardo Mondlane, 1977), p. 94.

support of the limited urban superstructure; receipts from transit trade fluctuated in response to conditions in neighbouring countries; the demand for mine labour was fully controlled by the South African Chamber of Mines and their recruiting offices in Mozambique; and, finally, the entire management of the economy, including its key sectors, remained in the hands of expatriates who had predominantly external interests.

CHANGES SINCE INDEPENDENCE

The economic situation

The different aspects of the vulnerability of the economy came into sharp focus at independence. Much of the basic infrastructure and productive facilities were damaged during the liberation struggle, in the closing years of colonial rule and subsequently in the Zimbabwe liberation war. Soon after independence there was an exodus of expatriate managerial, technical and skilled personnel and the majority of immigrant farmers and merchants, thus causing serious disruption in the administrative and productive systems. The economic dualism fostered during the colonial period took its toll. Earnings from transit trade fell drastically in 1976 with the closure of the Rhodesian border, in response to the Security Council resolution on the application of sanctions against that country. The strike in the port of Maputo offered a further setback to service sector revenues. The South African Government, unable to get out of its commitment to pay in gold for mine labour, reduced annual recruitment from Mozambique from 118,000 in 1975 to less than 40,000 in 1976; this resulted in a sudden decline in revenues from this source. Finally, economic difficulties were compounded by a series of natural disasters: floods in the major rivers (the Limpopo in 1975 and 1977; the Incomati in 1976, and the Zambezi in 1978); and prolonged drought in the northern region. A growing number of refugees also sought shelter in Mozambique.[10]

The post-independence economic crisis may be seen in terms of three major indicators: balance of payments, agricultural production and employment.

Balance of payments

Available information indicates that Mozambique has been facing a rapidly rising balance-of-payments deficit after independence (table 108), in part attributable, no doubt, to the fall in exports which, in turn, is explained by the decline in the output of cash crops. It is also partly attributable to increased food grain imports to meet urgent consumption requirements. The receipts from invisibles hide the fall in transit trade and miners' remittances.[11] Total revenues from railways, harbours and the transport system declined drastically from the end of 1974, reaching the lowest point in mid-1976 (see figures 5 and 6).

Continuous and rising deficits of this order have severely limited the ability of the Government to undertake the task of economic reconstruction, especially in the rural sector, and have also affected the choice of priorities.

Agricultural production

Food has become a major import since independence.[12] This is largely a result of disruption in production (table 109) due to a variety of factors. In the case of wheat, output declined because it was an item of urban consumption not related to rural demand. Sugar-cane was mainly a commercial-sector crop and its output fell when large farms were abandoned by expatriate farmers. Maize was the only major crop which had developed significantly in the traditional sector prior to independence; it was, however, affected partly by adverse pricing policies and partly by the breakdown of the rural trade and marketing network.

There was similarly a steep fall in the output of major cash crops (table 110), which led to a decline in export earnings (see table 108).

The fall in agricultural production had a number of serious repercussions on the economic situation. Besides accentuating the external vulnerability of the economy, it caused a major breakdown of the cash economy in the rural sector. There was a large element of circularity involved in this breakdown, especially through the collapse of the trading and marketing network *(cantineiros)*, and the commercial sector (plantations and *latifundios*). However, the cumulative effect of adverse production, pricing and labour policies was, perhaps, the most important factor in the agricultural crisis, generating a number of conflicts in agricultural and rural policies and priorities which will take time to resolve. Finally, it had a negative impact on the employment and incomes of a significant part of the rural labour force, particularly that which depended upon the commercial sector.

Table 108. Balance-of-payments estimates, 1973-78
(US$ million)

	1973	1974	1975	1976	1977	1978
Merchandise trade						
Imports	261	323	295	396	495	528
Exports	175	200	169	147	150	176
Balance	−86	−123	−126	−249	−345	−352
Invisible items						
Payments	71	80	92	96	90	83
Receipts	163	200	255	243	200	216
Balance	+92	+120	+163	+147	+110	+133
Current balance	+6	−3	+37	−102	−235	−219
Over-all balance	−6	−22	−25	−154	−185	−239

Source. United Nations: *Report of the Secretary-General on assistance to Mozambique*, General Assembly, 34th Session (New York, doc. A/34/377, 16 Aug. 1979).

Table 109. Output of principal food crops, selected years
('000 tons)

Crop	1970	1973	1975	1977	1978
Wheat	4.5	6.0	3.0	2.0	3.0
Rice	98.8	108.0	101.0	35.0	35.0
Sorghum	194.7	213.0	180.0	230.0	200.0
Millet	32.0	10.0	8.0	8.0	7.0
Maize	373.2	565.0	250.0	350.0	400.0
Sugar-cane	2 571.4	3 600.0	2 400.0	2 100.0	1 700.0
Cassava	2 547.4	.	2 300.0	2 150.0	2 150.0

. = not available.
Source. *Anuario Estatístico, 1972*; FAO: *Production Year Book*, various years.

Table 110. Output of major cash crops, selected years
(tons)

Crop	1970	1973	1975	1977	1978
Cashew nuts	200.0	206.5	95.0	180.0	150.0
Cotton	139.0	102.0	85.0	67.0	78.0
Sisal	29.8	26.7	19.0	18.0	18.0
Tea	18.7	18.7	13.0	14.0	14.0

Source. *Anuario Estatístico, 1972*; FAO: *Production Year Book*, various years.

Agrarian policies and rural poverty in Africa

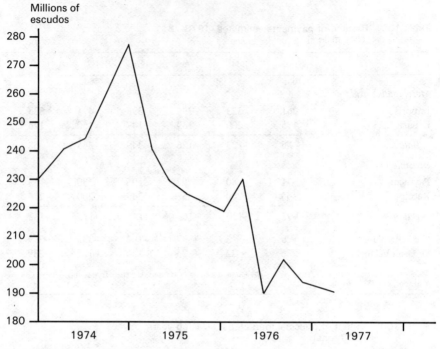

Figure 5. Total revenue of railways, harbours and transport

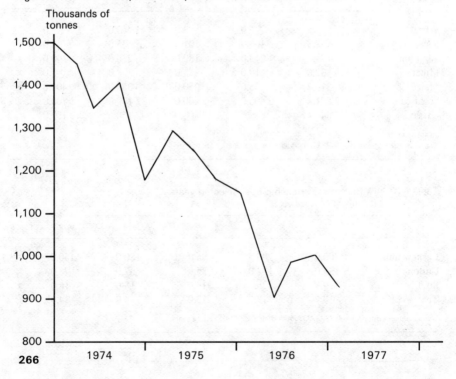

Figure 6. Traffic in the ports of Maputo, Beira, Quelimane and Nacala

Employment

It is difficult to assess the over-all impact of the economic crisis on the employment situation mainly on account of the shortage of concrete data. Information collected by various aid missions has, however, been pieced together to provide some insight into the situation (table 111). In comparison with the employment levels in 1970 (table 105), there was a steep fall in both activity and employment in all non-agricultural sectors by 1976; in most cases, employment appears to have fallen by almost half. In consequence, urban unemployment levels rose sharply. The unemployment situation was further complicated by two factors: one was the drastic curtailment of opportunities in the neighbouring countries, whether of mining employment in South Africa or of plantation work in Southern Rhodesia. In addition, some migration took place from the rural areas to urban centres in search of wage employment.

Although the nationalisation of abandoned farms and agro-industrial complexes in the rural areas led to some stabilisation of employment in the commercial sector, the bulk of the rural population had to fall back on subsistence farming. Given the livelihood configurations in the different regions of the country, the search for urban wage work was an inevitable result of the breakdown of the rural cash economy.

It is easy to exaggerate the dimensions of the economic crisis. While the fall in agricultural and industrial production has been serious, it has sometimes been argued that the subsistence sector, through remaining isolated, has not been affected to any appreciable degree; thus, isolation is said to have as it were cushioned it from the effects of the general economic crisis. This view is hardly acceptable. In the short term, the economic difficulties severely restricted the Government's capacity to undertake programmes in the rural sector for the direct benefit of the mass of the poor farmers; in the long run, they would appear to have swayed the order of priorities away from the family sector.

Socio-economic objectives and priorities

The economic difficulties no doubt contributed very largely to the formulation of socio-economic objectives and priorities by the Third Congress of FRELIMO in February 1977. The Congress, which appears to have recognised the practical imperatives of the situation, stressed the need to increase production and productivity in all economic sectors. In addition, its approach was influenced by some major social and political factors.

It will be appropriate to note that the socio-economic goals of Mozambique were fashioned mainly out of the experience gained during the liberation struggle. The political context of the struggle appears to have influenced the trend towards a socialist approach and ideology and the need to "extend

Table 111. Estimates of employment in non-agricultural sectors, 1973 and 1976 (thousands)

Sector	Total employment	
	1973	1976
Mining	.	6.0
Manufacturing	161.0	79.0
of which		
Construction industries	20.6	9.0
Metal works and light industries	6.0	4.4
Export industries	30.0	22.5
Construction		
Civil construction and public works	28.0	23.0
Trade		
Commerce and banking	.	39.0
Transport		
Ports, railways, cargo handling	36.0	23.0
Services		
Public administration	.	59.0
Public services	.	38.0
Personal services	.	32.0

. = not available.
Sources. Some of these estimates are based on SIDA: *PM Om Moçambiques ekonomi och biståndsläge* (Nov. 1976), quoted in *Mozambique: Food and agriculture sector preliminary study* (Uppsala, Swedish University of Agriculture, Nov. 1976). The estimate for mining sector is for domestic employment only; employment in South African gold mines was 44,100, as shown in table 106.

and reinforce the democratic power of the people". Throughout the liberation struggle, however, it was the rural peasantry which was in the forefront of FRELIMO's activities. This has, no doubt, largely contributed to the subsequent emphasis on the development of agriculture, as well as on the importance of peasant-worker association in all later ideological formulations.

A further important factor was the ethnic composition of the country and the need to overcome tribal divisions and conflicts (one major theme of FRELIMO's political campaign was "to die a tribe and be born a nation"). These rivalries had been fully exploited by the colonial Government. The subsequent emphasis on reducing regional inequalities, particularly in the social sectors and in the access to political structures and decision-making processes, can be attributed to this factor. However, regional economic diversity, in terms of both natural resources and economic infrastructure, necessarily led to the recognition of the need for regional specialisation in both agriculture and industry. In addition, the rural strategy has differed in important respects in different regions largely in response to past livelihood patterns.

Finally, FRELIMO looked upon the need to break away completely from the colonial pattern of development as a paramount consideration in its socio-economic strategy. The transformation of colonial *aldeamentos* (fortified hamlets) into *aldeias communais* (communal villages) in the northern region was an expression of the need to replace colonial modes of life. Indeed, the concept of collective production and communal living within the framework of co-operatives or *aldeias communais* appeared to be the right answer both in terms of ideology and by way of contrast to earlier colonial patterns.[13]

The main "directives" adopted at the Third Congress derived from the basic premise of "agriculture as the base and industry as the dynamic and decisive factor", and required the Party and the Government to promote the increasing socialisation of agriculture, to accelerate the process of industrialisation and promote the creation of heavy industry, to develop and consolidate the role of the State in the economy, and to guide the process of development through over-all economic planning. The principal sectoral objectives were stated to be as follows: "In agriculture, we must give priority to the production of the main foods that our people need; in industry, we must increase and diversify the production of the main goods that the people need; and we must organise the commercial network that ensures the distribution of produce to the people and guarantees the storing of surpluses, linking it with the transport sector."[14] Within these broad objectives, sectoral priorities were indicated and quantitative targets were formulated for the period 1977-80.

Later developments have, however, shown that socio-economic policies were largely tempered by the exigencies of the economic situation and the relatively slow process of national reconstruction. There is now more and more reliance on selective interventions and market incentives in the production centres. Explicit reliance on the private sector in agriculture, industry, trade and business, albeit to a small degree, appears to have become an accepted part of the economic system. Considerable efforts have also been made towards developing management systems with increasing accountability. In addition, a rural institutional network is being established.

Rural institutions

When one considers the development of rural institutions in Mozambique, it is perhaps crucial to underline the importance of FRELIMO's views on leadership, organisation and socio-political mobilisation. During the period 1972-76, in particular, these inter-related themes were central concerns in all political training, especially at the FRELIMO schools for the training of party cadres. This is hardly surprising. In a country where the large majority of the people were forced to work for extremely low wages, the notion of work had become almost synonymous with the notion of exploitation. To change this approach into a participative process of decision-making in which work is linked to the workers' interest and responsibility and to the well-being of the society as a whole was an immense task.

Two innovative and participative organisational devices were introduced for this purpose. The first was that of the "dynamising group" *(groupo dinamizador)*. The main function of these groups was to raise the people's political consciousness and bring about a heightened awareness of the purpose of the liberation struggle. They were to involve the people in participatory decision-making and in electing local leaders. The groups were organised as an experimental measure during the period of the transitional Government in 1974-75 and were initially seen as temporary links between the party and the people. Members of the group were selected by the local population or the workers concerned, primarily on the basis of ability, leadership qualities and willingness to work. They were established in all places of work, study or residence ranging from villages to government ministries.

A typical "dynamising group" was composed of six to eight persons—not necessarily members of FRELIMO—with distinct responsibilities.[15] These included the organisation of discussions with the people regarding party policies, the development process, production (or work) targets, and collective effort. Special attention was given to organising activities for women in co-operation with the Organisation of Mozambican Women. Among the many changes promoted by these groups, free and open discussion of issues was one of the most important.

The second innovative form of organisation was the *responsavél*. The members of dynamising groups were originally called *responsavél*; the term was later extended to all heads of production or work units, and it is currently used to indicate the decision-making level of leadership throughout the hierarchy, especially in the provinces. Thus, it has now become the generic term for leader.

The term originated during the liberation struggle. It applied to ordinary men and women who had never before had any authority—farmers and miners, both men and women—who, in response to the imperatives of war, became part of a highly disciplined organisation. Shouldering responsibility within the liberation organisation transformed these individuals into leaders.[16]

The notion of *responsavél* in Mozambique seems to carry with it an understanding of leadership that is distinct in two aspects. First, the leader is not the "boss", that is, the one who makes all the decisions, but rather the one who is responsible [for ensuring] that the decisions made by all are carried out. Thus, the concept of *responsavél* seems to connote responsibility more than power. Second, the development process seems to be understood by Mozambique's leadership as the welding together and the realisation of plans originating both at the top levels of authority and at the bottom—that is, from the national leadership as well as from the people who selected the leadership. The authority of the *responsavél* rests in being accountable to all levels of decision-making.[17]

These organisational innovations have, no doubt, undergone changes as more experience has been accumulated in the working of the party and the Government. The "dynamising groups" have been replaced by party cells and workers' councils in most cases. There is also a perceptible change in the *responsavél* concept, perhaps in response to the slow progress made by many enterprises and farms taken over by the Government. These organisational changes are likely to continue even as the nature of the challenges and tasks before the Government is modified in the course of time[18] and better-defined hierarchical structures emerge.

However, the participative approach has also led to significant institutional changes in the rural sector. The transformation of rural production relationships was high on the agenda of FRELIMO's socio-economic programme. The resultant "socialisation" of agriculture took three main forms. These are briefly discussed below.

State farms

It is convenient to group plantations, *latifundios* and *colonos* together under this category. The state farms were formed as a result of the Government's take-over of all such rural enterprises abandoned by their expatriate owners.[19] Thus the farms vary greatly in size, workforce, products and efficiency.

In line with the political ideology of the Government, the nationalisation of abandoned expatriate farms appeared to be a logical step. Moreover, this could be justified on the strictly economic ground of recovering and rehabilitating valuable assets threatened with disintegration. The farms were run by the resident workers until "dynamising groups" were established and *responsavéls* appointed. The former have given way to workers' councils. Workers are organised in brigades, and each brigade selects a representative to sit on the council overseeing the day-to-day operations of the state farm.

In the initial stages, as production on these farms declined, the first charge on their output was the fulfilment of the workers' food needs and the main goal was to keep the enterprise operational, even though it was uneconomic. A special office was established to rehabilitate the state farms sector. The next step in development was the establishment of production targets with a view to reaching pre-independence output levels. More recently, the need to develop the farms as efficient and modern production units has been stressed, and steps are being taken to introduce a high degree of mechanisation in their operation. They are expected to produce food for meeting urban needs and cash crops for export. It has further been recommended that "state farms assume their role as centres for the propagation of agricultural techniques as well as commercial distribution centres for the products of co-operatives and the family sector, making an effort to guarantee the supply of inputs ... the structures of the regional communal villages take part in the planning and control of state farm activities so as to ensure full employment for rural labour and the proper distribution of state support to the different production sectors."[20] Meanwhile, the Government has been moving towards the reorganisation of the larger plantations and farms into autonomous parastatal units.

Communal villages

The communal village was conceived as the basic political, social and economic unit which should, in due course, develop as an autonomous administrative and political entity with its own administration, security, justice, finance and basic services. This objective is, however, far from being achieved. Early in 1980 about 1,060 communal villages were reported to be in existence with a total population of approximately 1 million. They were at different stages of development.

The concept of the communal village arose from the experience of transforming the colonial *aldeamentos* and reorganising them for collective activity. In the second place, communal villages were formed when large-scale resettlement of the population took place after floods in the Zambezi and the Limpopo. Finally, the Government has actively encouraged the formation of communal villages in other areas through political mobilisation as well as through the provision of basic services.

The principles for establishing communal villages, their functions, organisation of production and work, and other related aspects, were set down in a resolution adopted by FRELIMO in 1976.[21] They stress the need for political mobilisation and for the establishment of a production base as the essential elements in the formation of the village. It is further emphasised that work will be organised in a collective manner: "Salaried work comes to an end and production becomes organised in two manners: family production, and collective production with a co-operative orientation." Private ownership is not allowed in the communal villages; however, each family is permitted to own and cultivate "a family plot not exceeding half a hectare in irrigated areas or 1 hectare in dry-land farming regions. The size of this family property can vary within these limits on account of different factors, such as the number of family members ... Each family may also hold as personal property the hand tools necessary for their family production and a limited number of animals. There is no limitation upon the number of small stock." The decisions on all questions of family property are taken by the communal village collectively.

The internal organisation of the communal village consists of a general assembly and an executive committee. Often the latter is synonymous with the party cell, and is responsible for planning and guiding the daily tasks. All members of the communal village are obliged to participate in the tasks assigned to them by the committee—production tasks as well as political, social and cultural work.

The collective production is utilised for *(a)* meeting the food needs of the people; *(b)* reserves (seed) for the following season; *(c)* financing the social services; and *(d)* repayment of loans and other production expenses. Of any surplus production that may be achieved, one-fifth is to be placed in a fund for the upkeep, maintenance and purchase of production inputs, and for the promotion of political, social and cultural activities.

The Government does not provide any services or inputs to the communal village free of charge except some technical advice, particularly for the construction of buildings, roads and irrigation facilities and for agricultural extension, as well as support for education, political and cultural services. A communal village is expected to become self-sufficient within two to three years.

Policies and programmes relating to the development of communal villages are co-ordinated at the national level by the National Commission for Communal Villages; there are also regional commissions and district committees.

The communal village approach is a challenging and ambitious one in the Mozambican context. The transformation of the rural peasantry into disciplined participants in organised collective endeavour is bound to take time, and many difficulties must be overcome before any significant progress can be made. Conflicts are reportedly arising in time allocation between work on collective farms and family plots in several areas. Low productivity of land and labour have combined to make food self-sufficiency a distant goal for most communal villages. Given the limited crop mix and the absence of irrigation in the overwhelming majority of cases, labour use is markedly seasonal. There is hardly any opportunity for off-season work outside the village and thus for earning cash to meet non-food needs. The question of motivation and participation on the part of the people is, clearly, critical to the success of the whole experiment. While the communal villages in the northern region derived their inspiration from the liberation struggle and its attendant historical circumstances, and in the southern region from the compulsion of natural disasters, it is difficult to mobilise the peasantry in the regions with no special circumstances. Moreover, such mobilisation depends upon effective local leadership. To mount an economic as well as a political programme of such magnitude throughout the country has strained available human and material resources to the maximum.

Co-operatives

This is nowhere more evident than in the co-operatives. The People's Assembly recommended that "village co-operatives be clearly defined as the body to integrate all the collective economic activities of the villages. The co-operative must therefore be polyvalent, embracing agricultural production, trade, building, artisanal or industrial production, etc."[22]

Public Law 9 of 1979 defined seven types of co-operatives according to their main activity: agricultural, fisheries, industrial, handicraft, consumer, housing, service. With the establishment of communal villages, the co-operative movement has gained ground rapidly. By the end of 1979 there were

314 agricultural production co-operatives in the country; of these, 44 were defined as collectives, 121 as pre-cooperatives and 149 as co-operatives. In most communal villages production co-operatives are in their initial stages; their organisational and management problems are still to be resolved, and their production base has yet to be firmly established.

Each agricultural co-operative has a general assembly, a controlling committee and a board of directors. The formation of these bodies is a prerequisite for the registration of the co-operative and for receipt of government assistance.

At the national level, an office for the organisation and development of agricultural co-operatives was established in March 1979 with wide terms of reference;[23] many regional and district offices were also set up in the following months. However, perhaps the most serious problems being faced by the co-operative organisation are the lack of trained personnel and inadequate supply of agricultural inputs. The slow progress in the development of these co-operatives has led to a revision of priorities: it has thus been decided that one pilot co-operative should be selected in each district for concentrating technical support, as well as resources, so that it may serve as a model for agricultural co-operatives as a whole.

A related problem which is likely to arise before long is that of co-ordination of the activities of the various government organisations which are being established in relation to the communal villages: the district committee for communal villages, the district office of agricultural co-operatives, and the district office of consumer co-operatives.

The consumer co-operatives are developing as independent entities, possibly in response to the more urgent need of replacing the supply and distribution network in the rural sector. Soon after independence the Government established "People's Shops" *(Lojos de Povo)* to take over the functions earlier performed by the *cantineiros*. Many consumer co-operatives have recently been formed in place of People's Shops. About 700 consumer co-operatives are reported to have been formed during the period 1977-79. The over-all control and direction of these co-operatives rests with the Co-ordinating Commission of Consumer Co-operatives of the Ministry of Internal Commerce, and its provincial and district offices.

Consumer co-operatives face the challenging task of replacing the conventional supply distribution network in the rural areas. Their main problem has been inadequate or delayed supply of essential consumer goods. Their limited success has been perhaps the most important factor in the government decision to allow private shops to be established in rural areas.

GROWTH AND DISTRIBUTION

It is rather early to make an assessment of the impact of various government policies in the rural sector; hardly five years have elapsed since the

Third Congress of FRELIMO at which policies, priorities and production targets were laid down. However, some clear patterns are emerging from the experience of developments so far. These may be considered in the context of the twin aims of growth and distribution being pursued by the Government.

To begin with, we may look at the Government's capital budget and the share of different sectors as reflected in it (table 112).[24] The figures presented in the table are the shares proposed; actual expenditure has always fallen short of the budgeted amounts for two main reasons: first, the implementation capacity has been severely constrained by the shortage of trained manpower; and, second, the element of external financing assumed in successive years has been high. External aid and borrowings have not matched the estimated demand.

Agriculture shows a falling share of the investment budget. Most of the programmes included here relate to state farms and large agricultural projects, and to their accompanying high degree of mechanisation. The share of communal villages and co-operatives is extremely small, while no investment is envisaged in the family sector. The increasing share of industry and energy, transport and communication, and public works shows that investment is going largely to the urban sector and to the development of the transit and export trade network. The colonial pattern of economic dualism appears to be persisting in Mozambique, both in the economy as a whole and in the rural sector.

The factors which explain this investment bias are not difficult to identify. Soon after independence, urban food requirement was the most urgent priority before the Government.[25] It appeared that this demand could most easily be fulfilled by strengthening the capacity of the state farm sector and by the rapid mechanisation of large farms. Later experience has, however, not borne out these assumptions. Indeed, the question of allocating investment between large farms and the family sector needs to be re-examined carefully, even from the point of view of urban food needs.

The case for the mechanisation of state farms appears to be equally weak, particularly in the face of growing unemployment in both urban and rural sectors, and the urgent need to provide wage work to a significant part of the rural peasantry. The trend in rural-urban migration since independence should provide a clue to the magnitude of the problem. Such migration places a greater strain on available resources not only by reducing labour input into agriculture but also by drawing upon the limited urban food supply.[26]

What is urgently required is a comprehensive review of the total strategy for rural development. Several elements of this strategy already exist, including, inter alia, the establishment of agricultural production co-operatives and communal villages, the replacement of the rural supply and distribution network, programmes of rehabilitating the rural infrastructure, and social investment, particularly in education and health. In this strategy, if the right production and labour utilisation relationships can be established, the state

Table 112. Composition of investment budget, 1976-80[1] (percentages)

Sector	1976-77	1978-79	1980-81
Agriculture	22	19	18
Industry and energy	6	10	12
Transport and communication	23	24	26
Public works	20	23	23
Social sectors	12	7	7
Others	17	16	14

[1] The figures are two-year averages; those for 1980-81 are estimates. (See also note [24].)

farms sector has a vital role to play. Clearly, there is a need to keep the state farms going and to make them a productive and surplus-producing sector. However, it is now becoming apparent that this cannot be achieved either by resorting to the exploitative labour relationship of the colonial period or by rapid mechanisation.

A critical element in this strategy would seem to be the role of the family sector. It has been noted earlier that, throughout the colonial period, this sector provided most of the agricultural output, even under negative conditions for survival. This should be evidence enough of the adaptability of the family sector and of its capacity to respond to modest investment and favourable terms of trade.

Another urgent need is for the systematic examination of labour and production relationships in the rural sector with a view to establishing a long-term balance between rural institutions and avoiding conflicts of interest between them. At first sight, it would seem as if scarce resources (capital, inputs and manpower) are being channelled between competing institutions which have not yet developed interdependent and mutually reinforcing relationships.

This may not be the intention of policy but may be the result, in the short term, of absorptive capacity. A more even spread of investments over all production sectors, including the family sector, appears to be necessary within the framework of an integrative relationship.

Equally important will be an attempt to co-ordinate the labour requirements of different components of the agricultural sector in order to even out the impact of seasonality, reduce the need for large-scale mechanisation and provide some opportunities for the peasantry to engage in cash farming.

The balancing of investments between food crops and export crops is also a necessary element in the rural development strategy. Their foreign-exchange-saving and foreign-exchange-earning roles can be made complementary. If the historical trade balance is any indication, it is perhaps the foreign-exchange-saving role which may prove to be more important in the medium term. This again points to the need for greater support to be given to the family sector.

It is central to the Government's long-term strategy that the livelihood patterns of the colonial period should be completely transformed. Limited experimentation with alternatives has so far concerned itself mainly with organisational form rather than productive content. While growth has been emphasised in all sectors of economic activity, rural development will often succeed only when its distributive effects are felt by the majority of the population. And while the expansion of literacy, education and health services do ease the problems of distribution, in the particular situation of Mozambique a marginal rise in the production and incomes of the peasantry at large could be the turning-point for both rapid growth and improved distribution. For this to be achieved, the dualistic pattern of economic development will need to be altered.

CONCLUSION

As mentioned at the outset of this chapter, Mozambique's development experience is too recent to allow a full assessment to be made at this stage. It emerges from the preceding analysis, however, that since independence an effort has been made to promote equitable rural development through major institutional change. While it may be too early to provide a full appraisal of the experience, we may draw a few lessons for the future from the preceding analysis.

Given Mozambique's natural and human potential, the long-term economic prospects can be expected to be good. However, there is an immense challenge to be met in the immediate future; seven years after independence, considerable problems remain to be overcome.

The dual structure inherited from Mozambique's colonial past is perhaps the most formidable danger for future development. This dualism between large modern estates and small family farms appears to continue even under the new institutions. There is clearly a need to balance the developments in the rural sector within an integrative framework: for example, there may be more to lose in terms of distribution than to gain in terms of growth if the state farms sector and large agricultural projects run too far ahead of the family farm sector. Moreover, given the colonial framework of livelihood patterns, it is perhaps more important to provide minimum benefits to the peasantry at large through improved terms of trade in order to promote growth with distribution. This is necessary also in order to relieve the acute unemployment which prevails in the towns. In other words, a strategy of rural development for the future which aims at reducing the existing flagrant dualism should give greater weight to the family farm sector.

Related to this point is the whole issue of food supply. The food problem of urban areas should not be looked at in isolation. There is a need to review the various components of the food production, supply and distribution system. Current reliance on large farms to meet urban food needs (and exports) does not appear to have produced the expected results. This is a pattern which was established during the colonial period under very different systems of labour utilisation; it is facile to assume that those systems can be overcome by resorting to mechanisation. It is necessary to explore other alternatives for reaching national and regional food self-sufficiency. With its natural resources, Mozambique has the potential of becoming a net food exporter in the foreseeable future.

Another feature of Mozambique's inherited economic structure was its integration into international markets through exports. Recent developments point to an increasing dependence on exports of cash crops and limited agro-processing at home. The dangers of such a pattern of dependence are obvious: not only does it imply external vulnerability of the economy but it also generates conflicts in the rural sector between food crops and cash crops. Experience elsewhere in Africa has shown that food security must be an important component of an equitable rural development strategy.

It is also natural in such circumstances that within the over-all investment strategy there should be competition for resources between production and service sectors. Here again, a balance should be kept between the need to provide the population with social services and investment requirements for future growth.

In its colonial period Mozambique was further characterised by under-development of its human resources. Perhaps this will continue to be the most formidable constraint on rural development in the foreseeable future. There are two aspects of the problem which need to be handled in co-ordination. First, there is a shortage of trained personnel in all sectors of activity. Careful and comprehensive planning, together with the implementation of manpower development and training programmes, are essential for this purpose. Some progress has been made in this direction, but there is a need to develop an over-all and well co-ordinated framework for dealing with short-term and medium-term manpower problems. Second, to a large extent, the shortage of trained personnel and the consequent weakening of government services can be made good by promoting and developing participatory institutions. Much progress has been made in Mozambique within a short period in this respect. There is now an urgent need to consolidate this gain, especially as regards training, organisation and management of co-operatives, and developing a solid base of economic activity in the communal villages.

In formulating the socio-economic strategy for the 1980s Mozambique would thus need to consider both the problems emanating from the major constraints on growth and those created by its dualistic structure. Various policy options should be examined with a view to mitigating the constraints and reducing the persisting dualism.

The fact that there is a continuous and pragmatic review of policies and programmes at the highest levels in the Government gives grounds for hope that Mozambique will emerge in the relatively near future from its post-independence problems.

Notes

[1] Among these the more important were: Niassa Company, Mozambique Company, Zambeze Company, Boror Company, Madal Society and Sena Sugar Estates. See Allen Isaacman: *The Zambesi prazos: The Africanisation of a European institution, 1750-1902* (Madison, University of Wisconsin Press, 1972); and R. Palmer and N. Parsons: *The roots of rural poverty in central and southern Africa* (London, Heinemann, 1977). The Mozambique Company, incorporated in 1891 for a term of 50 years, exercised sovereign rights over the territories of Manica and Sofala. It had control of agriculture, commerce, industry, mining, communications, transport, taxation and customs, and issued its own currency and postage stamps.

[2] Part of the area was cultivated directly by the *latifundios* while peasants cultivated the remaining as tenants paying rent in kind or in labour in a semi-feudal relationship. Some Mozambicans also participated in the land concessions system; they were drawn from among *assimilados*.

[3] Under these decrees the *indigenas* were subject to several regulations which affected their freedom of movement, wages and working conditions. They had to carry a passbook for identity, obtain travel permits *(guias)* and take permission for sale of crops or slaughter of cattle. Under the decrees almost 99 per cent of Mozambique's population was legally defined as *indigenas* and could be forced to work for the expatriates or for the Government.

[4] A circular (566/D-7) laid down seven items which could exempt an *indigena* from compulsory labour. These were: *(a)* self-employment in a profession, commerce or industry; *(b)* permanent service under the State, administrative corps or private persons; *(c)* work for at least six months as a day labourer under the State, administrative corps or private persons; *(d)* to be within six months of return from South Africa or Rhodesia under a legal contract of work; *(e)* cattle-raiser with at least 50 head of cattle; *(f)* to be registered as an *agricultor africano* under the statute of "african agriculturist" (Diploma Legislation 919 of 5 Aug. 1944); and *(g)* to have completed military service and be in the first year of "reserve" status.

[5] Some estimates place the number of Mozambican agricultural (mainly plantation) workers in east Transvaal at as high a figure as 80,000 and in Southern Rhodesia at 50,000. This peak was reached around 1970-72. A good proportion of these workers (about 30 per cent) were illegal migrants.

[6] On this subject see especially the detailed study by Ruth First et al.: *The Mozambican miner* (Maputo, Centro de Estudos Africanos, Universidade Eduardo Mondlane, 1977). The deferred wages (37.5 per cent) were paid in gold at the official price and credited to Portugal's account. The subsequent sale of gold by Portugal at much higher world gold prices yielded substantial earnings for the Portuguese Government. In 1973 Mozambique received £25 million from this system while Portugal received about £40 million from gold sales at free market prices. The workers received deferred payments in Mozambican escudos less commission.

[7] Average salaries (in 1970 escudos; US$1 = 28.75 escudos) per annum in different sectors were reported to be as follows:

Sector	1960	1970
Agriculture	1 821	3 629
Mining	2 215	4 640
Processing industry	2 885	5 662

According to *Portugal e capital multinacional em Moçambique, 1500-1973,* Vol. II, these averages were low because of the very low wages of African workers; Europeans working on farms earned 55 times more than Africans, while Asians received about one-third of the European salary.

[8] ILO: *Poverty and employment in rural areas of the developing countries,* Report II, Advisory Committee on Rural Development, Ninth Session, Geneva, 1979, table IV, p. 13.

⁹ These schools were heavily concentrated in urban and semi-urban centres, and for a very long time the scattered rural settlements were outside the reach of these facilities.

¹⁰ During 1977 the number of refugees entering Mozambique averaged over 1,000 per month; by mid-1978 the total number was reported to be over 70,000. See United Nations: *Report of the Secretary-General on assistance to Mozambique,* General Assembly, 33rd Session (New York, doc. A/33/173, 12 July 1978).

¹¹ ibid. Net receipts on invisibles and capital account transactions include grants and long-term loans which were expected to meet about half the trade deficit. The direct costs to Mozambique of applying sanctions was estimated in 1976 to be US$139-165 million in the first year, US$108-134 million in the second year and US$106-132 million annually thereafter.

¹² The food import bill in 1978 was US$51.8 million and in 1979 US$64 million. Its main components were maize, 81,000 and 117,000 tonnes; rice, 103,000 and 66,000; and wheat, 53,000, and 148,000 tonnes respectively.

¹³ These arguments were included in the resolution on communal villages adopted at the Eighth Meeting of the FRELIMO Central Committee in February 1976.

¹⁴ See Third Congress of FRELIMO, 3-7 Feb. 1977, Documento informativo no. 6 (1.6.1978), Serie E (Maputo, Centro Nacional de Documentação e Informação de Moçambique, 1978).

¹⁵ See Allen Isaacman: *A luta continua* (Binghamton, NY, University of the State of New York, 1978), pp. 36 ff.

¹⁶ Quoted from Frances Moore Lappé and Adele Beccar-Varela: *Mozambique and Tanzania: Asking the big questions* (San Francisco, Institute of Food and Development Policy, 1980).

¹⁷ ibid., pp. 71-72; see also pp. 76-79.

¹⁸ The most notable evidence of these changes is contained in a recent speech of President Machel (4 Dec. 1979): "It is not the workers' collective that takes decisions; it is the Director that decides ... the hammer in the FRELIMO emblem represents, in addition to the primacy of the working class, its power and decisiveness. Once decisions are taken in consultation with those affected, they must be carried out decisively. But just as the head of the hammer concentrates its force, so should the head of an agency or operation." (Unofficial translation.)

¹⁹ A small number of expatriate farmers have, however, remained after independence. Under the Land Act of July 1979 these private farms are permitted to continue, regardless of their size; the main criteria determining their continued existence are economic efficiency and contribution to production and export.

²⁰ Resolution on agriculture and communal villages adopted at the Fourth Regular Session, 18-22 June 1979, by the People's Assembly.

²¹ Resolution on communal villages, 1976 (see note ¹³ above). The question was subsequently discussed at the Third Congress and finally approved by the Council of Ministers as the Document of Nacala in July 1977. The National Commission for Communal Villages was established in the same month.

²² Resolution on communal villages.

²³ The Gabinete de Organização e Desenvolvimento das Cooperativas Agricolas (GODCA) was established by Decree No. 41/79 of 31 Mar. 1979.

²⁴ The figures have been taken from the reports to the United Nations General Assembly quoted above. Estimates for 1980-81 have been pieced together on the basis of information available from recent government sources. A considerable degree of reallocation between sectors has been required in preparing the figures. Investment in large agricultural projects has been taken out of "public works" and put under agriculture, while construction expenditure from transport has been taken to public works. Sectors such as internal commerce, water supply and drainage, mining, and so on, are grouped under others. The table therefore shows variations from figures published elsewhere.

²⁵ The highest priority in the agricultural sector, according to the economic and social directives of the Third Congress, was "to guarantee the supply of the main agricultural products and give special attention to the supply of basic necessities to the urban centres". See Documento informativo no. 6, op. cit.

²⁶ These arguments have been fully brought out in a recent paper. See Marc Wyuts: *On the question of mechanisation of Mozambican agriculture today: Some theoretical comments* (Maputo, Universidade Eduardo Mondlane, Centro de Estudos Africanos, Dec. 1979), and "The mechanisation of present-day Mozambican agriculture", in *Development and Change* (London, Sage Publications), Jan. 1981, pp. 1-27.

NOMADS AND FARMERS: INCOMES AND POVERTY IN RURAL SOMALIA[1]

10

Vali Jamal

The Somali economy provides an interesting contrast with the other African economies analysed in this volume. It is an economy based on nomadic pastoralism, on which as much as three-fifths of the population depend for their incomes and subsistence. It is, moreover, an economy in which the main growth-generating force in recent years has been provided by livestock exports to Saudia Arabia and other countries of the Gulf. Despite this the economy remains largely subsistence-oriented, with most nomads and farmers still untouched by the market economy. Finally, it is an economy about which relatively little has been written.

In this chapter we shall document the recent progress of the economy, paying particular attention to the extent of poverty amongst the nomads and farmers, the mechanisms responsible for it and the impact on it of the export boom. The chapter is organised as follows: in the first section we take a detailed look at the structure of the economy, especially that of the livestock and rural sectors; in the next section we present a profile of poverty and incomes for rural Somalia based on regional figures of livestock and land-ownership, and using an appropriately defined poverty line; in the third section the trends in inequality are analysed and the mechanisms responsible for them examined; the following section provides a brief look at government investment and pricing policies as they have affected the two sectors; and, finally, the conclusions of our investigation are presented.

More than the customary word of caution is required concerning the data used in this chapter. Statistics on Somalia are relatively sparse and unreliable, which no doubt accounts for the dearth of writings on its economy. Recently, estimates of the national income have been published, along with much background data.[2] We have made extensive use of these to draw a picture of the Somalia economy. As might be expected in a subsistence-oriented economy, problems arise in estimating production and these are apparent in the national accounts figures. In such cases we have derived our own estimates, based on survey data and other information about the working of the Somali economy. It is considered that the picture that emerges throws interesting light on the rural economy and on rural incomes and poverty. It is hoped that, as more statistics become available, the picture we have drawn will gain sharper focus.

STRUCTURE OF THE ECONOMY

In view of the great preponderance of the subsistence sector in the Somali economy, especially the subsistence livestock sector, we shall confine our analysis of the structure of the economy to monetary GDP only in the first instance. The problem is twofold: one related to the interpretation of GDP figures and the other to the quality of the data. The first problem is familiar in most developing countries, but is especially crucial in Somalia, where a great part of the subsistence output consists, on the one hand, of high-valued products such as milk and meat and, on the other, of comparatively low-valued goods such as maize and sorghum. Thus, if the livestock sector is valued to include own consumption it would unrealistically dominate the total GDP, and nomads' incomes too would loom large over other incomes, whereas the situation in terms of real welfare would be quite different. The problem with regard to the quality of the data is that a considerable part of the subsistence output (in the form of milk consumption) has been underestimated, to the extent of as much as one-fifth. In subsequent analysis we shall make amends for this to derive comparable food-consumption profiles and incomes.

A second point with respect to income figures is that the national accounts, by definition, estimate the GDP: namely, incomes accruing to factors of production from activities undertaken in the domestic economy. The recent upsurge of Somali workers emigrating to Arab countries and the money they repatriate to Somalia certainly suggests that the GDP figures considerably underestimate the national income. It is reckoned that between 100,000 and 150,000 Somalis now work in Arab countries, earning three to four times the local wages (estimated at around 960 Somali Shillings per month). If we adhere to median values, these figures imply that the Somalis abroad earn Sh 5,000 million per annum, or some 28 per cent in excess of the total income generated in the monetary economy in 1978, and some 87 per cent more than the income generated in the urban economy. This figure, even if we conservatively estimate that 20 per cent is repatriated one way or the other, would amount to three-quarters of the total wage bill in the country (where the total number of wage earners is 120,000) and 37 per cent of the modern sector income.

However, even with all the underestimation inherent in the nature of the economy and in the supplements received from nationals working abroad, the basic fact remains that Somalia is a poor country, and one moreover with

Table 113. Estimates of total and monetary gross domestic product of Somalia, selected sectors, 1978, and real growth rate, 1970-78

Sector	GDP in 1978 (Million shillings and % share)		Average real growth rate, 1970-78 (% per annum)	
	Total (%)	Monetary (%)	Total	Monetary
Agricultural	4 207 (61.1)[2]	1 032 (27.8)[3]	+ 3.0	+ 2.1
Livestock	3 173 (46.1)	571 (15.4)	+ 3.3	+ 2.1
Livestock change	1 125 (16.3)	– (–)	+ 10.9	–
Crops	621 (9.0)	324 (8.7)	– 0.5	+ 0.1
Other[1]	413 (6.0)	138 (3.7)	+ 7.1	+ 7.1
Industrial	706 (10.3)	706 (19.0)	– 0.2	– 0.2
Manufacturing	433 (6.3)	433 (11.7)	– 0.2	– 0.2
Services	1 972 (28.6)	1 972 (53.1)	+ 8.0	+ 8.0
Government services	608 (8.9)	608 (16.4)	+ 9.1	+ 9.1
GDP at factor cost	6 885 (100.0)	3 711 (100.0)	+ 3.3	+ 2.8
(Excluding livestock change)	(5 760)	– (–)	(+ 2.2)	–

– = not applicable.
[1] Fishing, forestry and poultry. In the original source, poultry is included in the livestock sector; here it has been placed with crops on the grounds that poultry is mostly reared by the non-nomadic population. [2] For agricultural GDP, valuation placed on agricultural products at producer prices has been used, rather than agricultural value added. The difference is slight (Sh 290 million in 1978). This means our total GDP figure would be higher than the figure quoted in the source by Sh 290 million in 1978 (and Sh 48 million in 1970). We have resorted to this approximation to facilitate later analysis. [3] Division between monetary and non-monetary agricultural GDP is made as follows: (a) for "Livestock", based on background data in Central Statistics Department, State Planning Commission: *Estimated aggregates*, op. cit.; (b) for "Crops", also on background data, using purchases by the Agricultural Development Corporation as a basis and some "topping up" to account for crops sold outside the official channels; (c) for "Other", a division in the ratio of 1:2.

Source. Central Statistics Department, State Planning Commission: *Estimated aggregates of national accounts at current and constant prices and economic indicators, 1970-79* (Mogadishu, Dec. 1979), tables 1 and 2 and tables relating to livestock and crop sectors.

an economic structure rooted in traditional activities. Table 113 shows estimates of the structure and growth of the GDP over the period 1970 to 1978. The implied income per head of Sh 1,400 would increase by two-thirds if we were to correct for underestimation of milk production. Similarly, the share of livestock would rise from 46 per cent to over two-thirds of the revised estimate. Even without this adjustment, the dominance of the livestock sector in the total GDP is quite striking. If one considers the monetary economy alone, the service sector accounts for an unusually large proportion of recorded output. Over the period 1970-78 the growth in GDP—in great part due to a rapid expansion in the service sector and to change in livestock valuation—barely kept pace with population growth.[3] The output of crops stagnated while that of the manufacturing sector declined.

We now look at the salient features of the livestock and agricultural sectors.

Livestock

There are two main traditional systems of livestock production in Somalia: *(a)* nomadic pastoralism, and *(b)* livestock production by settled farmers. In addition, urban dwellers also keep cows for milk and smaller animals for meat. Livestock is thus a constant feature of the Somali economy and, in the rural areas especially, the dominant feature, with the category of the pure agriculturalist a rarity. Nomads constitute about 60 per cent of the population; but even among the settled farmers (19 per cent of the population) perhaps no more than 10 per cent (i.e. 2 per cent of the total) can be counted as pure agriculturalists.

The settled farmers maintain a permanent homestead where most of their family and animals may be found at any time, and they obtain a major share of their food supplies from their land. These farmer/pastoralists are commonly found in the Juba and Shabelle river valleys, the Bay region, and the north-west. Animals are kept mostly for milk, which supplements the calories derived from grains. Nearer Mogadishu the cattle assume added importance as a source of cash from milk sales.

Nomadic pastoralism—the main system of livestock production in Somalia—supports the majority of the Somali population, providing it with its major source of subsistence in the form of milk. In recent years the herd has also become an important source of cash through the growth of urban demand and livestock exports to Arab countries. Production is carried out as a household enterprise, with the senior male of the composite family as herd manager. Most families keep a mixture of livestock species (camels, cattle, sheep, and goats) to take maximum advantage of species differences in terms of drought resistance, feeding habits and economic uses (camels provide milk and transportation, sheep and goats produce subsistence meat and milk and small amounts of cash, and cattle yield milk and cash income). A composite herd thus constitutes a rational strategy for balancing the objectives of drought security, subsistence needs and cash income.

Some statistics pertaining to livestock and nomadic population are given in table 114. The average nomadic family owns 10 cattle, 65 sheep and goats, and 12 camels, which, costed at average prices prevailing around 1977, represents a livestock wealth of Sh 23,900 per family and, converted into livestock units, a herd of 33 cattle-equivalents. The average of livestock units per family in the central regions is nearly twice the level in the north. In terms of total livestock units, the south accounts for about 52 per cent.

It may also be noted that, even in terms of the proportion of nomadic families, the south is no less pastoral than the rest of the country, 71 per cent of its population being nomadic compared with 79 per cent in the north and 81 per cent in the central regions. What distinguishes the south is that it has a majority of the settled agriculturalists: 63 per cent of the total, as against 28 per cent in the north and 9 per cent in the central regions.

Total livestock population is estimated to have grown by around 3 per cent per annum between 1969 and 1979 in terms of livestock units. The

Somalia

Table 114. Livestock population, livestock units and nomadic families, 1977

Region	Cattle[1]	Sheep and goats[1]	Camels[1]	Livestock units		Nomadic families ('000)	Ratio of nomadic to farm families
				No.[1]	Per nomadic family		
North	364	17 147	1 674	4 600	26.9	157.4	3.8
Central	700	8 198	1 257	3 296	51.3	59.5	4.2
South	3 891	7 363	3 024	8 591	32.4	233.0	2.4
Total	4 954	32 708	5 956	16 488	33.0	450.0	3.0

[1] Thousand head.

Sources. Central Statistics Department, State Planning Commission: *Estimated aggregates...*, op. cit., table 25 for total livestock population. Its division between regions is assumed to be on a proportionate basis to the 1975 livestock census as reported in Somalia: *Three-Year Plan, 1979-1981*, tables 19-3 and 19-4. Livestock units are in terms of cattle-equivalents, with cattle = 1, sheep and goats = 0.125 and camels = 1.25. These are the usual FAO norms. Total population is taken to be 4.77 million, of whom 56.6 per cent are nomads, these figures being official estimates derived from Somalia: *Special statistical issue for the tenth anniversary of 21st October revolution*. Regional distribution of population is based on Somalia: *Three-Year Plan, 1971-1981*, op. cit. table 19-1. Family size is assumed to be six members. Average figures per nomadic family assume that nomads own three times as much livestock as settled agriculturists.

growth of the herd and its total population at present are generally believed to have led to serious overgrazing and deterioration of the rangelands.[4] This is attributed to the traditional system of production, in which, while the animals are privately owned, the range which supports them is communally owned. The interest of the individual motivates him to build up as large and varied a herd as possible, with as many lactating female animals as the labour force can manage, and to move over wide areas to take advantage of favourable conditions for the animals. He has no incentive to limit his herd or restrict its movement since he cannot reap the benefits from any consequential improvement in the range. The end result is overstocking and its attendant ills—decreased productivity of pasture and animals.

Despite discouraging productivity trends in the livestock sector, its financial position now is much stronger than a decade ago because of the export boom. This has brought a substantial amount of cash into the hands of the pastoralists, increasing their purchasing power and their terms of trade vis-à-vis the rest of the rural, as well as the urban, sector.

The nature of the boom may be discerned from the figures given in table 115. It will be seen that, after the initial increase in numbers in the 1960s, physical quantities more or less stagnated in the 1970s, after which the total numbers fluctuated wildly. In the meantime, prices increased almost continuously, especially after 1972, the year by which the physical boom had practically collapsed. As a result, although fewer animals were exported in 1978 compared with 1972, the country earned 3.7 times as much from exports as six years previously.

This performance had a phenomenal impact on the producers' income, as may be seen in tables 116 and 117: in just eight years the total income in their hands increased fivefold, most of it coming from export sales. This implies

Table 115. Exports of live animals from Somalia, 1960, 1964, 1967 amd 1970-78

Year	Cattle			Camels			Sheep and goats		
	No. ('000)	Price (Sh per head)	Value (million Sh)	No. ('000)	Price (Sh per head)	Value (million Sh)	No. ('000)	Price (Sh per head)	Value (million Sh)
1960	12.4	239	3.0	6.3	337	2.1	574.5	58	33.1
1964	56.6	348	19.7	17.7	549	9.7	1 015.5	56	57.3
1967	35.7	344	12.3	36.7	585	21.5	933.4	64	59.8
1970	45.4	341	15.5	25.5	745	19.0	1 150.9	74	84.9
1971	56.1	330	18.5	23.7	700	16.6	1 184.6	74	88.2
1972	77.1	289	22.3	21.2	694	14.7	1 617.2	76	123.5
1973	69.6	504	35.1	27.9	835	23.3	1 322.7	105	138.3
1974	27.4	788	21.6	23.7	1 275	30.2	1 211.2	141	170.6
1975	38.7	866	33.5	33.4	1 421	47.4	2 304.4	131	301.1
1976	76.2	943	71.9	36.6	1 349	49.4	747.6	214	159.9
1977	54.4	767	41.7	34.6	1 393	48.2	902.1	210	189.6
1978	73.9	1 365	100.9	21.0	1 932	40.5	1 450.7	308	447.3

Sources. Figures for 1960, 1964 and 1967 from Somalia: Development Programme, 1971-1973, table on p. 50; for 1970-78 from Central Statistics Department, State Planning Commission: Estimated aggregates..., op. cit., tables 54-56. Sheep and goats have been combined together.

Table 116. Export sales (producer value and f.o.b. value)
(million shillings and percentages)

Animal	1970		1978	
	Producer value	f.o.b. value	Producer value	f.o.b. value
Cattle	6.4	15.5	52.9	100.9
Sheep and goats	48.3	84.7	296.0	447.3
Camel	6.7	19.0	17.0	40.5
Total	61.4	119.2	365.9	588.7
	(51.5)	(100)	(62.2)	(100)

Source. Central Statistics Department, State Planning Commission: *Estimated aggregates...*, op. cit., table 27 for producer value of export sales, table 55 for f.o.b. value.

Table 117. Producer income, 1970 and 1978
(million shillings and percentages)

Source of income	1970		1978	
	Total	%	Total	%
Export sales	61.4	55.1	365.9	64.1
Meat factory	6.5	5.8	·	·
Internal sales	38.1	34.2	182.0	31.9
Hides	5.4	4.8	23.0	4.0
Total	111.4	100.0	570.9	100.0

· = not available.
Source. Central Statistics Department, State Planning Commission: *Estimated aggregates...*, op. cit., tables 27 and 32. The following assumptions were made: slaughterhouse figures (internal sales) were adjusted in 1978 to bring urban consumption per head to 80 per day; 75 per cent of hides income was assumed to be for cash.

considerable commercialisation of the nomadic sector, a point that may be underlined by contrasting the change in the 1970s with that in the 1960s. Thus, extrapolating backwards from the 1970 figures by using assumptions about population growth and producer prices, we estimate that livestock income in 1960 should have been of the following order of magnitude: export receipts, Sh 19 million; slaughter receipts, Sh 22 million.[5] Between 1960 and 1970, therefore, the cash income of the nomads probably increased from Sh 41 million to Sh 111 million—that is, 2.7 times. In contrast, between 1970 and 1978 it increased fivefold, or by Sh 460 million.

While the export boom has had a smaller impact on the south than on the north, the internal market for the south was far more important. It is worth emphasising, however, that for both the northern and southern nomads the cash income earned from export or internal sales is only a small part of their "total income", represented by subsistence production as well as cash. While it has enabled them to buy grains and consumer goods on the market, the basis of their economy has remained the products they obtain from their animals for self-consumption and barter.

Agriculture

The majority of Somalis engaged in farming, who constitute 19 per cent of the total population, do so as dryland farmers in the area between the Juba and Shabelle rivers, particularly in the Bay region. Thus, of the 700,000 hectares estimated to be under cultivation,[6] 540,000 are cultivated as dry land—or rain-fed—farming and support 75-80 per cent of the settled agriculturalists. Another type of farming associated with smallholder agriculture is that based on flood irrigation, while a third type (controlled irrigation farming) is associated with large-scale units. An estimate of the number of families dependent on each type of agricultural system and the area they cultivate is shown in table 118.

The table further serves to show the geographical concentration of farming in Somalia. Over three-quarters of all farm families, and nearly 90 per cent of the cultivated land, occur in the four areas mentioned therein: Juba, Shabelle, Bay and Waqooyi Galbeed. With the exception of the last, which is in the north, all the others are in the south, so that the south, with three-fifths of all farm families and four-fifths of the agricultural land, including all the irrigated land, is truly the agricultural zone of Somalia.

Rain-fed farms are on average 4 to 5 hectares in size, with the total area cultivated at any time divided into several scattered plots in order to ensure the likelihood of at least some plots coming to harvest in the face of an uncertain rainfall pattern. Sorghum is the main crop, sometimes interplanted with pulses. Yields are low (300 to 400 kg per hectare is the usual range) on account of the poor resource base and poor farming practices. Field preparation is minimal and no insecticides, herbicides or fertilisers are used. Technology is based on the hoe, and the peak period of labour demand during weeding provides the main constraint to acreage expansion. However, larger families are as likely to invest in animals as in land, since the former bring better cash returns than food crops. In fact, many subsistence farmers carry on farming as a secondary activity to animal husbandry, even though they might be obtaining a majority of their subsistence needs from their land.

Flood irrigation farming is another traditional method of smallholder farming in Somalia. The cultivated area again varies according to the extent of the rainfall and flood; when these are sufficient, cultivation might be extended into dryland areas. The main system involves the successive planting of the soils along the rivers as the flood recedes. To optimise water use, fields are smaller and contiguous, and they have higher planting densities than in rain-fed areas. Maize is the staple crop, but it appears to be losing ground to sorghum, which yields better returns to labour. Beans, fruit, cotton

Somalia

Table 118. Estimated number of agricultural families and area, by type of farming, c. 1977

Type of farming	Families ('000)	Area ('000 hectares)
Controlled irrigation	10+35-50 workers	50
Flood irrigation	20-30	110
Shabelle river	15-20	80
Juba river	5-10	30
Rain-fed farming	120	485
Shabelle/Juba	35	263
Bay	26	100
Waqooyi/Galbeed	22	40
Others	36	85
Total	150	645

Source. Based on Somalia: *Three-Year Plan, 1979-1981*, op. cit., tables 19-3, 19-6, and 3-1. The difference between tables 19-6 and 3-1 is reconciled by assuming that rain-fed farming is overestimated in table 3-1 and that all the acreage in table 19-6 relates to rain-fed farming.

and sugar-cane are also grown. Technology is similar to that in the dryland areas. Although labour demands are higher on account of the generally heavier soils, yields are also higher, which fact offers the farmers sufficient incentive to invest extra labour in land rather than livestock, the extensive ownership of which is in any case constrained by the prevalence of the tsetse fly along the rivers.

The third major system of farming (controlled irrigation farming) is associated with large-scale enterprises, either private plantations growing bananas and sugar-cane—but also cotton, maize, fruits, and vegetables—or state farms based on the co-operative principle. Only a few families now depend on this type of agriculture, either as owner-farmers, as workers on plantations, or as volunteers on the co-operatives. In the past, these lands were claimed by class or village groups, although they made little use of them on a continuous or intensive basis. With the arrival of the colonial Power large tracts were requisitioned for plantations of bananas and sugar-cane. Successive governments have continued this practice, recently in the form of state farms, causing the permanent alienation of lands from their traditional owners. It is even believed that some lands under flood irrigation, too, have been appropriated by the expansion of controlled irrigation farming.

The structure of the agricultural sector in terms of output and growth performance is shown in table 119. The most striking and disturbing feature of the figures is the decline in total production, amounting in terms of production per head to 28 per cent, or 3.6 per cent per annum. It was not just the smallholders who suffered declines: both banana and sugar-cane planters registered declines too, which were even steeper than those sustained by the smallholders.

Table 119. Agricultural output and agricultural income, selected items and years

Crop	Thousand tons			Value (million Sh)		
	1970	1978	1979	1970	1978	1979
Food crops	294	271	272	120.4	239.1	246.0
Maize	122	108	108	54.9	81.0	81.0
Sorghum	158	141	140	55.3	105.7	105.0
Fruits and vegetables	225	246	265	93.8	258.9	307.9
Bananas	130	76	90	39.0	63.4	85.5
Industrial crops	511	376	396	70.6	122.9	123.8
Sugar-cane	461	330	350	11.5	15.5	16.5
Sesame	43	40	40	51.6	96.0	56.0
Total	1 030	893	933	284.8	620.9	677.8

Source. Central Statistics Department, State Planning Commission: *Estimated aggregates*..., op. cit., tables 18 and 19. Note that the figures under "Value" refer to value of the output (whether self-consumed or sold) at producer prices.

The reasons for this decline are many and varied. In the first place, one should note the poverty of the natural resource base and the technology used. In the smallholder sector, as has been pointed out, most farmers use practically no fertilisers or other biochemical agents. To combat decline in fertility they resort to shifting cultivation, which is likely to lead to declining yields. In the commercial sector this decline in yields has been especially marked in banana and sugar-cane production, and is attributed in part to management problems arising from the nationalisation of private plantations. The state farms which have taken over, and others which have been created to grow food crops, have experienced additional problems in attracting farmers to cultivate the land on a co-operative basis, and have resorted to mechanisation in an attempt to solve them. This has generally proved unprofitable and, with the passing years, the mounting maintenance problems have resulted in ever more inefficient production and lower yields.

However, not all the poor performance can be blamed on physical factors or on the use of inappropriate technology. Equally important, and perhaps more so, is the lack of economic incentives. As outlined in the second section of this chapter, prices have generally been kept low and have declined considerably in real terms. In the case of bananas, the price has been too low even to cover the costs of private plantations, and for the smallholders too low in relation to livestock prices. Such price trends have done nothing to "wean" the farmers away from livestock; on the contrary, they have reinforced their traditional preference for livestock, and are thought to have caused a partial abandonment of farming in favour of livestock-rearing. This in its turn has led to competition for land from animals, especially in the south, where much fertile land has in consequence become unusable.

Contribution of nomads and farmers

We conclude this section by taking account, in table 120, of the contributions of nomads and farmers in terms of *(a)* employment generation; *(b)* income generation; *(c)* foreign exchange earnings; and *(d)* food self-sufficiency.

In order to take account of the fact that farmers also own livestock, we have differentiated between the contribution of nomads and livestock on the one hand, and farmers/pastoralists and crops on the other. Nomads and the livestock sector easily "win" in terms of their contribution to employment, income and exports. They do well in the important category of food self-sufficiency, being practically self-sufficient. However, the crop sector does better, producing a surplus to feed half as many people as depend on that sector. Along with their livestock production, the farmers/pastoralists produce almost twice as much food as their own requirements. This should adequately disprove the assertion that the crop sector in Somalia is unproductive: for 19 per cent of the population to produce a surplus big enough to feed around 30 per cent of the whole population may be regarded as quite an achievement.

Our findings concerning Somalia's food situation are thus in sharp contrast with what has hitherto been believed. In the period 1975-77, for example, FAO was reporting *total* calorie availability (from home production as well as imports) to be 2,129 per head, or 97 per cent of the requirement.[7] The current estimate of the population[8] would reduce this to only 1,499, or 68 per cent of the requirement, and would imply that the country by itself produced only half of this. We do not believe this to be the case either for the period cited or for the period we have been considering.

The problem arises because of a gross underestimation of milk production by both FAO and the estimators of Somalia's GDP. Our own estimates show that the country produces around 93 per cent of its needs, of which it exports some (in the form of bananas and live animals), so that in the end 15 per cent of the country's requirements have to be imported.[9]

The foregoing facts needed to be corrected in order to show that the food situation in Somalia and Somalia's dependence on imports are not as serious as is sometimes implied. However, they must be seen within their proper dynamic context, for the fact remains that yields in both the subsectors have been falling for a long time. This fall is difficult to quantify in the case of the livestock sector, but is glaringly true of the crop sector. In 1970 the country produced 43 per cent more crops in terms per head than in 1979. Imports of cereals were half of those in 1977, and banana exports were double those of eight years later.

In the meantime the livestock sector began to make its own special contribution to food self-sufficiency, as a result of the livestock export boom, which enabled the country to export live animals at better prices and at a very favourable exchange rate in terms of calories; this may be seen in the fol-

Agrarian policies and rural poverty in Africa

Table 120. Relative contributions of nomads and farmers, c. 1977

	Nomads	Farmers/pastoralists	Livestock	Crops
Population (millions)	2.7	0.91	–	–
Percentage	56.6	19.0	–	–
Income (million Sh)	368	308	409	267
Percentage	13.6	11.4	15.1	9.9
Exports (million Sh)	–	–	433	89
Percentage	–	–	78.6	16.1
Food production (thousand million calories)	2 140	1 389[1]	2 432[2]	1 149
Percentage of:				
Own requirement	99	191	–	158
Country requirement	56	36	63	30

– = not applicable.
[1] 1,149 calories from crops plus 240 calories from livestock (equal to farmers' share of livestock-calories: 10.1 per cent of rural-produced livestock-calories). [2] 2,140 produced by nomads, 240 by farmers and 52 by urban livestock-owners.
Sources. Various tables in the text and working paper for food production. Most figures are averages for 1975-78, except food production which is for 1977.

lowing tabulation of the external prices and terms of trade of live animals against cereals in 1978 (1970 = 100):[10]

Live animal price (export)	391
Cereal price (import)	193
Terms of trade	197
Over-all terms of trade (1972 = 100)	141
Terms of exchange: cereal calories per livestock calories 1977	8.3
1978	12.2

The favourable price trends enabled the country to import twice as much cereal in 1978 for the same amount of livestock. At the same time it exported 28 per cent more livestock (in terms of livestock units), so that its ability to import (cereals as well as other goods) increased considerably. Livestock exports thus enabled the country to bridge the gap between food requirements and supply. Thus in 1977, the last year for which we have complete data, livestock exports amounted to Sh 302 million, whereas food imports amounted to Sh 318 million (22.2 per cent of total imports). The terms of exchange were very favourable to Somalia: 64,000 million livestock-calories purchased 527,000 million grain- and non-grain-calories.[11] Banana exports yielded a further Sh 55 million, so that it could be said that the agricultural sector as a whole produced enough to satisfy the country's food requirements.

INCOMES AND POVERTY

In this section we shall attempt to derive an income and poverty profile for Somalia. The task is rendered difficult by the comparative shortage of reliable surveys on which to check our estimates. The recently published livestock and acreage figures and national account estimates, however, provide useful material for an attempt to arrive at a tentative profile. Milk consumption remains a constant problem, and here we have used probable production coefficients, consumption data and some intuitive notions regarding nomadic diets in order to fill in the gaps.

The first step in our analysis is to define a poverty line. As has been consistently argued, income figures do not provide an answer because of the valuation problem. The best alternative (given that most rural producers work in a largely subsistence economy and consume different types of foods) is a poverty line in terms of calories. We shall define a consumption of 2,200 calories per head [12] as the threshold of poverty. To make this criterion operational, we shall relate it to livestock wealth, defined as the value of the stock of animals at current prices, and to land, the other asset of rural families. From this we shall obtain, using various coefficients, the income streams and calories flowing to the nomads and farmers. The procedure is shown in detail in Appendix A. What is important to note is that the poverty line depends in the case of the nomads on three factors: (a) the amount of livestock; (b) the amount of livestock products; and (c) the production sold on the market and/or bartered for grains.

This last is an important aspect of the Somali nomadic economy and has a great influence on food availability for the nomads since barter takes place at rates which are very favourable to them. Unfortunately, nothing concrete is known about its extent, although it can be inferred that far more of it takes place in the south than in the north, because of the existence of farmers with a grain surplus. Thus herds can be smaller in the south and yet satisfy the requirement of food. On these considerations we have placed the poverty line at a livestock wealth of Sh 16,000 in the south, Sh 19,000 in the north and Sh 17,450 nationwide. In physical terms these figures imply the poverty lines, compared with the average herds, shown in table 121.

The average family existing on a purely subsistence basis, that is to say, without any exchange, would need a herd equal to the national average to provide enough calories to feed itself; the milk/grain exchange enables its needs to be met out of much smaller herds because of the favourable rates at which this exchange takes place. [13] Its importance may be gauged from the following figures (calories per family per day):

	South	North	National
Produced	8 800	10 443	9 600
− Milk	775	775	775
+ Grain	5 200	3 500	4 375
Consumed	13 225	13 168	13 200

Table 121. Poverty line, average herds, livestock wealth, and livestock-calorie units, northern and southern Somalia, c. 1977

Area	Cattle	Sheep and goats	Camels	Wealth (Sh)	Livestock-calorie units
Poverty line					
South	12	22	9	16 000	42.6
North	3	72	8	19 000	52.2
National	7	48	9	17 450	47.3
Average herds					
South	15	28	11	19 978	53.2
North	4.5	108	12	28 207	77.3
National	10	65	12	23 851	64.7

Sources. Central Statistics Department, State Planning Commission: *Estimated aggregates...*, op. cit., various tables for average herds; poverty line and livestock wealth from Appendix A. Livestock-calorie units (a measure of livestock in terms of calorie yields) from G. Dahl and A. Hjort: *Having herds*, Stockholm Studies in Social Anthropology No. 2 (Stockholm, 1976), table 10.3, p. 229. Cattle = 1, camels = 2.4, sheep and goats = 0.4. Note that these units differ from FAO's livestock units, which are generally used for a comparison of livestock in terms of fodder requirements.

Taking the percentage of grain-calories in total consumption as an index of the importance of the milk/grain exchange, we find its level at 39 per cent in the south, 27 per cent in the north, and 33 per cent nationwide. The ratio, of course, also shows the importance of grains in the nomads' diet.

As for the farmers, their poverty line has to take account of their income from their *two* assets—land as well as livestock. Most of their subsistence would have to come from their *land* rather than livestock, since grains are much more calorific than livestock products and the farmers would have no opportunity to barter their livestock for grains, they themselves being supposed to provide the grains in the grain/milk barter.

Assuming that the average yield is 300 kg per hectare, a farmer can expect to produce around 2,900 calories per day from 1 hectare of land planted to grains or pulses.[14] In addition, on the assumption that he owns one-third as many animals as the average nomad, he can expect to get around 4,000 calories from his herd. Thus, to satisfy his family's total needs of 13,200 calories, an average farmer owning livestock would have to cultivate around 3.5 hectares of land, allowing for seed and waste. Where the herds are smaller, the total area under cultivation would have to be higher, and vice versa.

The extent of poverty among the nomads and farmers, given the poverty lines we have just established, may be judged on the basis of regional figures given in table 122. On average, only the nomads in Togdheer in the north, and Lower Shabelle and Juba in the south, should be regarded as falling below the poverty line, although allowing for intra-regional variation would put most of the southern regions, with the exception of Gedo and Hiraan, on the borderline. Unfortunately, few concrete data are available on this subject although it is generally accepted that livestock ownership is quite unequal. Some surveys throw partial light on the matter, but their validity is doubtful.[15]

Table 122. Nomadic and farm families: livestock, wealth and cultivated area, c. 1977

Area	Number of families ('000)		Livestock wealth per family (Sh)		Cultivated area per farm family(ha)
	Nomadic	Farmers	Nomadic	Farmers	
North	216.9	55.7	28 207	9 402	.
Mudug (Central)	35.0	5.9	43 419	14 473	0.8
Bari (North-east)	23.9	5.1	35 769	11 923	0.2
Galguduud (Central)	24.5	8.2	34 194	11 398	0.6
Sanag (North-east)	23.2	4.1	26 957	8 986	0.9
Waqooyi Galbeed (North-west)	55.8	21.9	25 401	8 467	2.0
Nugal (North-east)	13.5	2.7	20 690	6 897	0.8
Togdheer (North-west)	40.7	7.8	14 870	4 957	1.2
South	233.0	95.2	19 978	6 659	.
Gedo (Juba river)	37.2	4.1	31 969	10 656	4.2
Hiraan (Shabelle river)	23.9	4.1	30 912	10 304	7.1
Bay (Inter-riverine)	20.6	26.3	17 350	5 783	3.9
Bakool (Inter-riverine)	16.2	2.7	16 679	5 560	4.7
Middle Shabelle (Shabelle)	34.2	12.7	17 668	5 889	2.7
Juba (Juba river)	61.0	18.7	15 387	5 129	4.3
Lower Shabelle (Shabelle)	39.8	26.6	12 419	4 140	5.7
Total	450.0	151.0	23 851	7 950	3.3

. = not available.

Sources. Population figures assumed as follows: total population 4.77 million, projected from revised 1975 census figure of 4.5 million. Of this, shares of various groups were: urban, 22.6 per cent; settled farming, 19.0 per cent; fisheries, 1.8 per cent; nomads (as a residual), 56.6 per cent. This breakdown is given in International Fund for Agricultural Development (IFAD): *Report of the special programming mission to Somalia* (1979), table 1.1, p. 5. A family size of six members is assumed. Regional distribution of population from Somalia: *Three-Year Plan, 1979-1981,* op. cit., table 19-1. Livestock wealth figures are based on livestock population figures given in tables 19-3 and 19-4 of the Plan (undated to 1977 using figures given in Appendix A and producer prices prevailing in 1975-78). The division of livestock wealth between nomads and farmers is obtained assuming that the nomads on the average own three times as much livestock as farmers. Area figures are from table 19-6 of the Plan; per family figures are derived using population estimates updated to 1977.

Thus we can give only rough orders of magnitude of the number of nomads living in poverty. In the north we should expect to find them in the regions where the livestock wealth is below Sh 30,000, and in the south in all the regions except Gedo and Hiraan. Taking account of intra-regional variation we may assume the following proportions for nomads below the poverty line: Mudug, Bari, Galguduud, Gedo and Hiraan, 5 per cent; Sanag and Waqooyi Galbeed, 25 per cent; Nugal, 50 per cent; Togdheer, 75 per cent; Bay, Bakool, Middle Shabelle and Juba, 40 per cent; and Lower Shabelle, 75 per cent. The total number is thus 146,900 families out of 450,000, or 33 per cent, the proportions being 28 per cent in the north and 37 per cent in the south.

Turning now to the farmers, we see that in all the regions their livestock wealth by itself is insufficient to afford them an adequate supply of calories.[16] However, it is from their land that they obtain most of their sustenance and, based on the 3.5 hectare/Sh 7,950 livestock wealth criterion established earlier, we can infer from the fourth and fifth columns of table 122 that the southern regions which are solidly agricultural (namely Lower Shabelle, Juba and Bay) lie above the poverty line. Three other southern regions (Gedo, Hiraan and Bakool) do not have such large concentrations of farmers, but they also, along with their livestock holdings, fall above the poverty line. Middle Shabelle thus remains the only important agricultural region below the poverty line.

It is in the north that we should look for poverty among farmers, since most of them cultivate small tracts of land. However, many of them own considerable herds, which afford them the possibility of augmenting their energy requirements. Of the farmers who fail to produce sufficient crops, those of Mudug, Bari and Galguduud should be considered rather as pastoralists than as agriculturalists and they might just be able to obtain sufficient calories from their herds and land to feed themselves. This leaves Sanag, Waqooyi Galbeed, Nugal and Togdheer as those regions where most farmers fall below the poverty line, the total number of families involved being 36,500. In addition, 10,000 families in Bay and 5,000 in Middle Shabelle may be counted as being in poverty, giving a figure of 51,000 families, or just about 34 per cent of the total.

Thus, altogether a clear north-south pattern emerges. In the north we have rich nomads but poor farmers; in the south rich farmers but poor nomads who, however, have the possibility of bartering some of their milk and meat for grains to make good their calorie deficit. In sum, poverty in terms of hunger is much more likely to be prevalent in the north than in the south, especially among farmers, but also among nomads owning small herds.

Finally, three important qualifications should be noted. Our calorie-availability figures for nomads have been based on calories which are self-produced as well as those obtained on the market for cash and through the barter of milk for grains. This exchange takes place at very favourable terms for the nomads, so that the assumption about its extent has a great impact on calorie-availability figures. If more is bartered than we have assumed, there should be fewer nomads in poverty. Indeed, if a grain surplus were available everywhere and at any time, most nomads should be able to satisfy their calorie needs by bartering milk for grains.

Second, we have based our poverty line on calorie availability, with a calorie supply of 2,200 calories per head taken as the threshold of poverty. This has been done because of the subsistence nature of the economy; clearly, however, non-food items have been excluded in the calculation of the poverty line. With the mass of the nomads and farmers still peripheral participants in the cash economy, the inclusion of these items would certainly place most of them below a more broadly defined poverty line.[17]

The third qualification is of a more fundamental nature. What we have derived is the extent of poverty under normal conditions, caused by lack of productive assets—referred to as "endemic poverty" by the IFAD mission.[18]

Table 123. Output, price and income of some rural products and their price and income terms of trade, 1970-78

Product	1970	1978	Terms of trade, 1978 (1970=100)
Sorghum/maize			
Output ('000 tons)[1]	280	249	–
Price (shillings per ton)	390	750	93.8
Income (million shillings)	110	187	82.7
Bananas			
Output ('000 tons)	130	76	–
Price (shillings per ton)	300	834	135.6
Income (million shillings)	39	63	79.3
Livestock			
Output (index)	100	115	–
Price (index)	100	451	220.0
Income (million shillings)	106	548	252.0
Consumer price index	100	205	–

– = not applicable.
Source. Central Statistics Department, State Planning Commission: *Estimated aggregates...*, op. cit., various tables.
[1] In the case of sorghum/maize, output figures refer to total production, whereas strictly speaking we should be using figures of output actually marketed. These are not available for both the years. If the proportion of crops marketed remained the same, there should be no difference to the calculation of the income terms of trade.

We find that the extent of this is not so great as sometimes believed.[19] However, whenever poverty in Somalia is mentioned it is also understood in the sense of poverty caused by droughts, locusts and other natural calamities ("sporadic poverty" in IFAD's terminology) and, recently, of poverty associated with refugees ("disguised poverty", so called from the fact that the refugees live in settlement areas and depend on food aid and relief). It is important to bear these concepts in mind if one is to be fully aware of the highly volatile nature of poverty in Somalia.

TRENDS IN INEQUALITY

The extent of commercialisation in the rural sector has certainly been the most important factor affecting trends in inequality in that sector. As shown in table 123, the underlying market factors (output produced, its price and the price of consumer goods) have brought about a considerable worsening in the fortunes of the crop sector and a substantial improvement in those of the livestock sector.

The fall in the real price of crops, combined with the fall in their output, has meant that the farmers' purchasing power (or their income terms of trade) fell by around 20 per cent between 1970 and 1978. This is as true for the smallholders who dominate the marketing of grains as for the large-scale farmers who cultivate bananas and sugar-cane. These figures imply a narrowing of income differentials, which can be conceptualised as follows. During both periods the mass of the farmers remained outside the market, while a minority produced a grain surplus and another group produced crops for export and local industries. These groups have become poorer in an absolute sense in relation to their situation at the start of the decade, the smallholders because their price as well as quantity fell, the large-scale farmers through a precipitous fall in production which wiped out the improvement in their prices. Although one would normally welcome such a narrowing of inequality, it is not possible to do so here, for what has in fact come about is an increase in absolute poverty (which is especially disturbing for the smallholders) without any redistribution from the richer groups to the poorer groups. If any redistribution has occurred, it has taken place partly to the advantage of the urban population in terms of low prices of food.

The situation in the nomadic sector was quite different. As a result of the tremendous increase in export prices and some increase in production, the real income of that sector increased 2.5 times. An increase in inequality is thus implied, for it is the large herd-owners who have gained the most from the export boom, since they are able to relinquish a greater proportion of their animals for sale.[20] In the 1960s the nomadic economy was much more hermetic, most nomads obtaining their subsistence from their own herds and from barter with the farmers. This is still true for a majority of the nomads; however, for the minority who trade in the market, the export boom has meant the introduction of a new dimension of income—that of cash income—to the traditional dimensions of subsistence production and barter. The increase in income differentials thus implied could be quite substantial.

Again, one cannot decry this change: it is almost a natural outcome for an economy emerging from self-subsistence. Moreover, the spread of commercialisation and the accompanying improvement in the terms of trade has also bestowed some benefits on the small herd-owners by opening up to them the prospect of selling one or two animals on the market and making up their calorie deficit by buying grains. Unfortunately, while commercialisation has been rapid and widespread in terms of the evacuation of animals, it has lagged behind in terms of the distribution of consumer goods and food, so that the nomads have not been able to reap the full benefits of the favourable price trends.

What can be decried is that the distribution of gains from the export boom (just as in the case of the deterioration in the terms of trade of the farmers) has operated in great measure to the advantage of urban-based livestock dealers, or *dilaals*. In most years the nomads obtain around two-thirds of the livestock income, the Government takes one-tenth in the form of taxes, while the middlemen receive 10-15 per cent as commission, depending on the market.[21] When prices rise they pass on some of the increase to the producers, and retain the rest. Given the relatively small number of traders, this clearly results in a much higher increase in their incomes as compared with those of the more numerous producers.

In recent years the traders have gained even more with the introduction of a new system for the repatriation of export proceeds. Under this system the Government sets a minimum price for exports which the traders must repatriate at the official exchange rate. Now, with the operation of the so-called *franco valuta* system, this rate has been effectively declining as compared with the market rate, so that the traders have been realising a higher and higher price in shillings from their sale of cattle. While some of the benefit of this *de facto* devaluation could be passed on to the producers, there is no evidence that this has is fact happened. In effect the producers have been getting a smaller and smaller share of the export parity price.

GOVERNMENT POLICY

In this section we shall look at government policies with respect to the livestock and agricultural sectors as they have affected rural poverty and inequality.

Livestock

The policy of the Government with respect to the livestock sector has changed on an ad hoc basis, according to its perception of the constraints facing that sector. Thus in the first Five-Year Development Plan, as well as in the two development programmes which followed—namely up to the Second Five-Year Development Plan launched in 1974—stress was laid on the intensification of animal health and the organisation of livestock marketing. This concentration was based on the assumption that there was no problem regarding the number of animals which could be utilised, and that the sole objectives were to minimise the losses incurred as a result of animal diseases and to improve both external and internal marketing processes.

In the Five-Year Plan covering the years 1974 to 1978, although it was recognised that modern techniques of animal production and management of the rangelands were the more important factors affecting livestock productivity, disease control and marketing continued to receive considerable allocations. It was intended to raise productivity through comprehensive development farms which would experiment with modern techniques of crossbreeding to improve the production of meat, milk and other products. The much-publicised need for rangeland development received practical expression only halfway through the Plan period, and that also as a result of the 1973-74 drought. A large-scale project was initiated covering the three northern regions worst affected by the drought (Togdheer, Nugal and Sanag) which sought to redress some of the damage and suffering caused by the drought and, more importantly, for the first time tackled on a serious basis the problems of range conservation and management.

Further recognition to range development was given in the recent Three-Year Plan, for the period 1979-81, in the form of a project for the central region—covering Mudug, Galguduud and Hiraan—similar to the northern project. Including the northern project, some 70 per cent of the country's arid areas have now come under the control of the National Range Agency, and some 40 per cent of the pastoralists stand to benefit from the project. Altogether, range development has been allocated nearly 40 per cent of the total amount to be spent on the livestock sector.

However, despite these changes the priority given to livestock within over-all development remains low, its allocation of expenditure being merely 9 per cent, if poultry is taken into account, or 7 per cent if poultry is excluded. This is shown in the following tabulation along with the allocation for agriculture.

	Percentage of total expenditure		
	1971-73	*1974-78*	*1979-81*
Livestock	5.9	4.9	8.9
Agriculture	14.7	15.0	22.6
Total expenditure (million Sh)	1 000	4 308	7 104

Although the figures imply a considerable improvement over the previous Plans, especially when account is taken of the absolute amounts involved, the allocations can only be considered meagre in relation to the problems faced by the sector.

Agriculture

The problems facing the agricultural sector are similar to those of the livestock sector; namely, how to increase productivity and enhance equity. The Government has initiated a number of projects in response to these problems. Up to the time of the last Development Plan, the major emphasis was given to extending the area under full water control in the Juba and Shabelle river areas. There is great potential in this direction, as may be gauged from the fact that the present area under irrigation is only 50,000 hectares, whereas the potential is estimated at five times as much, most of it on the Juba river. The Plan document had estimated that the proposed irrigation projects would create around 30,000 new jobs between 1977 and 1983. In the event, because of various implementation problems, the actual performance fell far short of this target, so that the projects failed to make the anticipated contribution to the employment capacity of the agricultural sector.

In the current Three-Year Plan emphasis has shifted to "production projects" and support services. The basic aim of the former is to meet the country's objective of self-sufficiency in staple foods, while the latter seek to rectify the current lack of research in Somalia. These programmes, especially the support services, have the potential of more widespread impact than irrigation projects, and thus should meet the Government's twin aims of increasing productivity and making development equitable.

Another policy adopted by the Government in an endeavour to reach the masses is based on the creation of agricultural co-operatives. Their attraction for the farmers is that more and more they are the channels through which technical assistance, initial grants, credit and land are distributed. But despite this, the co-operative programme has had only mixed success. Thus, during the Five-Year Plan ending in 1978, 1,273 co-operatives were planned (150 multi-purpose, 1,051 group farms, 75 co-operative societies,[22] but only 271 were created, of which 47 were multi-purpose and 224 group farms. (No co-operative societies were formed.) However, membership reached the Plan target (27,251 against 28,250), while the acreage exceeded the target (63,029 against 38,250), although it is estimated that at any one time only 60 per cent of the acreage is actually cultivated. In the current Three-Year Plan the aim is to consolidate the present co-operative structure in order to provide a strong foundation for future expansion. Accordingly, only a modest increase in co-operatives is planned: 70 co-operatives (of which 50 are group farms) and 5,150 members. It may be seen that the co-operatives affect only a small number of farmers, so that the Government has not altogether succeeded in realising its desire to make development as widespread as possible.

It must be stated that government policies do not adequately confront the fundamental problems facing the agricultural sector, which may be described, with regard to the nomadic sector and the cropping sector respectively, as: *(a)* how to stabilise and eventually reduce the nomadic population; and *(b)* how to create a stable and controlled agriculture. There is general agreement among all experts that the continuing increase in the livestock population has exerted severe pressure on the natural resources required for the survival of the national herd. The visible signs of this process are increasing erosion in the drier areas, bush encroachment in the wetter regions and general deterioration of the range everywhere. The current situation in effect signifies a breakdown of the natural equilibrium, in which the traditional patterns of water use led to a fairly evenly distributed pattern of grazing, with grass and trees far from the wells grazed during the wet season, and the areas close to the wells during the dry season—a practice which limited the number of animals that could be supported on the range. This natural equilibrium has been modified by the building of government boreholes and private cement tanks (built principally by *dilaals*), as well as by the success of veterinary campaigns, especially the JP15 campaign against rinderpest and a change in the composition of the herds in favour of cattle.

The outcome has been a fall in productivity. Its ultimate cause, as has already been suggested, is the contradiction between the private ownership of animals and communal ownership of the range. The resolution of this contradiction thus requires the creation of new institutions and tenure arrangements, but so far no solutions have been devised by the Government to this end. In the medium term the solution must include not only a freeze in the nomadic population but even its eventual reduction. While this has been accepted by the Government as part of its policy, relatively little action has so far been taken on the matter.

In agriculture irrigation offers the best prospects; so far in Somalia irrigation projects have been confined to large-scale farms, without benefiting the masses of the farmers; co-operatives are an alternative, but they too have involved only a limited section of the farming community. Nevertheless, the Government has continued to devote the major share of its investment to these types of development, while the existing institutions for their part have concentrated their services on farmers relying on these types of agriculture—effectively, only one-fifth of the total number of farmers. Such critical services as input distribution, tractor hire and credit are practically not available to the predominantly subsistence-type smallholders. The only service they are able to obtain is from the Agricultural Development Corporation in connection with marketing.

In the face of this and of the perpetual struggle against an inhospitable environment, it is not surprising that most subsistence farmers regard farming as a secondary activity to be carried out in conjunction with animal husbandry. Farming in Somalia is a far more risky activity than pastoralism, so it would always remain a poor alternative to nomadism. The creation of a stable agriculture therefore requires that farmers have control over water through irrigation. Since clearly the present large-scale and costly schemes do not provide a solution, irrigation projects which reach the masses of the population will need to be devised. Fortunately, the bulk of the subsistence farmers live in or near areas where such schemes are practicable: along the main water arteries, particularly in Lower Shabelle, in dryland/irrigated areas currently under settlement schemes, and in the north-east where the resource base can be upgraded through the controlled utilisation of flood water. Only after such a modification of the farmers' environment can they be expected to respond to the introduction of new husbandry practices and inputs, and to make farming an attractive alternative to the nomadic way of life.

CONCLUSION

This chapter has concentrated on drawing a quantitative picture of the Somali economy based on some newly available data. Within the limits of their availability and trustworthiness, we have been able to shed some interesting light on the workings of the Somali economy. The situation with regard to the availability of food appears to be much healthier than hitherto believed, and there would also appear to be much less poverty (in terms of hunger) than commonly thought. These facts are admittedly based on an estimate of milk production, but we have shown that it is a plausible estimate, given the livestock population, composition of the herds and milk yields. Contrary estimates have contributed to the image of Somalia as a country where most nomads do not even meet critical calorie needs for mere body maintenance. We have shown that such estimates cannot be reconciled with any of the magnitudes relating to the Somali economy. We have also shown, using a similar criterion for poverty (food intake) that there is much less poverty amongst the farmers than hitherto believed.

The implications of the nature of the economy have also been brought out, as have those of the subsistence consumption of different population groups for comparisons of real welfare. It has been shown that, both internally and internationally, these render income figures that are inappropriate because of the high value placed by the market on the diet of the nomads.

While finding the situation, in terms of poverty, to be much healthier than hitherto believed, we have cautioned that this comparatively optimistic picture must be seen against the following three basic facts of the Somali economy: *(a)* the ever-present threat of natural disasters—droughts, floods, locusts; *(b)* the long-term decline in agricultural productivity; and *(c)* the lack of coherent programmes to solve the fundamental problems of the agricultural sector. The first would always colour any single-snapshot analysis of the Somali economy: the next year could bring drought, and if not drought, floods. One cannot account for them in the analysis, but one has always to bear them in mind. The other factors certainly have to be accounted for, for they show that the present situation has been reached at the end of long-term productivity declines in the rural sector. In both the subsectors the technology remains primitive, and the increasing population has intensified the inherent tendency towards decline in productivity. The solution of half the problem requires a stabilisation, or even reduction, of the nomadic population and its absorption in the cropping sector. The solution of the other half requires a frontal attack to be launched on poverty and on the decline in productivity which has been afflicting the farm sector.

APPENDIX A

Poverty line for the nomads

A poverty line for the nomads has to be based on their livestock wealth, the production of meat and milk and its distribution between production sold on the market or bartered, on the one hand, and, on the other, that retained for self-consumption. Thus we relate livestock wealth to income streams, in terms of livestock production, which yield calories through either self-consumption or exchange. The data needed relate to livestock wealth (defined as livestock numbers multiplied by livestock prices), production of meat and milk and the proportion of it sold for cash, a division of livestock wealth between farmers and nomads, and a division in terms of northern and southern Somalia.

Figures for meat income and meat consumption are derived from data in table A1, which are based on national accounts with the exception of one adjustment to allow for urban slaughter outside official channels on a seller-to-buyer basis, known to be especially prevalent in the case of the smaller animals. We have adjusted the urban meat consumption figure so as to bring it up to 80 grammes per head per day, a figure of the order generally found in consumption surveys. Then rural slaughter is obtained as a residual. Given the estimate of livestock population in line (8), the offtake ratios are 7.0 per cent for cattle, 10.2 per cent for sheep and goats, and 2.4 per cent for camels, which would seem to be within the expected limits.[23] In terms of value the livestock wealth amounted to Sh11,934 million and the income (meat income—cash as well as subsistence) ratio to 7.0 per cent, of which one-half was cash and one-half subsistence.

Regarding livestock ownership, we shall assume that agricultural families own one-third as much livestock per family as the nomads. Thus, the 25 per cent of the rural population who are agriculturalists would get 10 per cent of the livestock. We shall then assume that they have a proportionate share in meat and milk income, and in production.

Milk income figures have to be estimated, since the national account figures are quite unreliable.[24] Thus it is shown that in 1977 total milk production amounted to 443 million litres, of which the nomads took 358 million. This implies that the nomads consumed only 0.42 litre daily per head, which is a clear underestimate since it would give them only around 300 calories per day, leaving a deficit of as much as 1,900 calories to be made up from other sources. In 1954 a FAO survey showed that nomads consumed 4 litres of milk per day during the eight months of the wet season from March to October, and 0.5 litre during the four dry-season months from November to February.[25] An average intake is thus implied of 2.83 litres per day over the year. We shall be somewhat conservative and assume that the average consumption is 2.5 litres per day, enough to provide just over 1,800 calories, and that the agriculturalists consume one-third as much, on the assumption that they own one-third as much livestock as the nomads. To this must be added the milk consumption in urban areas. On the basis of these figures, total milk production would amount to 8 million litres per day. We have arrived at this figure from the consumption side. From the production side the most conservative estimates of milk animals, yield, and so forth, certainly indicate that the estimated production is feasible. Of the total milk production we shall assume that the amount represented by urban consumption is sold for cash.[26]

Finally we have to divide our figures on a north/south basis to allow for the greater chances the southerners have for bartering milk for grains. This enables them to satisfy their needs from smaller herds, exchanging milk for high-calorie cereals. Livestock wealth figures for north and south can be derived from national figures (see table 122 above). Meat and milk *production* (or consumption) are then assumed to divide on a proportionate basis. For meat income we shall assume that 50 per cent of cattle,[27] 90 per cent of sheep and goats and all camel exports[28] come from the north, that all the meat-factory animals (cattle only) come from the south[29] and that 80 per cent of the urban sale of animals originates in the south, reflecting the south's share in the urban population. For all these magnitudes, the nomads' share is assumed to be 90 per cent of the total. The resulting figures are shown in table A2.

An explanation of the figures might be in order. Around 1977, as we have shown in table A1, the total livestock wealth amounted to Sh11,934 million. Of this the nomads had by assumption 90 per cent, or Sh10,733 million, which, spread over some 450,000 families, implies an average wealth of Sh23,851 per family. Out of this figure, total meat income would be Sh368 million for the nomadic families (90 per cent of total), milk production 2,464 million litres (2.5 litres daily per head), and meat consumption 54 million kg (90 per cent of total rural slaughter). The north/south division is obtained as has already been explained.

In the last part of the table the total production of milk and meat is converted into

Somalia

Table A1. Livestock production, population and incomes, c. 1977 (average for 1975-78) (lines (1)-(8) in thousand heads, other figures as indicated)

Line	item	Cattle	Sheep and goats	Camels
(1)	Recorded urban slaughter[1]	96	320[3]	37
(2)	Unrecorded urban slaughter[2]	10	100	5
(3)	Meat factory slaughter	50	–	–
(4)	Rural slaughter = (5)-(1)-(2)-(3)	122	1 894	68
(5)	Total slaughter	278	2 314	110
(6)	Export	68	–	30
(7)	Total offtake = (5)+(6)	346	3 348	140
(8)	Livestock population, 1977	4 954	32 708	5 956
(9)	Offtake ratio = (7)÷(8)	7.0	10.2	2.4
(10)	Price, 1975-78 (sh/head)	550	165	640
(11)	Livestock wealth (million Sh) = (8)×(10)	2 725	5 397	3 811
(12)	Offtake value (million Sh) = (7)×(10)	190.3	552.4	89.6
(13)	Internal sales value (million Sh) = [(1)+(2)+(3)]×(10)	85.8	69.3	26.9
(14)	Export value (million Sh) = (6)×(10)	37.4	170.6	19.2
(15)	Cash value (million Sh)	123.2	239.9	46.1
(16)	Carcass weight (kg/head)	140	15	220

– = not applicable.
[1] It is assumed that figures given in the source under "Slaughterhouses" all relate to urban slaughter; they are entered here as "Recorded urban slaughter". [2] Unrecorded urban slaughter is estimated as explained in the text. [3] Average for 1974-78.
Source. Central Statistics Department, State Planning Commission: *Estimated aggregates...*, op. cit., table 26 (for lines (1)-(7)); table 25 (for line (8)); table 30 (for line (10)).

calories in order to illustrate what the average herd would mean in terms of calories if there were no exchange. An average nomadic family would be able to obtain 13,053 calories from its own production, which for practical purposes may be counted as its full requirements, the more so since we have omitted from our analysis income from hides, as well as blood and ghee production and consumption. While in the north, with larger herds, it would be able to produce up to 16.4 per cent in excess of its needs, in the south the subsistence production would amount to only 83 per cent of requirements.

The possibility of buying and/or bartering grain places the nomads in a far more advantageous position, especially in the south, where the farmers have a surplus of grains to sell. On the basis of table A2, and assuming that all coefficients maintain proportionality, a livestock wealth of around Sh 16,000 should provide sufficient calories in the south. Without exchange, this implies the availability of 8,800 calories. Reckoning exchange on the basis of 1 litre of milk for 1.5 kg of grains[30] (775 milk-calories for 5,200 grain-calories), and assuming that on the average a nomadic family exchanges 1 litre of milk, implies that the stipulated livestock wealth should suffice to meet calorie needs. In the north, exchange (or purchase) most probably takes place at a less favourable rate—for example 1 litre of milk for 1 kg of grains—and this would imply a livestock wealth of Sh 19,000 as the poverty line. Nationwide, the poverty line would be at a livestock wealth of Sh 17,450.

Table A 2. Livestock production and income in northern and southern Somalia, c. 1977

Item	Total (millions)			Per family (units)			Per family per day (units)			Equivalent calories per family per day		
	Total	North	South	Total	North	South	Total	North	South	Total	North	South
Livestock wealth (Sh)[1]	10 733	6 118	4 655	23 851	28 207	19 978	–	–	–	–	–	–
Meat income (Sh)[2]	368	201	167	818	927	717	2.24	2.54	1.96	768	870	671
Exports	204	172	32	453	793	137	–	–	–	–	–	–
Internal (including factory)	164	28	136	364	129	584	–	–	–	–	–	–
Meat consumption (kg)[3]	54	30.8	23.4	120	142	100	0.33	0.39	0.27	660	780	540
Milk consumption (litres)	2 464	1 405	1 069	5 475	6 478	4 588	15.0	17.7	12.6	11 625	13 718	9 765

– = not applicable.
[1] Livestock wealth figures based on tables 121 and A 1, scaled down to 90 per cent to reflect nomads' share. [2] Meat income total figure from table A 1, also scaled down to 90 per cent. Division of meat income is as explained in the text. [3] Meat consumption is set so as to yield 300 grammes per family per day and milk consumption to yield 2.5 litres per head per day; north/south division is then in a proportionate basis to livestock.

Note. This table derives from table A 1 and hence relates to average figures for 1975-78. In contrast table 117 relates to 1978 only.

APPENDIX B
Previous studies of income distribution

There have been three previous studies of income distribution in Somalia. Two have been based on an expenditure survey in Middle Shabelle among nomads, farmers and townsmen, and the other on the livestock census of 1975 and crop acreage data.

The Middle Shabelle survey was "discovered" by ILO/JASPA during an employment mission to Somalia in 1976.[31] Expenditure classes were duly reported and Lorenz curves drawn. They revealed a large open-ended class in all the sectors, as follows: in the nomadic sector, comprising 21 per cent of the households and 58 per cent of expenditure; in the urban sector, 40 per cent of households and 72 per cent of expenditure; and in the rural sector, 20 per cent of households and 68 per cent of expenditure. Nevertheless, Middle Shabelle was taken to be representative of the whole of Somalia and the conclusion was reached that income distribution in Somalia was highly skewed, especially in the rural areas. (The Middle Shabelle survey showed less inequality in urban areas than among the nomads and farmers.)

The ILO/JASPA mission also calculated a poverty line for Somalia. To provide 2,200 calories (assumed to be for an adult-equivalent),[32] it was calculated that a daily expenditure of Sh 2.57 would be needed, and for a family of five (said to be equivalent to three-and-a-half adults), a monthly expenditure of Sh 270. This was topped up by one-third to account for non-food needs, giving a figure of Sh 360 as a basic needs income (pp. 235-236).

JASPA did not attempt to combine the Middle Shabelle survey with its poverty line to derive estimates of the population in poverty. This was first done by Michael Hopkins.[33] He took the JASPA poverty line in urban areas, reduced it by 10 per cent for inflation to bring it down from 1976 to 1975[34] (the year of the expenditure survey), assumed that prices of the JASPA basket were 75 per cent of urban prices for farmers and 50 per cent for nomads, and arrived at the following figures for the poverty line and poverty incidence:

	Urban	Rural	Nomads
Poverty line (Sh per annum)	3 888	2 916	1 944
Percentage below	42	67	49
Average income (Sh per annum)	5 835	3 136	3 413

The JASPA basket contained such items as flour, rice, sugar, bananas and tea, which are as likely to be *more* expensive in rural areas as less expensive. But the main point is: What is the relevance for the nomads—or even farmers—of an urban diet, in which milk provided only 60 calories (less than 0.1 litre per day for an *adult*) and bananas as much as 120? And what do the income figures for nomads and farmers mean compared with those for urban dwellers? Was subsistence consumption counted? Apparently not. There are aspects of the "valuation fallacy" that we discussed in Appendix A in the approach adopted here.

Norman Hicks[35] also capitalised on the JASPA poverty line and the Middle Shabelle survey to estimate the incidence of poverty in Somalia. He arrived, however, at a figure of Sh 300 per month per family as the poverty line, allowing for "more than one income earner per family".[36] In terms per head this equalled the average supposed GDP at that time (US$ 110), and from this, assuming that Somalia's income distribution was similar to that of other low-income countries, Hicks concluded that it would be "safe to say that 65 per cent of the population live below the poverty line" (p. 3).[37] He found confirmation for this in the Middle Shabelle survey, finding that 70 per cent of both the rural and the nomadic families fell below the poverty line, compared with 42 per cent of the urban families. Hicks also commented that the finding in the Middle Shabelle survey that nomads and farmers have approximately equal average incomes contradicted the common impression that nomads are disadvantaged.

With all the conceptual problems attending these studies, one could dispense with a critique of the basic raw material used, namely the Middle Shabelle survey. However, since one is bound to encounter it in future studies, "income-distribution data" being scarce, some basic flaws should be pointed out. In the first place, there is the problem of the open-ended class. Second, the survey seems to have been confined to cash expenditure and thus ignores the greater part of the rural income represented by subsistence consumption. Third, as has been ably analysed by the IFAD mission,[38] some of the underlying data are totally implausible. The nomads sold or slaughtered a great deal of livestock (200 per cent of the herd if continued on an annual basis (p. 38)), yet meat production amounted to the equivalent of one goat (p. 40). The nomads lost, on an annual basis, 40 per cent of sheep and goats (p. 39). The price of hides rose steeply from Sh 3 per unit in the first round to Sh 40 (p. 41).

It is small wonder that the mission concluded that the surveys were of "little immediate use" for a study of income distribution, practically discarded them and made their own assumptions about reasonable offtake rates, incomes, and the like (pp. 39-43) for their own analysis of income distribution based on the newly available livestock and acreage figures. This is the most interesting study of income distribution in Somalia so far and bears spelling out in detail.

The poverty line is based on productive assets—livestock (in terms of wealth) and land—and the expected income from these. Livestock wealth is calculated as in this chapter,[39] and 10 per cent is assumed to belong to farmers.[40] Livestock wealth is then converted to income streams using some coefficients: initially said to be 9.5 to 13.7 per cent for rural households and 11 to 14.7 per cent for nomadic households. These figures are said to come from the Middle Shabelle survey, but this is really not so, for, as we have seen and as the authors themselves point out, the Middle Shabelle survey gives implausible figures. Actually, in the end, despite the earlier coefficients suggested, those used are 9 to 13 per cent for both nomads and farmers. These yield an annual income for the nomads of Sh 741 to 1,071, or Sh 900 per head—or around Sh 5,000 per annum per family. One is not given any information concerning what these mean in terms of production of milk, meat, etc., nor on the amount sold for cash or consumed domestically, but it is obvious that the information refers to all income. All the income, whether valuation of food self-consumed, or sold for cash, is treated as equal in terms of so many shillings per head. While this would be acceptable if the estimate were used for comparing nomadic income of different regions, the figure is used to compare income from land, and strong inferences are drawn. Thus cultivation is said to be 0.6 hectare per head; yields are 350 to 450 kg per hectare, the price of the composite crop is given as Sh 0.80 per kg, and hence average income is Sh 190 per head. The two income figures are set against each other and it is concluded that "the settled rural population depending on crop cultivation is clearly worse off than the nomadic population depending on livestock..." (p. 23). This is a clear case of the valuation fallacy.

The study then moves on to discuss over-all inequality. Surprisingly, after devoting a great deal of discussion to livestock wealth and calculating the ensuing income streams, these concepts are set aside in defining the poverty line. This is introduced quite abruptly by a sentence at the end of a paragraph with which it has no apparent connection, thus: "In this respect it is relevant to note that the minimum nomadic livestock holding of 60 to 70 sheep yields an income of just about equal to the poverty line" (p. 24). We have not previously been told what the poverty line is (in fact we are not told this in the whole chapter), nor how the figure of 60 to 70 sheep has been arrived at. An identical sentence occurs in another chapter (p. 64), where a figure of 60 sheep and goats is mentioned and attributed to I. M. Lewis, an authority on Somali nomads. One does not know what to make of this figure. If one were to relate it to the

Somalia

estimation procedure of the authors, it should give a family income of Sh 1,655 as the poverty line: 70 sheep/goats × Sh 215 price × 11 per cent offtake. Yet the only explicit statement of what the postulated poverty line implies, mentions a figure of Sh 3,500, made up of Sh 2,000 from the sale of animals and Sh 1,500 implied value of milk consumption (p. 65). This gives Sh 700 per head as the minimum income. By comparison, the average national nomadic income is Sh 900, and from the closeness of these two figures it is concluded that "over-all distribution of incomes is unlikely to be highly skewed" (p. 65) and further from this that "Nearly all of the nomads may ... be expected to enjoy incomes above the poverty line" (p. 24). Their own table 2.2 shows that a majority of the nomads in Togdheer (average income Sh 425 to 615), Nugal (Sh 591 to 854), Middle Shabelle (Sh 555 to 801), Lower Shabelle (Sh 482 to 697), and Juba (Sh 534 to 771)—altogether 42 per cent of total nomads—should be expected to fall below the poverty line.

The authors then move on to the farmers. A poverty line of 2 hectares per head is laid down (p. 24); on what basis is never made clear, but evidently it relates to the poverty line of Sh 3,500 per family. Thus family land would amount to 10 hectares. With yield at 400 kg per hectare and crop price at Sh 0.80 per kg, an "income" of Sh 3,200 is obtained. This again confirms that the authors consider the valuation of crops and livestock products for self-consumption to be comparable. Two hectares of land per person would yield 2×400 kg/hectare yield $\times 3,500$ calories per kg $\div 365 = 7,670$ calories per day, enough to feed nearly three-quarters of the postulated family of five. Nevertheless, the poverty line is put at 2 hectares per head and, since the average holding is 0.6 hectare, it is concluded that there must be considerable poverty amongst the farmers—a great deal, indeed, to judge from the figures (given in thousands of persons):[41]

	Poor	Out of	Percentage
Waqooyi Galbeed	100	118	85
Middle Shabelle	50	68	74
Lower Shabelle	100	143	70
Juba	75	100	75
Bay	100	141	71
Total	425	570	75

The authors do not list the poor farmers in such marginal farming regions as Togdheer (cultivated land per head, 0.38 hectare), Sanaag (0.14), Bari (0.37), Nugal (0.13), Mudug (0.13), Galguduud (0.09), Gedo (0.68) and Bakool (0.87). No doubt 75 to 100 per cent of these farmers would also be found to be poor according to the criterion laid down, so that in the nation as a whole at least 75 per cent of all farmers should be considered to fall within the poverty category.

Notes

[1] This chapter is derived from an ILO/JASPA (Jobs and Skills Programme for Africa) study on the rural-urban gap in African countries.

[2] See Central Statistics Department, State Planning Commission: *Estimated aggregates of national accounts at current and constant prices and economic indicators, 1970-79* (Mogadishu, Dec. 1979).

[3] It should be noted that we are omitting the increase in population resulting from the influx of refugees after 1977. According to one estimate, by 1978 there were 1 million refugees in Somalia.

[4] See ILO/JASPA: *Economic transformation in a socialist framework: An employment and basic-needs-oriented development strategy for Somalia* (Addis Ababa, 1977), Ch. 5, pp. 71-91 and technical paper 1, pp. 259-278. This theme is also taken up in International Fund for Agricultural Development (IFAD): *Report of the special programming mission to Somalia* (Rome, 1979), Ch. 3, Section 3.2, pp. 55-70.

Agrarian policies and rural poverty in Africa

⁵ The following assumptions were used: producers' receipts from exports were 50 per cent of f.o.b. value; total slaughter in terms of tonnage was 70 per cent (reflecting urban population growth rate of 3.5 per cent per annum), and meat prices two-thirds of its value in 1970.

⁶ This figure is obtained from Somalia: *Three-Year Plan, 1979-81*, p. 85. Another estimate is also given in the same document (p. 282) which puts the total area cultivated at 485,000 hectares. Quite likely the second estimate is of the area under rain-fed farming that is actually cultivated, which in the first estimate is put at 540,000 hectares. This could be land that is appropriated for farming rather than actually cultivated.

⁷ Food and Agriculture Organisation (FAO): *Production Yearbook, 1979* (Rome, 1979).

⁸ In 1976, 4.63 million, compared with FAO's estimate of 3.26 million.

⁹ See Vali Jamal: *Nomads and farmers: Incomes and poverty in rural Somalia* (Addis Ababa, ILO/JASPA, 1981; working paper).

¹⁰ Cereal price from Central Statistics Department, State Planning Commission: *Estimated aggregates ...*, tables 57 and 58, live animal export price from table 61 and over-all terms of trade from World Bank: *Memorandum on the economy of Somalia*, Report no. 3284-SO (Washington, DC, Mar. 1981), table 9, p. 20.

¹¹ In 1978 the terms were even more favourable, at 197 as against 133 in 1977.

¹² This is generally the norm recommended by FAO. It should be noted that this figure is in per head, not per adult-equivalent, terms. An adult male required 2,700 calories, an adult female 2,200, children 1,300, etc. Averaged for a typical population (in terms of age/sex composition), this generally yields the figure of 2,200 calories.

¹³ One litre of milk (775 calories) for 1 kg of grains (3,500 calories) in the north and 1.5 kg in the south.

¹⁴ 300 kg of grains × 3,500 calories per kg ÷ 365 days.

¹⁵ See *Pilot survey of nomadic population using hiloes as a source of water, Bardere district* (Aug. 1973) and *Pilot survey of nomadic households in Burau district* (Feb. 1974). There is also an expenditure survey among nomads, farmers and urban population in Middle Shabelle: *Multi-purpose household pilot survey in Middle Shabelle region* (First Round, Nov. 1975; Second Round, June-July 1976). This survey has been used in at least two studies of income distribution in Somalia. A detailed critique of this is available in IFAD: *Report of the special programming mission to Somalia*, op. cit., Annex to Ch. 2.

¹⁶ The fact that they do not have the opportunity of "bartering up" should be borne in mind; this makes their livestock much less valuable in "real terms" than in the case of nomads, especially those in the south.

¹⁷ One such estimate puts the percentage of the Somali population in absolute poverty in the 1970s as 75 per cent. See J. P. Grant: *Disparity reduction rates in social indicators*, p. 59. Such a figure might well be of the right order of magnitude.

¹⁸ IFAD: *Report of the special programming mission to Somalia*, op. cit., figure 2.1, p. 35.

¹⁹ See Appendix B, where we have discussed some past studies of income distribution in Somalia.

²⁰ It should be pointed out that the price of milk, in which the small herd-owners are more likely to have a stake, has not increased anything like the price of live animals, the index in 1978 being 173 (1970 = 100).

²¹ Based on Hunting Technical Services and Associates: *Livestock sector review and project identification* (Washington, DC, USAID, 1980), Vol. 3.

²² These three types of co-operative represent three stages of the co-operative movement, with multi-purpose co-operatives being the lowest form in terms of collectivisation and co-operative societies the highest.

²³ In 1975 Hunting Technical Services estimated the respective offtake ratios at 5.81, 16.21 and 5.1 per cent. However, they did not have the livestock census figures and had to estimate the total livestock population. On the basis of the census figures the ratios would be 3.9, 9.6 and 2.4 per cent respectively. See Hunting Technical Services: *Livestock sector review ...*, op. cit., Vol. 3, p. 152.

²⁴ We should also include hides and skins income, but the magnitude is small compared with other incomes. Thus in 1978 hides and skins income (cash plus subsistence) amounted to only 11 per cent of livestock *cash* income.

[25] Quoted in Hunting Technical Services: *Livestock sector review...*, op. cit., Vol. 3, table D 1, p. 103.

[26] For fuller details, see Jamal: *Nomads and farmers...*, op. cit.

[27] In 1975 it was reported that 83 per cent of cattle exports went through Berbera; however, this does not mean that all these cattle come from the north since southerners find it profitable to trek their cattle to the north for export rather than to ship them from Kismayo. Also, the southerners have larger cattle herds than the northerners, so that again they are likely to have a bigger share of the total exports.

[28] This should be read in conjunction not only with the preponderance of exports through Berbera but also the composition of herds. It is known (see table 114) that northerners have much bigger herds of sheep and goats than the southerners.

[29] The meat factory is located in the south in Kismayo. It derives most of its cattle from the southern regions of Gedo, Hiraan and other nearby regions, which are all cattle regions.

[30] The producer price of 1 litre of milk is 2.5 times that of 1 kg of grain, so that exchange could be taking place at even more favourable terms than assumed.

[31] See ILO/JASPA: *Economic transformation...*, op. cit., technical paper 5, pp. 327-340. It may be pointed out that the figures analysed therein come from the first round of the survey done in November 1975; there was also a second round seven months later (*Multipurpose household pilot survey...*, op. cit.).

[32] This is an error that is often committed. The 2,200 calorie figure is *per head,* assuming a certain age distribution for the population. An adult male requires 2,700 calories per day. One should either multiply 2,200 by 5 to get family requirements, or 2,700 by adult-equivalent family units. In other words, JASPA underestimated the food requirement: it should have been Sh 386 per month.

[33] M. Hopkins: *Somalia and basic needs: Some issues* (Geneva, ILO, 1978; mimeographed World Employment Programme research working paper; restricted).

[34] There is a slight arithmetical error here, which is worth pointing out since it might bother the reader, as it did the author. In correcting for inflation (p. 19) Hopkins evidently multiplies by 0.9, which is incorrect: he should divide by 1.1. This would give a poverty line of Sh 3,924 per 360 days or Sh 3,978 per 365 days. If a correction had been made for the confusion between *per head* and *per adult-equivalent,* the poverty line for urban areas would be Sh 5,600 and correspondingly for the farmers and nomads. This would increase the incidence of poverty: urban to at least 60 per cent, rural 71 per cent, nomads 59 per cent.

[35] N. Hicks: *Poverty and basic needs in Somalia* (Washington, DC, World Bank, 1978; mimeographed).

[36] ibid., p. 2. Actually, this is a JASPA assumption, counting 1.2 earning persons per family (ILO/JASPA, op. cit., p. 236). The number of income earners per family should make no difference to the poverty line, or to the estimate of family income in a family budget survey, since by definition such a survey measures family income. Also, since the procedure is applied to all groups by Hicks, what does "more than one income earner per family" mean for a nomadic family?

[37] The newly available GDP figures would have given an income per head of US$ 180, using the estimate of population then current, or US$ 159 according to new population estimates. Correcting for milk underestimation would push up the average income to nearer US$ 250; where would the argument based on a comparison of the poverty line and average income then leave us? The point is not about figures as such, but about income figures in a subsistence economy such as Somalia's. Somalia would always look richer than non-pastoral countries and any attempt to apply inferences from those countries to Somalia would yield spurious results.

[38] IFAD: *Report of the special programming mission to Somalia,* op. cit., Annex to Ch. 2, pp. 36-49.

[39] It should be noted that the producer prices assumed (camels, Sh 1,850; cattle, Sh 1,300; goats and sheep, Sh 315) are much higher than anything prevailing then (averages for 1975-78 were: camels, Sh 640 (in 1978 Sh 809); cattle, Sh 550 (Sh 715); sheep and goats, Sh 215 (Sh 204)).

[40] This is done across the board so that in some regions where the farm population is much lower than the national average — Gedo prominently, and Hiraan and Bakook to some extent — farmers end up with unrealistically high livestock income compared with that of nomads.

[41] Figures of poor from p. 24, total population from table 2.1, p. 32.